Discovering the Cosmos with Small Spacecraft
The American Explorer Program

Brian Harvey

Discovering the Cosmos with Small Spacecraft

The American Explorer Program

Published in association with
Praxis Publishing
Chichester, UK

Brian Harvey
Templeogue
Dublin
Ireland

SPRINGER-PRAXIS BOOKS IN SPACE EXPLORATION

Springer Praxis Books
ISBN 978-3-319-68138-2   ISBN 978-3-319-68140-5   (eBook)
https://doi.org/10.1007/978-3-319-68140-5

Library of Congress Control Number: 2017955838

© Springer International Publishing AG 2018
This work is subject to copyright. All rights are reserved by the Publisher, whether the whole or part of the material is concerned, specifically the rights of translation, reprinting, reuse of illustrations, recitation, broadcasting, reproduction on microfilms or in any other physical way, and transmission or information storage and retrieval, electronic adaptation, computer software, or by similar or dissimilar methodology now known or hereafter developed.
The use of general descriptive names, registered names, trademarks, service marks, etc. in this publication does not imply, even in the absence of a specific statement, that such names are exempt from the relevant protective laws and regulations and therefore free for general use.
The publisher, the authors and the editors are safe to assume that the advice and information in this book are believed to be true and accurate at the date of publication. Neither the publisher nor the authors or the editors give a warranty, express or implied, with respect to the material contained herein or for any errors or omissions that may have been made. The publisher remains neutral with regard to jurisdictional claims in published maps and institutional affiliations.

Cover design: Jim Wilkie
Project Editor: Michael D. Shayler

Printed on acid-free paper

This Springer imprint is published by Springer Nature
The registered company is Springer International Publishing AG
The registered company address is: Gewerbestrasse 11, 6330 Cham, Switzerland

Contents

Acknowledgements	vi
Dedication	vii
About the Author	viii
Author's Introduction	ix
1 Foundations	1
2 Early Explorers	51
3 Explorer and the Crisis in Space Science	138
4 Faster, Better, Cheaper?	174
5 Future Missions and Conclusions	254
Bibliographical Note	271
Appendix 1: List of Explorer Missions	273
Index	276

Acknowledgements

I wish to thank all those who assisted with the provision of information, photographs, articles, ideas and suggestions, especially Dave Shayler; Bert Vis; Gurbir Singh; Dr Nathan Schwadron, University of New Hampshire; William Barry (NASA); and Suzanne Parry, Mary Todd and Ben Jones of the British Interplanetary Society (BIS) for use of the library and photographs.

Photographs are courtesy of NASA unless otherwise stated.

To
Judith, Valerie, Alistair and Charlie

About the Author

Brian Harvey is a writer and broadcaster on space flight who lives in Dublin, Ireland. He has a degree in history and political science from Dublin University (Trinity College) and a MA from University College Dublin. His first book was *Race into space – the Soviet space programme* (Ellis Horwood, 1988), followed by further books on the Russian, Chinese, European, Indian and Japanese space programmes. His books and chapters have been translated into Russian, Chinese and Korean.

Author's Introduction

Explorer 1 was the first American satellite to reach orbit, on 31st January 1958, salvaging American space pride after the first two Soviet Sputniks had orbited the Earth the previous year. Explorer 1 also achieved a significant scientific result, as its instruments enabled James Van Allen to present the case that Earth was circled by radiation belts.

What is less well known is that the Explorer program, in its various forms, continues to the present day, making it the world's longest continually running program of scientific space exploration. Explorers have been the principal instrument for unmanned American space discovery in near Earth orbit (although four also reached the Moon). Up to the present time (2017), 94 Explorer-class satellites have been launched and more are promised for the coming years. The importance of the humble Explorer program may have been overlooked, thanks to high-profile, large scientific missions (Hubble, for example) and by extraordinary American successes with planetary probes right across the solar system, but so far, no one has chronicled this series, its scientific results over six decades and the value of such small spacecraft as instruments of scientific discovery. This book, *Discovering the cosmos with small spacecraft – the American Explorer program*, is designed to mark the 60th anniversary of Explorer 1 (2018).

The core of the book is a description of the Explorer missions to date: their purpose, mission, instruments, results and discoveries. This is set in the broader context of the infrastructure built up around the missions (such as the Goddard Space Flight Center), ground control, management, operations, dissemination, methods and launchers used, from the large Deltas to the diminutive Pegasus. It will also look at the politics of space science, the priority given to that discipline and, within that, this form of exploration by small spacecraft. At one stage, for example, missions in the Explorer program got bigger in size but fewer in number (there was only one mission between 1982 and 1988) and the series almost died out, only to rediscover its true vocation in the period of 'faster, better, cheaper' during the 1990s.

Dividing the Explorer program into distinct phases in a manner that was manageable for the reader was a challenge and there was no perfect way by which this could be done. The American space program as a whole has a number of natural dividing lines: the establishment of the space program as a reaction to Soviet achievements (1950s); the heroic period of the 1960s, culminating in the landing on the Moon; the radical revision of the program under the Nixon presidency, associated with the introduction of the Shuttle (1970s); reconsideration in the late 1980s following the *Challenger* accident; and the current period of the space station, commercialization and robotic conquest of the solar system (2000s onwards). This book broadly follows these natural dividing lines.

Accordingly, Chapter 1, *Foundations*, outlines the establishment of Explorer, set against the creation of NASA and the field centre most associated with Explorer, the Goddard Space Flight Center. The extensive Chapter 2, *Early Explorers*, outlines the heyday of the early Explorer program, with its numerous subsets for aeronomy, astronomy, meteorites and so on, whose natural course ran to Explorer 55 in the mid-1970s. Chapter 3, *Explorer and the crisis in space science*, outlines the most difficult period of the program from the mid-1970s to the 1990s, in which there were few missions (only eleven) and which is now regarded as having been a period of crisis in space science. Chapter 4, *Faster, better, cheaper?* examines the recovery of the program with new approaches and rules, one of which has led to extraordinary scientific outcomes, starting with SAMPEX. Chapter 5, *Future missions and conclusions*, looks at new missions in prospect, looks back over the program and provides conclusions.

Nomenclature

How to best identify and name spacecraft has been a problem in virtually all the world's space programs. Unlike its Soviet equivalent, the Cosmos program, Explorer was never originally designed as a stand-alone program of small scientific spacecraft. The term 'Explorer' was given to it only once the first was in orbit and was the quickly-made choice of the incumbent President. Explorer was one of two rival programs – the other being Vanguard – to get a satellite into orbit for the United States, with what is now known as Explorer being run by the army while Vanguard was associated with the navy. The first five Explorers were all army projects and it was not until NASA was established that Explorer began to emerge as the scientific program it later became, starting with Explorer 6. Early Explorers were given 'S' designations ('S' for 'scientific'), normally with a number; indeed, this was long the designator of choice in the Goddard Space Flight Center and, to complicate things, might be followed by 'a', 'b' or 'c' designators which might be upper or lower case.

Not long afterwards, various subsets within the Explorer program acquired specific designations according to their role, such as 'Atmospheric Explorer' (AE) or 'Beacon Explorer' (BE) and these are used here alongside the numerical title. The 'Interplanetary Monitoring Platforms' (IMP), which had their own alphanumeric identifications (A, B, C etc) were also Explorers and there was a period of overlap between the 'S' series and these subsets. To complicate things further, there were a number of intrusions into the program by individual, well-established satellites that came from other institutions (University of Iowa, Naval Research Laboratory) but which acquired Explorer status (Injun, Solrad). From the 1990s, the subsets were re-invented according to size and scope, such as SMEX (Small Explorer), MIDEX (Mid-sized Explorer) and UNEX ('University Explorers').

Even the 'S' designators were not unique to Explorer, also being applied to the orbiting astronomical and geophysical observatories as well as to British satellites. At one stage, Goddard even issued a listing of designators to address the confusion.

The problem was exacerbated after Explorer 55 in 1975, the point at which NASA itself stopped using the Explorer numerical designator (officially, there is no such thing as 'Explorer 58', for example). Goddard, though, still occasionally assigned numbers to Explorer long after this point. The Explorer program continued to have an identity, however, normally reflected in the 'E' or 'EX' for Explorer in its title, such as Cosmic Background Explorer (COBE) or Interplanetary Boundary Explorer (IBEX), though this was not true in all cases (such as NuSTAR and Swift). Furthermore, the term 'Explorer' in a mission's initials normally means that it is part of the Explorer program, but again there are some exceptions. At the time of writing, 94 missions had been launched that could reasonably be considered part of the Explorer family and to guide the reader, the number in the sequence is also given for each mission. TERRIERS, for example, is referred to as '(Explorer 76)' in brackets to help identify where it fits in the sequence, but is not otherwise described as Explorer 76, since NASA never used that designator and that decision is respected. Hopefully, this will all become clearer as the reader progresses.

There are other such anomalies, including missions that may not have formally started out as part of the Explorer program but subsequently acquired the title Explorer and over time became considered part of the program, such as the International Sun Earth Explorer (ISEE). Some dispute whether COBE was 'really' part of the program, but it was an integral part of Goddard's program and is thus included. The formal Explorer program also covered some *instruments* flown on other spacecraft, rather than actual satellites, (Japan's *Hitomi*, for example, or NICER on the ISS): these are mentioned, but the book focuses on stand-alone satellite missions.

In the early days of the Explorer program, Roman numeral designators were normally used (e.g. Explorer XVII) and these survived to the mid-30s (e.g. Explorer XXXIII). These tended to be scarcer as the program passed 40 (XL) and 50 (L); less familiar territory to non-Latin scholars. Arabic numerals came to be used over time and these are applied retrospectively (e.g. Explorer 17, 33). The final note on metrics is that this book follows the provisions of the Congressional Metric Act, 1866 and subsequent Metric Conversion Act, 1975. It is the system utilized in most of the world (except Myanmar and Liberia), so it is used here.

For the launch centres, 'Cape Canaveral' is used to refer to both the civilian and military set of launch pads, although technically the civilian part is the Kennedy Space Center (KSC), the military is Cape Canaveral Air Force Station and they are also called the Eastern Test Range. Vandenberg Air Force Station is used to refer to the launch centre in California, even though the civilian part is the Western Space and Missile Center, Lompoc and collectively they are called the Western Test Range. The term Pacific Missile Range has also been used. For orbital parameters in both the tables and text, the principal sources used are NASA documentation; to 1989, the Royal Aeronautical Establishment *Table of Earth satellites*; and *Spaceflight* magazine. Note that many accounts may give slightly differing figures of orbital parameters, depending on the moment of determination and further complicated by conversion to metric, so this is an attempt to use as standard and authoritative sources as possible. There may occasionally also be differences in the dates, due to the time zone applied.

This book intends to give balanced coverage to each mission: its purpose, the spacecraft, instrumentation and results. I have drawn substantially on NASA-based sources. Where a series of Explorers is involved, descriptions are normally given only for the first of the series, rather than be repeated each time thereafter. Each description, though, is far from equal length. Some missions failed and little more can be said about them, while others brought back a treasure trove of science and led to thousands of published papers. In the case of some scientific missions, however, little information is readily available, though some must exist in pre-digital libraries. For the Interplanetary Monitoring Platform satellites, a series history was compiled, giving us a narrative, but this was unusual. Toward the end of the 20th century, mission results were dispersed across hundreds or thousands of papers in dozens of journals. Few had a single mission report that could provide an overview.

This book is intended for the general reader, so technical terms are kept to a minimum. To facilitate reading, some are introduced here. The capital 'R' is used to indicate the distance from Earth of a satellite as a function of the Earth's radius (thus a satellite at R30 is orbiting at a distance out equivalent to 30 times the radius of the Earth). 'Period' is the time it takes for a satellite to orbit Earth. The degree symbol (°) is used to express the angle at which a satellite crossed the equator, also called the angle of inclination (90° would be a polar orbit). The abbreviation 'Δv' is sometimes used for the thrust applied for a manoeuvre. The term 'PI' is used frequently, meaning the Principal Investigator on a mission.

Finally, this is only a start. Investigating the Explorers makes one ever more aware that each mission has many more stories, anecdotes, personalities and outcomes than could be covered in this introduction. There are many missions about which little is still known, gaps to be filled in and discoveries to be made, so this book is an invitation to others to take up the baton and lead on.

Brian Harvey
Dublin, Ireland, 2017

1

Foundations

The icy wastes of Antarctica are the unlikely starting point for Earth's history of scientific space exploration. On 5th January 1922, the great Antarctic explorer Ernest Shackleton died at Grytviken, South Georgia, bringing to an end what is now called the heroic age of Antarctic exploration. The few expeditions that followed were interrupted by the Second World War and Antarctica returned to its long slumber, undisturbed by human presence – until 1950. Then, two eminent scientists, Sidney Chapman and James Van Allen, moved for a worldwide effort to bring scientists across the planet together for a new, noble endeavour in international scientific collaboration, the Third International Polar Year. This was intended as a worthy successor to the earlier International Polar Years of 1882-3 and 1932-3. In the course of proposing the idea to colleagues, the view was taken that it should be a much more ambitious year, not restricted to polar and should instead study the natural phenomena of our whole planet; so it evolved into the 'International Geophysical Year' or IGY. It was a concerted effort to re-build science in the post-war world and turned out to be the most successful international collaborative scientific venture of the middle of the 20th century. Countries the world over warmed to the idea and by the time it was under way, the IGY had 66 nations involved and 60,000 scientists participating, their international links enduring long after the IGY was over. The present-day bases in Antarctica are among its many legacies.

Chapman set up an organizing committee in 1952, to which scientific committees worldwide quickly subscribed, with Van Allen participating on its panel for rocketry, cosmic rays and aurorae. In both the U.S. and USSR, scientists quickly seized upon the idea of a satellite as the most practical and memorable way of marking the year, with the USSR indicating as early as 1953 that an artificial Earth satellite would be its objective for the year. In November 1954, Van Allen wrote a four-page memo, *Outline of a proposed cosmic ray experiment for use in a satellite (preliminary)* and began assembling a prototype in his laboratory. The special committee for the IGY issued a statement on 4th October 1954, endorsing the idea of an Earth satellite to mark the year and encouraging participating countries to launch them; this was in turn approved by the American national committee at its meeting on 18th May 1955. So that was it, the Third International Polar Year had become the IGY and the landmark event of the IGY would be an Earth satellite.

2 Foundations

The idea of a satellite was not new, the best-known designs being those of the RAND Corporation (*Preliminary design of an experimental world-circling spaceship*, 1946) and the U.S. Navy Bureau of Aeronautics (*Structural design study for a high-altitude test vehicle*, 1946).[1] Studies of a minimum satellite were also conducted in Britain, notably by the British Interplanetary Society (*Minimum satellite vehicles,* 1948 and *Earth satellite vehicles*, 1950), imagining a satellite in the order of 5kg, but the Americans were largely unaware of these efforts until Alexander Satin of the Office of Naval Research visited London in 1952.[2] Lacking direction or support from higher authorities, these studies failed to progress amidst the large number of post-war research and defence projects. Now, with the IGY, these concepts had a real chance to progress. But first, they needed a rocket.

The Rocket: The Redstone

The arrival in the United States of rocket scientists from the wartime German A-4 (or V-2) program, led by Wernher von Braun, meant that there were practical prospects for building a rocket sufficiently powerful to put such a satellite into orbit. The Germans exfiltrated from Germany during *Operation Paperclip* had first been brought to Fort Bliss, Texas, in September 1945, a desolate part of the state near El Paso, where they were put under close U.S. Army supervision; 'prisoners of peace', as they called themselves. In April 1946, they fired the first A-4 from American soil, the launches taking place at White Sands in the neighbouring state of New Mexico not far from Roswell (where, some believe, aliens landed the following year).

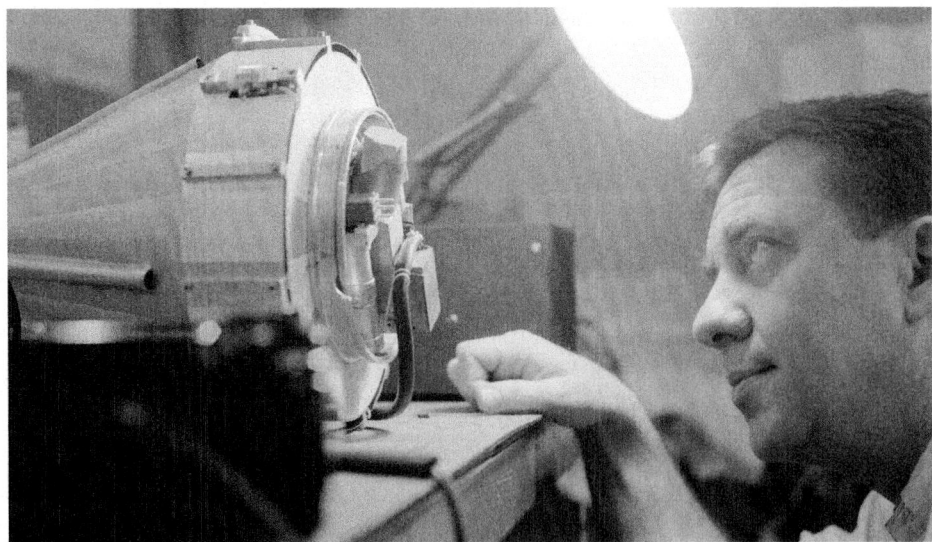

James Van Allen with space probe

Back in Washington DC, James Van Allen was a physics postgraduate in the Directorate of Terrestrial Magnetism (DTM) where, during the war, he had developed the proximity fuse that played a distinct role in the allied air victory in both Europe and the Pacific. The project became too big for DTM so it moved out to a new institution established for the purpose, the Applied Physics Laboratory (APL) of Johns Hopkins University in Baltimore, Maryland (much later to play in important role in the Explorer story). One of the organizers of *Operation Paperclip* was Ernst Krause, a nuclear physicist of the Naval Research Laboratory (NRL). As he began to ship A-4s back to the United States, Krause wondered whether instead of fitting a warhead, it would be possible to fit scientific instruments. Accordingly, he convened what was called the Upper Atmosphere Rocket Research Panel, to which he invited James Van Allen. Other later panel members were Homer Newell and William Pickering (the panel went through several iterations: V-2 Upper Atmosphere Research Panel (to 1948); then without the V-2 (to 1957); then the Rocket and Satellite Research Panel). When Krause left the panel, Van Allen became its chairperson in January 1947. This panel decided on the instrumentation to be carried on the A-4s. Van Allen fitted a simple cosmic ray detector to the first rocket.

The U.S. Army fired its first A-4 from White Sands on 16th April 1946, one of 67 between then and 1952. Like the first A-4 launch from wartime Peenemünde, Germany, the initial launch on American soil was not a success, with von Braun having to blow it up only seconds into its mission. When the first successful launch did take place, Van Allen's cosmic ray detector quickly registered an increase in cosmic rays as the rocket ascended. The first picture of Earth from space, at an altitude of 100km, was taken by an APL camera on 24th October 1946. The scientists frequently met von Braun and his German colleagues in the shed where the A-4s were assembled. One of von Braun's senior engineers was Ernst Stuhlinger, who had undertaken cosmic ray research as a student of Hans Geiger, the man after whom the 'Geiger counter' was named.

A-4 altitude tests

On 24th February 1949, a small upper stage called the WAC Corporal was fitted (WAC was named after the Women's Auxiliary Corps, because it was the little sister to the bigger rocket). It reached 393km, the seed of the idea that, with small upper stages, ballistic rockets could reach ever higher altitudes, or eventually orbit. When the supply of A-4s ran out, Van Allen moved his instrument to a small U.S. Navy sounding rocket called the Aerobee, a derivative of the small WAC. Scientists began to write up the findings of these missions, including Homer Newell (1915-83), an important figure in the development of early American space science. He held a PhD from the University of Wisconsin (1940) and then served as a theoretical physicist at the Naval Research Laboratory from 1944. Newell supervised the A-4 launchings, subsequently writing *Pressures, densities and temperatures in the upper atmosphere,* followed by the first of nine books, *High altitude rocket research,* the next year.[3]

In spring 1950, von Braun and his group moved to the Redstone Arsenal in Huntsville, Alabama, where their first task was to develop a new missile for the army. A development of the A-4, the new missile was called Hermes and its design was completed in 1950.[4] Von Braun then designed a successor with lightweight aluminium alloys, transistors and a 50% more powerful engine, able to fly up to 400km. When the Korean war broke out, the army was directed to adapt Hermes as a mobile field weapon with a range of 320km, using the engine of the Navaho missile. Von Braun named it 'Redstone', after his new home, on 8th April 1952. Its first flight was 20th August 1953 and 16 were built in the Redstone Arsenal, with later manufacturing by Chrysler in Michigan. The USAF formed a Redstone battalion in April 1956 which, after 36 tests, became operational on 1st June 1958 with the U.S. Army in Germany. Early testing had its fair share of explosions, but the missile became so successful that it became known as the 'ol' reliable'.

Although originally designed for a conventional warhead, it was also capable of flying a nuclear one; the 3.1 tonne W-39, capable of 100 Hiroshimas at 2.5 megatons. Redstone was indeed used for high-altitude nuclear tests. A Redstone launched *Teak* on 31st July 1958, when a W-39 exploded at 77km altitude over the Pacific, creating an 18km fireball that was visible in Hawaii some 1,300km away. A second test, *Orange*, exploded at 43km altitude on 11th August. Both tests were supposed to be secret, but the electromagnetic wave from *Teak* blacked out the Pacific from Canada to Australia for 11 hours, the explosions could be seen in Hawaii and they created aurorae over Samoa 3,200km away. Such spectaculars – which generated worldwide protest – were banned by the test ban treaty of 1963, although a small number of countries continued with them for some time after. Redstone was retired on 30th October 1964 and was replaced by the solid-fuelled Pershing.

The Redstone had an illustrious civilian history. It was chosen for America's first manned Mercury missions in 1958, although 800 changes were required to modify it. Redstone carried America's first two astronauts, Alan Shepard and Gus Grissom, into space in 1961, but most Americans were blissfully unaware that their first astronauts had launched on a rocket modelled on the wartime German V-2. The Saturn 1B, later used to launch Apollo 7 and 18 and the Skylab, also used Redstones but as tanks clustered around the first stage. The Redstone's last missions were at the other end of the Earth, with the missile being used for SPARTA tests by Britain and Australia for anti-missile research across nine tests from Woomera over 1966–7. Finally, a Redstone was used for WRESAT on 29th November 1967 to put Australia's first and only satellite into orbit, the rocket's last known launch.

First Redstone

The one missing part of the equation was a launch base that posed less danger to civilians than White Sands, where there was a risk of off-course missiles falling both within the state of New Mexico and on neighbouring Mexico. This was resolved with the selection on 9th July 1946 of the Joint Long Range Proving Ground Base, later named the Eastern Test Range, but now known to the world as Cape Canaveral. The first Redstone was first launched from there.[5]

Vanguard Versus Von Braun

Wernher von Braun was not just a missile builder and had always seen rocket-building as part of the much bigger project of space exploration. Like his opposite number in the Soviet Union, Sergei Korolev, he had found that only by working for the military could he make progress in the design, construction and firing of rockets and satellites. During the 1950s, von Braun had not only built rockets for the army, but had popularized the coming age of space in the mass media and outlets such as *Colliers Magazine*. In 1954, von Braun wrote *A minimum satellite vehicle based on components available from developments of the Army Ordnance Corps*, offering to launch it on the Redstone, then under development. In his article, von Braun described such a satellite as having 'an enormous value to science, especially upper atmosphere and radiological research,' offering 'a tremendous wealth of information, particularly about primary solar radiation effects on weather and radio communications.'

6 **Foundations**

Whatever the subsequent evolution of what became known as the space race, the initial focus was very much on a scientific project in line with the spirit of the IGY. The American Rocket Society's proposal to the National Science Foundation (24th November 1954), *On the utility of an artificial unmanned Earth satellite*, justified such a satellite on the grounds of the information it could provide in astronomy, astrophysics, geodesy and geophysics. In an appendix to this, *The satellite vehicle and physics of Earth's upper atmosphere*, Homer Newell of the Naval Research Laboratory wrote of how the satellite would provide a vantage point above the Earth that would enable the physicist to understand the energy and rays reaching Earth from the Sun and the galaxy. An influential study at this time was S.F. Singer's proposal for a Minimum Orbital Unmanned Satellite of the Earth (MOUSE), originally presented at the International Astronautical Federation in Zurich, Switzerland in 1953, then later in the United States at the American Rocket Society in Baltimore, Maryland on 20th April 1955. Singer gave detailed attention to the satellite's potential for geophysics, astrophysics, investigations of the upper atmosphere, studies of radiation and research into cosmic rays, micrometeorites and the ionosphere – very much the agenda of the subsequent Explorer program.

At the invitation of the Office of Naval Research, representatives of the services, industry and the research community, including von Braun, met in Washington DC on 25th June 1954. Von Braun calculated that, with a cluster of 37 Loki Surface-to-Air Missiles (SAMs)

Wernher von Braun

fitted on top, Redstone could launch a satellite in the order of 2.25kg to 6.75kg into a 260km orbit. This was formalized as a joint army-navy study project, called Orbiter, after a meeting at Redstone Arsenal on 3rd August. It was suggested as a candidate project for the IGY and was presented formally to the Department of Defense (DoD) in September 1954. From the perspective of the Explorer program, a point of interest is the instrument package suggested for this satellite, which was to measure air density, altitude and the ionization content of the atmosphere.

Von Braun and James Van Allen soon found themselves involved in the politics of who would build and launch the first satellite. Although the army, von Braun and his German rocketeers were the obvious candidates to lead the Earth satellite project, the Naval Research Laboratory designed a technically ambitious program for an artificial Earth satellite, adapting a naval sounding rocket called Viking to enable it to place a smaller satellite (only 1.5kg) into orbit. It was one of the smallest rockets ever developed. Although at the time the small size of American rockets was seen as a disadvantage, it forced the United States to miniaturize its instruments, equipment and satellite design in order to fit the rockets, which ultimately led to greater scientific value when those rockets became more powerful.

The question of which team should be given the go-ahead – von Braun or the NRL – was arbitrated by a committee with the less than eye-catching title of the Advisory Group on Special Capabilities, which came to a less than unanimous decision, the closest possible at 4-3. Accordingly, President Dwight D. Eisenhower's administration announced that the Viking-based proposal was the preferred choice for America's candidate for the IGY. Several grounds were given: based on a sounding rocket, it could be presented as a civilian project (even if it were run by the navy); it would promote technical development; and it would ensure that energies were not distracted from the ballistic missile program.[6] On 9th September 1955, the DoD instructed the navy to proceed. The term 'Vanguard' was not attributed to the project for another year, until 16th September 1956. Officially, this brought the original project Orbiter to an end.

Effectively, in the 'army vs navy' struggle, the navy had prevailed. Prejudice against von Braun and his German team may also have played a subliminal part. Had they realized how close the Russians were to launching a satellite (which the National Security Council did and ordinary folk could have been if they read the *New York Times*), the decision might have gone the other way. Throughout 1956, the advisory group held to its view that Vanguard was the better program, even if von Braun could reach orbit first. However disappointed he may have been with the decision to go for Vanguard, von Braun quickly and persistently promoted his rocket as an alternative should things go wrong.

When Vanguard was declared as the winner in July 1955, James Van Allen began adapting his instruments to the smaller Vanguard design and formally proposed them for inclusion in September 1955. The following month, he was appointed to the IGY Technical Panel on the Earth Satellite Program, responsible for the selection of its instruments. As Vanguard got ever smaller, now with room for only 1kg of experiments, the panel met on 1st June 1956 to whittle down the 25 instruments proposed to a short list of four, which included Van Allen's own cosmic ray experiment. He was well aware that von Braun's team continued to promote the Redstone as a launcher should Vanguard fail or the decision be overturned, however, so he visited Huntsville to make sure that his experiment could equally fly on any of von Braun's rockets.

Competition between Vanguard on the one hand and von Braun's team on the other became the subtext of the early American effort to develop an artificial Earth satellite. There were several who questioned whether the navy team had the capacity to carry out its mission and in August 1955, the Research and Development Policy Council (of the DoD) unanimously recommended that the time-risk factor in the scientific satellite program be brought to the attention of the Secretary of the Defense, to determine whether a Redstone backup program would be necessary. Von Braun continued to hover in the wings, reminding the authorities of his ability and preparedness to launch a satellite. General Bruce Mendaris formally established the Army Ballistic Missile Agency (ABMA) out of the Redstone Arsenal on 1st February 1956, as an institutional home for von Braun and his German colleagues.

In the meantime, von Braun's continued missile work brought him ever closer to launching a satellite. The army commissioned the Jupiter in 1955, as an Intermediate Range Ballistic Missile (IRBM) able to reach over 2,000km, giving the work to von Braun because of his record on the Redstone. Jupiter was a single stage vehicle 18.4m tall, weighing 29.9 tonnes and with an S03 engine of 68,040kg thrust. Von Braun modified the Redstone by lengthening the tank by 1.65m and increasing the burn time to 155 seconds. In effect, it was a Redstone with longer fuel tanks.[7] It was first tested as Jupiter A in March 1956.

It quickly became clear to von Braun that warheads could be delivered with greater range and accuracy if separated from the rocket, unlike the V-2 weapon which impacted as a single entity. One problem was that such warheads had a tendency to burn up or even explode during the friction of descent. To address this, von Braun persuaded the army that nose cone tests were important if the new missile were to be successfully fielded. To this end, he constructed the Jupiter C, ('C' for 'cone', though some say 'composite') for nosecone tests. Two solid fuel upper stages (called Sergeant) were attached, designed to ensure that the warhead reached both high altitude and sufficient velocity on descent to test it fully. He fitted 11 Sergeant rocket motors on the second stage, three on the third and one on the fourth. The Jupiter C was now 21.7m long and could orbit 11kg and von Braun was, in effect, disguising satellite work within the Jupiter program. He knew that his rocket could reach orbital altitude and velocity with these tests.

Unfortunately for him, so did the Pentagon, which allowed him to make the nose cone test on condition that he filled the top stage with sand so that it could not quite make it into orbit. The Pentagon was convinced, with some justification, that von Braun was trying to 'pull a fast one' and 'accidentally' get a satellite into orbit, hence the order to use the sand. Jupiter C launched on 20th September 1956, reaching 1,097km altitude and 5,366km downrange. Nobody except the Pentagon realized the significance of this test, which was little publicized. This Jupiter C had four stages, but the fourth was inert and there was no satellite.

The second launch was on 15th May 1957 (Redstone 34/Jupiter C1), but was not entirely successful because sharks ate the flotation bag after splashdown and it was not recovered. The third was on 8th August 1957, which was subsequently recovered and displayed by President Eisenhower as 'the first object to travel in space'. This Jupiter C soared 600km high, reached 19,300km/hr and parachuted the nosecone 2,140km downrange within a 400m circle. With this, the program was announced as concluded.[8]

Vanguard Versus Von Braun 9

Jupiter C

The first operational Jupiter missiles were delivered to the United States Air Force (USAF) in August 1958 and were stationed in Italy and Turkey two years later (the latter being subsequently withdrawn as the *quid pro quo* for the Russian withdrawal from Cuba).

Foiled in his subterfuge to reach orbit, General Mendaris, the chief of ABMA, ordered that a backup rocket be put in storage, just in case it might be needed. In fact, two were retained, one called 'missile 29'. He was allowed to keep the second rocket only for a 'long-term storage test', though he secretly hoped to have the opportunity to put it to use. The ABMA continued to press to have the opportunity to launch a satellite, renewing the proposal for a 7kg satellite in April 1957, to be launched that September. Von Braun also continued to argue publicly that he could and should be allowed orbit a satellite. This exasperated the Pentagon, which issued a directive on 29th July 1957 prohibiting the army, navy or air force from discussing space projects with the press, at a time when Soviet announcements of an upcoming satellite launch were becoming ever more frequent and specific.

Despite all the lobbying, it was not absolutely certain that Redstone had sufficient performance to achieve orbit anyway. According to Ernst Stuhlinger, it was 7% short of that capability. The Specific Impulse (SI) of its alcohol/lox combination was 284, but it needed 305 to reach orbit. General Mendaris went back to the builder, North American, in Downey, California, to ask for 'the best man they had' to bridge the gap. Instead, Downey offered the only woman engineer in the plant, who had no formal qualifications compared to her

900 graduate male colleagues, but whom the manager had no doubt was the best. Mary Sherman Morgan had been a chemist responsible for quality control in an ammunition factory making shells, so she knew a lot about fuels and explosives.[9] Now, although directed to improve the oxidizer, she instead felt that the solution lay in a more powerful fuel. She favoured, as its base, UDMH, a Russian formula, although it alone was not capable of reaching SI 305. Morgan also identified a denser chemical, diethylenetriamine, called DETA, which, when mixed 40% with UDMH 60%, could reach SI 310. Her new fuel was later called hydyne. The combination was tested at the Santa Susana Field Laboratory in the Simi Valley, California and, after a number of false starts, eventually made a full 155-second run and achieved the thrust hoped for.

Sputnik

By chance, the Secretary of Defense, Neil McElroy, happened to be visiting von Braun's plant in Huntsville on the evening of 4th October 1957. When partying there was interrupted by the news that the Soviet Union had just launched the world's first Earth satellite, the deathly silence that greeted this announcement soon gave way to raging frustration that it need not have been this way. Von Braun promptly volunteered that he could get an American satellite into orbit in 60 days with his Redstone rocket, but McElroy, unconvinced, gave him 90.

But what of the other key personality in putting together the American satellite project? James Van Allen was sailing in southern waters aboard the *USS Glacier*, a month into a four-month expedition in the Antarctic (where this story started), making cosmic ray experiments with small Loki sounding rockets. It was there that he learned of the launch of Sputnik, quickly picking up its signals as it flew overhead.

When the USSR launched its second satellite into orbit on 3rd November, with a dog on board, the pace picked up. Many Americans went out into their back yards to watch the two Sputniks tracking across the early winter skies of the United States, an ever-present visual reminder of apparent Soviet superiority in space travel. Five days later, on 8th November 1957, Neil McElroy directed the army to proceed with the launching of a satellite, providing funding for two Redstones to orbit an Earth satellite by March 1958, either after Vanguard, or in its place should it falter.

Technically, the Jupiter C was a three-stage rocket designed to test ablative nose cones for nuclear warheads. Explorer 1 was launched on the four-stage version, which acquired the fresh name of Juno 1.[10] Either way, they were Redstone derivatives. The rocket was 21.7m tall, 1.77m in diameter and weighed 29 tonnes. Its main stage was capable of a thrust of 37,640kg, its three upper stages of diminishing thrust of 7,845kg, 2,450kg and 816kg respectively. To ensure maximum possible accuracy, the entire upper stage was set spinning before lift-off. Close examination of the launch of Explorer 1 shows the exposed upper stages (2, 3 and 4) spinning round and round during the final moments of the countdown. Stage 1 was a lengthened Redstone able to fire for 155 seconds with hydyne. Stage 2 comprised 11 Sergeants, stage 3 was three Sergeants, while stage 4 was a single Sergeant at the base of the satellite, also designed to spin. It was shipped to Cape Canaveral on 20th December. These stages fired for remarkably short periods, as little as six seconds.

Meanwhile, the launch of Vanguard drew close. The original Vanguard satellite weighed 10kg, a 51cm sphere to study atmospheric density, but it was later decided that the first test launches would fly a small 1,470g sphere for tracking purposes, with two transmitters on 108MHz, one battery and one solar panel, rather than instruments as such. An important aspect of Vanguard, one which also later became important for Explorer, was that it needed a tracking system. Up to this point, the only tracking system in operation was the optical telescopes of the Smithsonian Institution. Vanguard enlisted the help of the Army Map Service to install ten electronic tracking sets, called Minitrack, around the world, a system with low noise and high accuracy of much less than an arc second.

Original Vanguard satellite type

Vanguard made slow progress and after a summer of aborted tests, the second, called Test Vehicle 2 (TV2), launched on 23rd October as the first stage only. Two more tests (TV3 and TV4) were scheduled before an orbital attempt would be made with TV5, likely in 1958, but TV3 was then moved up as an orbital attempt for December 1957. Contrary to popular belief, the Americans never adopted an open broadcasting policy for their launches: the press simply arrived, got little cooperation and was made feel unwelcome, but there was no law to keep the journalists out.

With the launch being broadcast on nationwide radio and television, Vanguard barely rose off the ground before its thrust failed and it fell over on its back, blowing up in a giant fireball of flame and black smoke. Amazingly, the tiny satellite was thrown free, ending up on a nearby beach still beeping. But that was not where it was supposed to be beeping and the press had a field day, calling it *Flopnik, Kaputnik* and even *Stayputnik*. Secretary of State John Foster Dulles appealed to the President never to announce a satellite launch in advance again and to 'wait until we are certain that it is in orbit.' Eisenhower agreed that in future, all such unfortunate publicity should be avoided.

Vanguard crash

Von Braun and Van Allen's Satellite

The authorities now swung fully behind the long-spurned von Braun, who took his old Jupiter C, number 29, out of storage and dusted it off. But what of its payload? Van Allen had been asked to include a cosmic ray detector on Vanguard, but had cleverly designed it in such a way that it would also fit a Redstone-based satellite. General Mendaris made William Pickering, director of the Jet Propulsion Laboratory (JPL) since 1955, responsible for such a satellite and he quickly decided that Van Allen's Geiger counters should be the instrument of choice for what they considered at the time to be the new 'project Orbiter' satellite.[11] Mendaris was able to issue such a command because at the time JPL was still part of the army (JPL later transferred to NASA when it was established). Pickering already had a concept of what a JPL satellite should be like and it was rumoured that JPL had already built a small satellite and was keeping it in a filing cabinet until such time as it could be launched.[12] JPL completed the task in December 1958 and the new satellite was fitted with an internal and external thermometer (JPL), meteorite erosion gauges and a meteorite impact microphone (USAF) and a Geiger-Mueller radiation counter (Van Allen). There was such a rush to put together the instrumentation that aircraft were made available for JPL engineers to fly to and from Van Allen in Iowa. Because it was part of the army, JPL scientists had to have prior security clearance, but they had to bring in external engineers who could not be told the purpose of the project (though they figured out anyway).

When it emerged, the 14kg Explorer was 1m long, 15.2cm diameter and weighed 4.8kg, but was kept attached to the fourth stage motor, making it 205cm long and 14kg. It was a bullet-nosed cylinder with four aerials at mid-point. It was made of steel, with the outer surfaces sandblasted and coated with aluminium oxide to ensure an even temperature because the designers were quite worried that it would fail through overheating or freezing. There were two transmitters – a high powered one and a low-power Microlock one – with two antennae in the satellite body and four with weights on the side.[13] It had 10mw and 60mw transmitters operating at 108MHz. Van Allen's instrument experienced numerous calibration problems and it was not installed until the day before the originally scheduled launch.

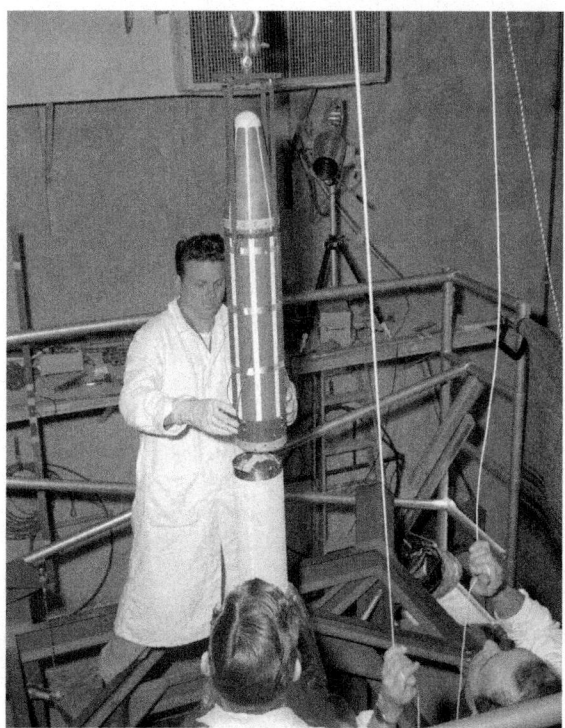

Explorer 1 lowered for installation

14   Foundations

Explorer 1 installed

Launching the Satellite

'Storage test' missile 29 was brought out to Cape Canaveral on 20th December 1957 and was immediately moved to the army's Hangar D, where integration and checking of the engines and upper stages were completed on 13th January 1958 (the next, rival Vanguard was 1,500m away in Hangar S). It was hoisted to pad 26 on 17th January. The upper stages were installed a week later and then the pyros a week after that, with all radios in the area turned off to avoid sparking an accidental explosion. Because the thrust of the solids was uneven, they were spun at 550rpm, creating 300G and taking power from electric aircraft motors. A feature of the launching system was that the upper stage firings had to be calculated for the moment of the apex of the trajectory, but commanded from the ground, with that task falling to Ernest Stuhlinger using his slide rule.

William Pickering's work on a prospective satellite at JPL in Pasadena had been known to the press since November 1957, but after the Vanguard failure the following month and in accordance with the President's wishes, its proposed launch date was classified. General Mendaris stressed that the rocket must be covered in tarpaulins and that secrecy violations would be handled severely. Decoy schemes were encouraged, the principal one being that Pickering was presenting a paper in New York. This was given a blaze of advance attention to distract the media, on the basis that he would surely be in Cape Canaveral had a launch been intended. General Mendaris issued those connected with the project with strict instructions not to talk, with staff warned not even to speak to their wives about it. They were instructed to identify the rocket only as a routine Redstone and that it must be moved to the pad under cover of darkness, so that it looked like just another Redstone test. The upcoming launch was logged as a test of 'missile 29', which it was. The only people who were given the launch date were members of the National Security Council and some senior Pentagon staff. This was a double irony, for the accepted narrative from this period was that Sputnik's launch was secret and that America's first satellite was launched in a glare of publicity: in fact, the Explorer launch was secret while there had been a crescendo of Russian publicity in advance of Sputnik's launch which nobody had taken seriously, with the honourable exception of the *New York Times*.

In reality, the story leaked and the *New York Times,* being even-handed in giving its readers advance publicity for both Soviet and American launch attempts, ran a preview. Some press – two film crews – did get to Cape Canaveral in time, but they had no facility to broadcast the launch live and get the story out as it happened. Eventually, the press was secretly informed on 24th January, but with the prior agreement and condition that no information be released until after launch, an agreement which the press kept.

It was still by no means certain that von Braun's satellite would be first into orbit because the Vanguard team was given another chance, this time with the backup vehicle from the December disaster, called TV3BU ('backup'). Cape Canaveral had allotted the Vanguard and Juno teams alternate slots for launch and tracking. Their two launch pads – 18 and 26 – were not far apart and, in keeping with the two-year rivalry between them, both teams eyed one other through binoculars. Vanguard was given the first slot of the new year, counting down on 23rd January and reaching T-9 minutes before postponement because of bad weather. On the 24th, the Vanguard count reached T-22 seconds before an umbilical cord stuck and forced a re-count, next time getting to T-14 seconds when a valve stuck. On the 26th, there was an acid leak and the count was called off. The Vanguard team was given a new slot of 3rd February.

Now the game of launch Russian roulette pointed to von Braun's team. The first attempt was scheduled for 29th January, but that day a jet stream moved over the cape, bringing with it a 300km/hr wind at high altitudes that could have blown the Juno off course or even broken it apart, so the launch was held for two days. Cape Canaveral's commander was President Eisenhower's forecaster, General Donald Yates, who had famously got it right in predicting that summer storms would briefly abate on the morning of 6th June 1944, facilitating the allied landings in Normandy, so no one dared challenge his legendary reputation. The army team wanted to try again on the 30th, but even though conditions at Cape Canaveral were calm, the hourly weather balloons sent up from there continued to report a raging high-altitude gale. The winds were no better on the morning of the 31st but, more in hope than confidence, they again scheduled launch for 10.30pm. The chief forecaster at the cape was a young 24-year old lieutenant who, in an echo

Explorer 1 on its rocket

of D-Day, predicted an abatement that evening, so Mendaris ordered the rocket to be fuelled. There was a dramatic moment when the gantry rolled back, the floodlights were turned on and, as Medaris dramatically recalled, 'the missile stood like a finger pointing to heaven.'

The 31st was probably the last chance to launch, because the cape's launch facilities had been booked for the next period for the rival Vanguard team. That evening, General Mendaris issued the order to go ahead and the service tower was rolled back 100m. At T-12 seconds, the upper stage was spun up to 550rpm. It took a full 15 seconds for the Jupiter to build up thrust on the pad, even longer than the later monstrous Saturn V Moon rocket (eight seconds), as flames lit up the pad. Launch took place at 10:40pm local time, the Jupiter soaring into the night sky and gradually turning over 40° to the horizontal. The Juno main stage burned out at T+157 seconds at an altitude of 100km where it began its coasting trajectory. Stuhlinger calculated the moment for stage 2 ignition at 404 seconds, with the second, third and fourth stages burning for six seconds each. At 6 minutes 52 seconds, there was a signal which indicated that the satellite had reached sufficient altitude and speed for orbit, but like Sergei Korolev the previous October, they would have to wait a full orbit until the satellite returned overhead above the United States. Those press who had not gone away were corralled into a theatre at Patrick AFB, so they did not actually watch the launch in any case.

Explorer 1 launch

In Orbit

Von Braun calculated that if the launch were successful then the first signals would be picked up in California 106 minutes later. Americans on the east coast would have to rely on hearing from stations on the west coast. When the Juno I disappeared into the eastern horizon at the cape, all they had to go on were doppler measurements, guidance telemetry and the observed time of disappearance which, although they appeared to be in order, did not guarantee that orbital velocity had been achieved. Signals from the satellite could be received as it headed east over Antigua, Nigeria and Singapore, but this was of no help to the anxious mission controllers because the Antiguan receiver failed due to rust. Although signals were picked up both in Nigeria and Singapore, the signal stations there had no long-distance telephone to pass the good news on to Cape Canaveral! Telephone communications capabilities at the time would now be considered antiquated.

Although he would much preferred to have been at the launch, von Braun had been sent ahead to Washington DC for a press conference and told to wait by the teletype. He recalled later that he wore a smart suit in case of success, but had brought a set of dark glasses so that he could make a quick, furtive getaway from the press if things went horribly wrong. As they waited, von Braun recalled how miserable he was, made worse by the Vanguard folk who had already hastened to commiserate. They waited in the war room of the

18 **Foundations**

Pentagon, with one of those present commenting that it felt like an election night, waiting for the results to come in.

The canny General Mendaris had made preparations for, but had not yet convened, a press conference. He had a colleague at the back of the room with an open line to the Microlock station in Earthquake Valley, California, which was waiting to pick up the first signals from Explorer as it approached the west coast. There would be no announcement if the launch failed and no one would be told. And fail it did, or so they thought, for nothing came in. Holding on a phone line to California, there were no signals. Some of the press team began to console them that 'they didn't make it this time,' while people began to ask von Braun what might have gone wrong. Suddenly, at the 114-minute point, when they had nearly given up, the tracking station reported good strong signals. Amateurs also picked up the signals, the San Gabriel Radio Club, Earthquake Valley. The only explanation for their lateness was that the satellite had gone into a much higher orbit than planned, which was indeed the case. The orbit reached was not the one intended, making the satellite eight minutes late.

Mendaris now called his press conference, giving an hour's notice for a 2am start in the National Academy of Sciences on The Mall. Pickering, Van Allen and von Braun were driven through deserted mid-winter streets from the Pentagon, with Van Allen wondering if anyone would be there. It was a wet night, but such was the crowd of press that they could

Explorer 1 press conference

Von Braun in Huntsville afterward

not get in through the front door and had to use a side entrance. They were greeted by a full auditorium as press bulbs flashed. The news took the press by storm and the 2am press conference turned to hysteria that lasted till 4am. Von Braun, Dr Pickering and Dr James Van Allen spontaneously lifted a model of the satellite aloft in an enduring photograph that forever symbolized their long-awaited night of triumph. Von Braun told the press that they had established a foothold in space and would never give it up. In staid Huntsville, there was dancing outside the courthouse. Telegrams poured in, von Braun was declared a national hero and children sent in their savings to help pay for his next satellite.

Like the Russians, the Americans had given no thought as to what to name their satellite. The Russians had simply called theirs an Artificial Earth Satellite (formally called AES in scientific papers), *Iskustvenniy Sputnik Zemli*, but the word Sputnik had stuck with the western press, so Sputnik it was. Of all the questions von Braun had expected, the one he never anticipated was the name of the satellite. Von Braun promised the reporters he would 'get back to them.' When the President was told of the launching, he thanked his informant, saying he did not want 'too much of a hullabaloo' about it. Eisenhower had the last word, though, for the naming of the satellite fell to him. Offered Missile 29, Deal, Highball and Topkick, he volunteered Explorer. It sounded perfect, offering even further distance from its military roots.

Explorer 1 was photographed from the ground on 18th March 1958, although this information and the accompanying photograph were not published until many years later. The photographer was Robert Cameron, then director of the optical tracking station at

Olifantsfontein, South Africa and who subsequently worked at Goddard. The Smithsonian had commissioned Baker Nunn tracking telescopes, with cameras, to follow early American satellites from three sites: South Africa, Spain and Iran, but the South African Baker Nunn did not arrive until a week after Explorer entered orbit. It took some weeks to set up and calibrate and, with the help of the nearby Minitrack radio station, obtained precise enough predictions to get the first photograph. Amazingly, they also spotted the later tiny Vanguard satellite.[14]

President Eisenhower

Explorer 1 transmitted for 112 days, until 23rd May 1958 when its batteries finally died. It also recorded 153 micrometeorite impacts. The satellite crashed into the Pacific Ocean on 31st March 1970 after 58,000 revolutions. As for von Braun, after Jupiter C he went on to design the Saturn rockets that took America's astronauts to the Moon over 1968–72. They remain the largest and most challenging rockets ever built in the United States to this day and, confirming his genius, none of the Saturn V's 13 launchings ever failed. In 1970, he was prevailed upon to work in NASA Headquarters, for him an unsuccessful assignment, before leaving to join the Fairchild company in 1972 and eventually dying of cancer in 1977 aged 65.

Explorer 2

With the USSR's achievements equalled and honour satisfied at last, the United States could now proceed less nervously with its space program. First, though, there were two reminders that 'launching rockets is hard.' The next Explorer, with an identical satellite, was unsuccessful on 5th March 1958, when the fourth stage failed to ignite and it splashed into the Atlantic. Extensive tests failed to reveal why, but the suspect was a faulty igniter. The next Vanguard attempt had also failed on 5th February, less than a week after Explorer 1's success. All went well at first as it soared into the night sky, but at 57 seconds a guidance failure caused the rocket to veer off course, break up and then explode.

The miserable but persistent Vanguard team finally achieved success on 17th March, the first of three to orbit Earth. Vanguard 1 is still in orbit some 50 years later and will stay aloft to 2197. Its battery powered beacon worked for 20 days, but its solar powered beacon worked for six years, giving useful information on the density of the upper atmosphere and the shape of the Earth's geoid. It was 16cm across, weighed 1470g, is the longest-standing spacecraft still in orbit and is arguably the first mini-satellite. After four further failures, the 10.8kg Vanguard 2 made it into orbit on 19th February 1959 and the heaviest, the 23.6kg Vanguard 3, on 18th September that year.[15] These were the only three successes from the 11 launches in the program.

Vanguard 2 was the first of the larger, original scientific satellites planned for the program.[16] It was intended to take the first pictures of Earth from orbit, but the final stage collided with it and sent it tumbling. The original Vanguard 3 was launched 13th April

Counting down in the blockhouse

1959, with two satellites, a sphere with a precision magnetometer to measure Earth's sphere and an uninstrumented balloon, but the second stage failed and the name 'Vanguard 3' was held back for the next launch. The next Vanguard 3 was an original satellite to measure Earth's radiation budget, but on 22nd June 1959 the second stage again failed. The third Vanguard 3 used the last rocket available and the original satellite was adapted to carry a magnetometer and other instruments. This version successfully reached orbit and provided a wealth of scientific information, a positive end to a troubled program. The upper stages were recycled to the Thor Able and Scout launchers used for subsequent Explorer missions.

Explorer 3

The same type of satellite as Explorer 1 and 2 was launched again as a U.S. Army satellite on 26th March 1958. Weighing 14kg and sometimes also called Gamma 1, its purpose was to record radiation using a Geiger-Mueller counter and micrometeoroid data using a detector. The insertion into orbit was not perfect, sending Explorer 3 as high as 2,801km but with a low perigee of 188km, enough for only 93 days on orbit. In a tumbling motion with a period of seven seconds, it lasted in orbit only until 27th June 1958. The microphone picked up only a few impacts, mostly during a meteor shower in May. The thermometers showed that the temperature control precautions were working well.

The big difference with Explorer 1, one whose importance was greatly underestimated at the time, was that Explorer 3 was the first satellite to carry a tape recorder. The humble tape recorder was actually one of the most important pieces of equipment for the new space age. Without one, tracking stations were dependent on a few minutes of signals collected whenever the satellite passed overhead and had no idea what the instruments were detecting during the rest of the orbit – in other words most of the time. Getting tape recorders right was one of the biggest challenges of the space age (those familiar with old spool tapes or cassettes will recall that they did not always work perfectly within Earth's gravity either). Essentially, space engineers had to adapt commercial industrial tape recorders and their tape, but these had never been designed for the heat or cold of space (tape can become damaged at quite low temperatures), radiation, or working in zero gravity, nor to function for many years without fixing, repairing or unjamming. With a tape recorder, Explorer 3 was able to play the tape when it passed over a tracking station and dump two hours of recordings on fast speed in just five seconds ('dumping' is the crude but technical term used). Explorer 3's tape recorder was 6.35cm wide and weighed 226g. Just as Explorer 1 had provided anxious moments as it entered orbit, so too did Explorer 3, the issue this time being the tape recorder. It was a full day before ground stations were successfully able to command the tape recorder to play back.[17] This was not the first hiccup with a tape recorder and in an attempt to get on top of the continuing problems, the Goddard Space Flight Center later convened a three-day conference of experts devoted entirely to getting recorders to work.

The radiation instrument on Explorer 1 had created something of a mystery. As it reached apogee, Van Allen's counter on Explorer 1 recorded radiation for parts of its orbit, increasing with altitude and then shutting down. An additional problem was that Explorer

1 data were limited to what was picked up over tracking stations and was not continuous. Given what they had unwittingly stumbled across, the scientists worked the data in a way that would seem leisurely now, but luckily for them time was on their side. The readings from Explorer's Geiger counter were sent on tapes by ordinary mail from the receiving stations to Van Allen in Iowa and a backlog soon built up, with each being tackled in turn by his students. Altogether, 694 tapes of Explorer 1 data were collected. It was originally thought that the Geiger counter was not working, as the instrument blanked out and counts dropped to zero. When the first tapes arrived in Iowa, it was the task of George Ludwig to listen to them. He logged the gaps in the tapes as 'no data'. However, there was clearly more to it than that, because the counts would rise and then stop, time after time, which was odd. Nobody could explain such a strange pattern. Ludwig argued that they needed complete, orbit-by-orbit data and the best way to collect that was with a tape recorder. He had been working on the idea of a recorder to store data since 1956 and he had it ready to go for Explorer 2 (which failed) and then 3.

Once again on Explorer 3, the cosmic ray experiment appeared to fail at 1,000km. The radiation counts rose, blanked out and then came on again from a lower base, leading Van Allen to suspect that such a recurrence might not be an equipment problem after all. On 2nd April, Van Allen picked up his first Explorer 3 tape recording download from the Naval Research Laboratory in Washington DC. By 3am that morning, with the pattern of 'rising counts-gap-falling counts' clearly established, he had come to the conclusion that there was nothing wrong with the instruments, but that there was a 'mysterious physical effect'. When the tape reached Iowa, it was analysed by three colleagues. On a hunch, they got out a spare radiation counter and passed it in front of an x-ray machine, replicating how high levels of radiation choked the counter. The clue was counter-intuitive: to assume not that the instrument had gone off-scale *low* but instead that it was overwhelmed – saturated – and had gone off-scale *high*. When Van Allen returned to Iowa, there was a message on his chair saying simply, but absolutely profoundly, 'space is radioactive', a real 'eureka moment'. Van Allen tested the theory in his own lab himself: saturation did indeed silence the readings.

They spread the sheets of readouts all over the physics laboratory to try piece them together. It became apparent that the satellite was entering and exiting radioactive regions that were shaped like a doughnut. Armed with graphs, Van Allen and his colleagues presented their findings at a conference in Washington DC on 1st May 1958, putting forward their conclusion that above 1,000km there was a zone of intense radiation held in place by magnetic force. Van Allen announced that he had found 'geomagnetically trapped corpuscular radiation' in a band. 'You mean like a belt?' asked a reporter. Something along these lines had been predicted as far back as 1907 by Norwegian mathematician Carl Størmer (whom Van Allen regarded as an inspiration).

The event was a sensation and led to Van Allen becoming the cover story on *Time* magazine. It was the main point of discussion at the fifth general assembly of the IGY from 30th July to 9th August that summer. In July 1959, he presented the results to a cosmic ray conference in Moscow, there meeting the USSR's top radiation scientist, Sergei Vernov. Van Allen's paper, eventually published in *Jet Propulsion 28* (September 1958), had the prosaic title of *Observation of high-intensity radiation by satellites 1958 alpha and gamma*, authored by himself, George Ludwig, E.C. Ray and C.E. McIlwain. Not long afterward, a NRL physicist was in Europe and made an offhand reference to the 'Van Allen radiation belt' and the name stuck. Within weeks, it was being called the 'Van Allen radiation belt' worldwide. It was the outstanding scientific discovery of the IGY. The United States might have started from behind in the space race, but had made the first great discovery of the new space age. Ironically, the earlier Russian Sputnik 2 had also detected the radiation belts, but lacking a worldwide tracking system and a tape recorder, the significance of the detection had not been appreciated.

Iowa improbably became the go-to place for the news media, military and visiting scientists, all amazed that such a discovery could come from such a modest place. James Van Allen, although best known for the radiation belts that bore his name, went on to a long and distinguished career in space science through the 1960s and 1970s including many Explorer missions, the last being Explorer 52, the *Hawkeye* mission (see chapter 2).

As for the famous tapes, they were piled high and lodged in the basement of the old physics building, constructed in 1910 and which had once been an abandoned particle accelerator chamber. Following bad floods in Iowa in 2008, they were beginning to accumulate humidity and the library decided to organize a rescue-and-restore mission with the help of funding from the Roy J. Carver Charitable Trust.[18]

Explorer 4

The Explorer 4 mission was concerned with nuclear tests in the atmosphere. From the *Trinity* test on 16th July 1945 onward, nuclear testing had been done largely at ground level, mainly in deserts (e.g. Nevada, Australia) or out of the way places (Novaya Zemlya). From the late 1950s, nuclear engineers were interested to learn about the effects of letting off atomic bombs high in the atmosphere. There had already been high-altitude tests over Johnson Island in the Pacific – *Teak* and *Orange* on the Redstone rocket, as noted earlier. Now the Americans planned a new series of tests, called *Argus,* the idea of elevator-engineer-turned-Livermore-laboratory-military-scientist, Nicholas Christofilos (1916-1972) (called 'the crazy Greek' in the press), who persuaded the Atomic Energy Commission that clouds of artificially created space radiation would disrupt the arming and fusing systems of incoming Soviet warheads and cause them to go off, long before they reached the ground. These tests were intended to be much higher, at 200, 240 and 540km respectively, far above the atmosphere.

Thanks to Explorer 1 and 3, the ability of satellites to measure radiation was now well known and this was an opportunity to use satellites to measure the impact of *human-made* radiation belts. When the *Argus* test was proposed, Van Allen was suggested as the man who could measure their radioactive impact. He agreed in advance, at a no-notes meeting, to build four detectors to monitor them, all to be constructed in 77 days and requiring more subtle calibration than their predecessors. The mission was put together quickly and it was not part of the original IGY program. The satellite would be put into a higher inclination orbit of 50°, compared to the 33° of earlier satellites. Explorer 4 was deliberately put into orbit in good time (26th July 1958) ahead of the blasts (due August), to measure the before- and after-effects. The 1kT *Argus* blasts duly took place in secret over the South Atlantic on 27th and 30th August and 6th September, but there was no hiding the sudden

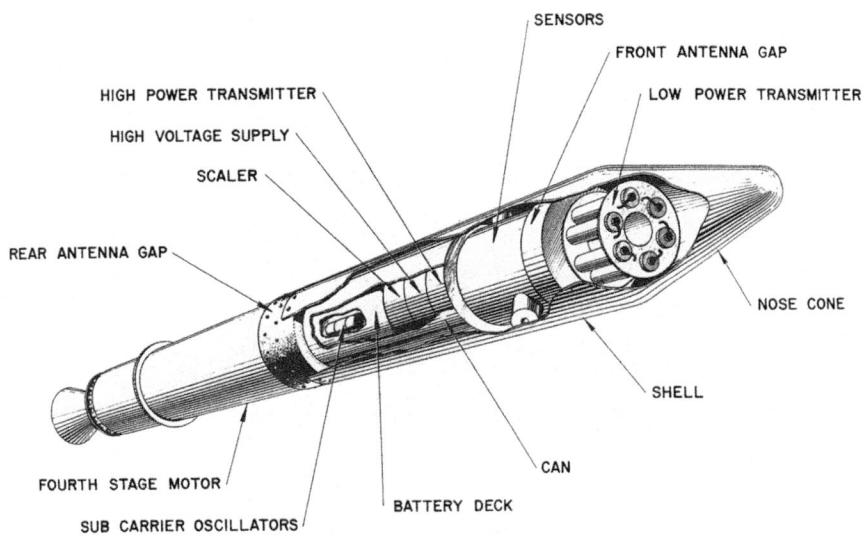

Explorer 4 cutaway

subsequent magnetic storms and auroral displays over the north and south poles. The dual nature of the mission was not revealed at the time, but when the Russians picked up the radiation from *Argus* the Americans made a virtue of necessity and de-classified the tests the following year.

Explorer 4 weighed 8kg and carried two Geiger and two scintillation counters. The counters were re-calibrated to allow for higher radiation intensities at higher altitudes, avoiding the saturation experienced on Explorers 1 and 3. The Explorer 4 tapes duly reached Iowa, where a team of 15–20 students managed by Annabelle Hudmon was ready to analyse them. Hudmon and Van Allen were aware of *Argus*, but the students were not told. Explorer 4 traversed the natural radiation belts several times a day and did the same with the new *Argus* belts. The data assembled by the students delineated the inner and outer radiation belts – and the *Argus* belt, though they did not know that. Explorer 4 quickly identified the inner radiation belt, a gap, the outer belt and the belts created by *Teak*, *Orange* and now *Argus*. The different nature of the artificial and natural radiation belts was quite evident.

Because of its higher inclination, Explorer 4 could map belts from 50°N to 50°S and was able to detect much higher radiation densities than Explorer 1 or 3.[19] Explorer 4 brought the first map of the spatial distribution of inner belt protons up to 30MeV and electrons up to 3MeV, showing complicated changes with latitude and longitude as well as local variations, for example due to the South Atlantic Magnetic Anomaly.[20] Moreover, it provided the first evidence of a larger outer belt dominated by protons, later confirmed by the Pioneer 3 Moon rocket that never reached the Moon but determined the outer belt between 14,000km and 20,000km. Van Allen and his colleagues found electrified particles – mainly electrons – trapped in Earth's magnetic field, spiralling to and between the poles along magnetic lines.

Explorer 4 launch

The first findings were presented to the American Astronautical Society on 27th December. A fortunate discovery was that Explorer revealed that the artificial radiation from the blasts decayed quite quickly, turning into a shower of particles in the upper atmosphere. The air force would have to set off atomic bomb blasts in space every month to renew them, which it knew would probably strain the limits of public support for this form of protection from the nuclear enemy (some of the concepts were re-invented as President Ronald Reagan's 'Star Wars' program in the 1980s).

Explorer 4's low-power transmitter and the plastic scintillator detector failed on 3rd September 1958. The two Geiger-Mueller tubes and the caesium iodide crystal detectors continued to operate normally until 19th September 1958. The high-power transmitter ceased sending signals on 5th October, probably due to battery failure and the satellite fell out of orbit on the 23rd.

At the end of Explorer 4, Presidential Science Advisor James Killian sent a memorandum to the President.[21] This summarized the early scientific results of the Explorer program. Killian told President Eisenhower that the principal results had been in the area of radiation, but that outcomes had also been achieved in micrometeorite impact as well as temperatures and air density. The principal surprise was the intensity of radiation above 990km, rising by a factor of several thousand from 0.002 roentgens/hr to more than 2/hr., up to 2,400km. This had implications for the need to shield space travellers, as well as for the heating of the high atmosphere, the amount of visible light, radio noise and ionization. As for meteorites, Explorer 1 registered only one in 32 days, too limited a sample for statistical analysis. Air density fell from $23,547 \times 10^{-14}$ kg/m^3 at 176km, to $1,409 \times 10^{-14}$ kg/m^3 at 370km.

Explorer 5

The army had two more Explorers left. Explorer 5, identical to 4 and announced as a 17kg magnetospheric satellite, launched on 24th August 1958, but the first stage collided with the upper stage and sent it spinning in the wrong direction. It was also supposed to monitor the three *Argus* blasts over the South Atlantic, so the air force had to rely solely on the data of Explorer 4. The last army launch, on 22nd October 1958, would have been called Explorer 6, the payload being a balloon, but the spinning upper stage vibrated to the point that it broke free. It was not given a name, for the new policy was not to give a name to failed launches. This was the last Juno 1 rocket and the 'real' Explorer 6 flew almost a year later. By then, however, the program was no longer U.S. Army, but NASA.

Space Science in the New NASA

Explorer was the American response to the challenge of Sputnik, but it did not take American political leaders long to realize that matching the Soviet achievement was only the beginning. Sputnik provoked a sense of crisis about America's leadership in the world, its technological supremacy and whether it was investing sufficiently in science and engineering. A properly organized space program was called for. From the point of view of this narrative, the critical issue was: what role would space science in general (and the nascent Explorer program in particular) play in this new organization?

A first issue was whether there should be a new organization responsible for space; and the second, connected issue was whether it should be military or civilian. Given that both Vanguard and Explorer had connections to the military (navy and army), this was an important decision. Van Allen – who had himself served in the navy – and his colleagues had argued for a full, civilian space agency as soon as the first Explorer reached orbit, the obvious option being to upgrade the National Advisory Committee for Aeronautics (NACA), which had been founded in 1915 to support the burgeoning aircraft industry. NACA's director, Dr Hugh Dryden, welcomed the opportunity to build on the organization. He brought a proposal directly to senate leader Lyndon Johnson in February 1958 which he quickly translated into the National Aeronautics and Space Act, 1958. Senate Bill 3609 was amended and signed by the President on 29th July 1958. The new organization would be called the National Aeronautics and Space Administration (NASA), retaining the aeronautics brief. It would become one of the most familiar acronyms not just in the United States, but worldwide.

Dr Hugh Dryden

Prizing the space program away from the grasp of the feuding military was the subtext. De-militarizing the space program accorded well with the instincts of the President who, though he was a military man himself (he was normally announced in the media as 'The President, General Eisenhower'), remained suspicious of over-weaning military power. Although other options for the space program were considered, including military supervision, no one in the Congress defended or preferred a military-led program. Eisenhower had always been wary of the 'military industrial complex' and it was ironic that it took a military man to civilianize space. In a second irony, the man who was not interested in space set up the infrastructure that later conquered the Moon and the solar system.[22]

Having established that the new space program would be civilian, the first decision was to address the importance of space science. The Rocket and Satellite Research panel, chaired by Van Allen, proposed immediately creating a 'national space establishment' whose aim was an intensified program for scientific and technical developments with small instrumented satellites of the Earth. This panel was the re-iteration of the Upper Atmosphere Research Panel that had overseen the A-4 launches from 1946. On 9th December 1957, W.W. Kellogg of the RAND Corporation sent his proposal *Basic objectives of a continuing program of scientific research in outer space* to members of the rocket and research panel.[23] Kellogg proposed the use of satellites in the 20–30kg range to study the ionosphere and to observe photons, particles, magnetic fields, Earth's radiation budget, meteorites, solar ultraviolet and x-ray radiation, the distribution of hydrogen in space, auroral particles and extragalactic and cosmic rays. Heavier and more advanced spacecraft could make astronomical spectrograms, test the theory of relativity and detect high frequency cosmic noise. Weeks later, the panel proposed *A national mission to explore space* with *A national space establishment* (27th December 1957), 'concerned with such questions as the properties of the upper atmosphere, the nature and intensity of electromagnetic and corpuscular radiations from the Sun, the character and distribution of matter in space and the electric, magnetic, and gravitational fields' with 'an intensified program of scientific and technical developments with small instrumented satellites of the Earth, immediately.' This, in effect, became the definition of the Explorer program.

Another early intimation of official thinking came from the President's Science Advisory Committee, convened at the end of 1957 to recommend the direction and pace of the space program. Called *An introduction to space – a statement by the President* (26th March 1958), their report listed the first objective of the space program as exploration and discovery; 'to go where no one has gone before' (years before *Star Trek* popularized it), coupled with scientific observations that would add to our knowledge of the universe. Scientific purpose came first; 'national prestige' came third. Top of the list were physics, geophysics and meteorology, with manned flight coming in last.

Priorities were then considered by NACA, which established a Special Committee on Space Technology under General James Dolittle in spring 1958. The committee included Wernher von Braun, Hugh Dryden, Abe Silverstein, William Pickering and James Van Allen; all the leaders of the early American space program. The committee issued its *Recommendations to NASA regarding a national civil space program* on 28th October 1958. This set the agency's major objectives as scientific research in physical and life sciences, technology, manned spaceflight and human benefit. Instruments would observe and measure the geophysical and environmental phenomena of the solar system, cosmic processes, the atmosphere and the space environment. The first objective, *Scientific research*, explained how it was now possible to get above the atmosphere to study radiation and particles, thanks to the excellent start that had been made during the IGY. Scientific excellence was now evaluated worldwide in terms of success in the exploration of space and the U.S. should now achieve and maintain an 'unselfish leadership'. The priorities of such research should be gravitational and magnetic fields, geodesy, geophysics, radiation, meteorites, the ionosphere, the effects of weightlessness and the upper atmosphere.

This document undoubtedly gave a high priority to space science: 'instruments mounted in space vehicles can observe and measure geophysical and environmental phenomena in the solar system, the results of cosmic processes in outer space and atmospheric phenomena, as well as the influence of the space environment on materials and living organisms. A vigorous, coordinated attack upon the problems of maintaining the performance capabilities of man in the space environment is prerequisite to sophisticated space exploration.'

It proposed a continuation of the work begun by the IGY. This it elaborated in a subsequent paragraph on the Sun-Earth system, radiation, geodetic and geophysical studies and deep space observations, proposing a series of satellites in the 14kg, 140kg and 1,400kg class. This was published as the *Report of the working group on space research objectives*, on 14th November 1958, now of course under the rubric of NASA. This is worth reproducing in detail, since it indirectly became the work program for the Explorer series:

- Non-directional monitoring of radio-frequency radiations, especially those wavelengths absorbed or reflected by the atmosphere.
- Observations in the ultra-violet region of the spectrum: wavelengths 1,000 to 3,000Å, the astrophysically important lines.
- Exploratory ultra-violet observations in the far ultra-violet region of the spectrum, wavelengths less than 1,000Å.
- Studies of auroral radiations and of the interplanetary plasma.
- Cosmic ray exploration.
- X-ray exploration.
- Gamma ray exploration.
- Extragalactic radiation.
- Observations on meteors, particularly meteor showers.
- Magnetic field measurements (scalar magnitude).
- Measurements of radiation energy balance of the Earth.
- Observations of cloud cover.
- Measurements of atmospheric density.
- Measurements of refraction of radio waves by the ionosphere.
- Experiments with powered communications repeaters (10 kc/sec band width).
- Biological experiments.

The launching of one satellite per month was suggested, with the number of experiments on each mission kept to a minimum in the interests of simplicity and reliability. The report added that it would be 'wise' to continue with a program of small satellites of this class even as larger ones became available.

Meanwhile, the technical panel formed for the IGY in July 1955 issued a progress report in 1957 under Homer Newell, listing future priorities as geodetic, air density, the ionosphere, cosmic rays, solar radiation in the Earth's environment, the magnetic field and radiation balance.[24] Then, the National Academy of Sciences created a Space Science Board (SSB) on 4th June 1958 to continue the satellite work of the IGY, directing it to assess priorities for future space research and meeting several times that summer in New York. Lloyd Berkner, President of the associated universities, sent out a telegram on 4th July 1958 on behalf of the SSB, inviting proposals on space science. It was circulated to 150 people, but more than 200 responded. Young scientists spoke later of how this unexpected invitation began, for them, a lifetime career in space exploration. It also had the indirect effect of mobilizing the space science community. The SSB sifted and evaluated the proposals and they became the basis for the recommendations made that autumn, presenting them on 1st December 1958 as *Recommendations of the Space Science Board for space experiments* and suggesting missions in the areas of astronomy, the theory of relativity, ionospheric physics, fields and particles and meteorology. In effect, it had turned the 200 proposals into 30 missions over a decade. These academic scientists became the lobby to the Congress for the space science program. In effect, the scientists took advantage of the discovery of the radiation belts to argue that the new NASA should be guided by the Academy of Sciences Space Science Board in a formal advisory role, giving scientists a direct role in the selection of instruments.

Space Science in Retreat?

NASA was established on 1st October 1958 and was based around what are called to this day 'field centres'. The first of these was William Pickering's JPL, taken from the army. The second was von Braun's rocket team in Huntsville, despite howls of protest from the army which delayed the transfer and only abated when it was renamed the Marshall Space Flight Center after General Marshall. As for the navy, the 160 scientists of the Naval Research Laboratory in Beltsville, Maryland, the home of Vanguard, formed what was later called the Robert Goddard Space Flight Center.

According to Homer Newell, space science occupied a favoured position at the formation point of NASA and this is obvious from the presentations made by various boards and panels over 1957-8.[25] The NASA Act even made space science one of its major activities. There had been enthusiasm for the IGY, Sputnik had shown up American under-investment in science and a civilian space science program fitted in well with President Eisenhower's preferences. Hugh Dryden's organigrams for NASA had an Office of Space Science, a senior (associate) administrator reporting to the administrator and a defined role for the scientific community as a whole and the SSB in particular. Dryden had been very much involved in the creation of the SSB, so it was reasonable to presume that, with their guidance, space science would be the top priority of NASA.

But the 'favoured position' of science quickly unravelled. NASA came under pressure to move fast to match the USSR, especially in the next anticipated goal of the space race, manned spaceflight. This set the scene, as Newell presciently analysed later, for the battle between NASA and the scientific community over the value of the small and less costly projects versus the large and expensive ones. Alert to this, there was a concern, in the National Academy of Sciences in general and the SSB in particular, that space science would be shut out by an agency focused entirely on manned flight; indeed, that was one of the reasons that prompted the formation of the SSB.

Although the NASA Act made space science one of its major activities, it also gave its Administrator considerable freedom of action to interpret its mandate and did not make provision for any formal advisory group or input from the science community. This made the choice of first Administrator crucial, since that would determine the culture, atmosphere and priorities of the new NASA, as the first such choice determines any new institution. On 8th August 1958, General Eisenhower appointed not NACA director Hugh Dryden, as anticipated, but Keith Glennan, a Republican, aggressively competitive with the USSR and a champion of manned flight. Positively, Glennan greatly respected Dryden and brought him in as his deputy (the loyal Dryden kept his disappointment to himself). In the new Glennan organigram issued before the end of August, Dryden's Office of Space Science had been replaced by Space Flight Programs, which went to Abe Silverstein, a propulsion engineer from NACA Lewis. There was now no liaison with the scientific community, something which Homer Newell pointed out was likely to cause distress therein.

Glennan made it clear that his first priority was manned spaceflight, which was signalled by NASA formally approving the Mercury program in its first full week, on 7th October 1958. Glennan abruptly changed the new organization from one focused on space science to one devoted to a massive engineering project, manned space flight.

Abe Silverstein

Moreover, Glennan also made it clear that NASA would decide the space science program, making the role of the SSB essentially advisory and distant. He also fought off the idea of any kind of broad advisory committee to NASA. The scientific community was very much put in its place, sparking a letter by John Simpson of the University of Chicago to the Associated Universities on 30th July 1959, questioning whether NASA's goals would effectively strengthen science and would make the most effective use of American scientists. When it came to experiments, he was unclear as to whether scientists would be doing their own experiments or purely be technicians supplying equipment. Possibly in response, Hugh Dryden wrote to the SSB on 20th October 1959, informing them that he would like to have its continuing input of thoughts, ideas and recommendations. This was a long way short of the relationship the scientific community would have liked and reflected poorly on the thoughtful suggestions it had put forward for prospective space science programs. Worried about what had happened to space science, NRL's Homer Newell went to Abe Silverstein.

Keith Glennan (left) with Lyndon Johnson

The largest group of scientists transferred directly to NASA was in what had been the NRL, so its fate was critical for the future of space science. Although the NRL was the *Naval* Research Laboratory and formally a military institution, its staff mainly comprised civilian researchers with university backgrounds and ambitions in scientific research. Homer Newell re-argued his case in his memorandum *The United States program in space research* on 27th October 1958, identifying the principal areas of scientific interest as being atmospheres, ionospheres, high-energy particles, fields, photons and astronomy. In the event, Silverstein responded positively, appointing Newell as assistant director for space science and some of his science colleagues as program chiefs. Between them, they smoothed the way for the transfer of other scientists into NASA. Silverstein was in a hurry and was not interested in waiting around for lengthy discussions with the SSB. His decision to appoint Newell may well have been critical in protecting the role of space science.

A year later, NASA effectively responded to Newell's memorandum in its *Long range plan* of 16th December 1959, which started with the objective of 'the expansion of human knowledge of phenomena in the atmosphere and space', but listed the role of the United States as a world leader in space exploration a lowly fifth. Space science was allocated a budget in 1960 of $82m, set to rise to $300m by 1969, the highest priority after manned space flight. Space science had effectively survived and had protected its position in the new NASA.

NASA now re-drew its relationship with the scientific community, at first sight downgrading its role. This critical period is examined in some detail here, as it set the agenda for the Explorer program for many years, arguably to the present day, as well as the critical issues of *Who decides?* and *How?* Silverstein set up a Space Science Steering Committee (SSSC), which first met in February 1960 and, with Newell, set down procedures whereby all soliciting, receiving and evaluation of proposals must go through the SSSC at NASA Headquarters. Within two months, the SSSC issued Technical Management Instruction 37-1-1, *Establishment and conduct of the space science program – selection of scientific experiments* on 15th April 1960. This effectively remains in place.

According to John Naugle's analysis in *First among equals – the selection of NASA space science experiments*, although 37-1-1 appeared long, cumbersome and unlikely to work, it was extremely important, a sound document and a road map through an uncertain future. This was the bible from now on. The director of space flight programs appointed a space sciences steering committee and its six scientific subcommittees. Scientists would make their proposals, which would go to the appropriate committees but also be sent to the appropriate field centre for comment (which for the moment meant JPL or Goddard). Their deliberations would then go to the steering committee for recommendation and the director would make the final decision over missions, experiments and experimenters. In other words, there was going to be strong central control, doubly ironic since the supposedly centralized competing Soviet space program was much looser.

Document 37-1-1 had the merit of ending the process, going back to the V-2 panel, whereby scientists adjudged their own proposals. It set down clear dividing lines, roles and responsibilities. This was not as easy as it sounded, as it challenged NASA to find scientists for these subcommittees who had sufficient knowledge to evaluate proposals, because most such scientists were likely active and potential competitors. 'Retired' space scientists must have been difficult to find in the early 1960s. One merit of 37-1-1 was that with all committee members identified, external critics could quickly pounce on any potential conflicts of interest, which kept the system honest. According to Naugle, from 1962 this selection process 'began to work'. Decisions that had made their way through the process were rarely, if ever, overturned or subsequently interfered with. Historian David Portree makes the point that for all the criticism of Eisenhower, he put in place the system that made NASA work in general and space science in particular, before he left office.[26]

Document 37-1-1 was refined over time. The practice arose whereby NASA formally announced plans for new missions, so that all scientists could present proposals from an equal playing field, later called a 'fairness doctrine'. Originally, they were called Announcements of Flight Opportunities (AFOs), but the 'F' part was later dropped. This process was not always followed at the start, with the experiments on the Interplanetary Monitoring Platform (IMP) being effectively presented as an in-house *fait accompli* with no AFO. The IMP series was a spectacular success, but it was the last one issued without an AFO. Over the years, the process became more bureaucratic, requirements to prevent potential conflicts of interest more rigorous (arguably self-defeating) and the process of preparing a proposal became almost more demanding than carrying out the mission itself. There was a further substantial refinement of the process during the period of 'faster, better, cheaper' (see chapter 4). The principles set down, though, were important.

But where did this leave the science community in general and the SSB in particular? The first discussion as to the respective roles of NASA and the SSB occurred early enough, during NASA's first month at the third meeting of the board on 25th October 1958. The board considered but did not approve a proposal that it should effectively plan NASA's science program, but one of its members warned of the conflict of interest involved if the scientific community were to arbitrate which of its members were to fly and who did not. NASA, for its part, barred its scientists from membership of the SSB to avoid such a conflict of interest.

In December, the trio of Newell, Silverstein and Glennan agreed between them that the space science program would be planned and executed by NASA. This was formally approved on 15th December. NASA would welcome recommendations from the SSB, but would set priorities and schedules and select the scientists and experiments. Newell did not want the SSB involved in day-to-day supervision of the program. Thoughts, ideas, recommendations and strategic suggestions were fine, but not guidance or participation. Any results would be published by NASA, or in open scientific literature in accordance with accepted scientific practice. Newell believed that Silverstein considered creating a NASA space science journal but this did not happen, although NASA did develop its own in-house special publications, regarded with disdain by some in the scientific community. Scientists generally published across a broad range of scientific journals, both domestic and international (e.g. IAF, COSPAR), which was good for openness and diversity but made keeping track of outcomes difficult in comparison to, say, the Soviet program, where the main outcomes could be found in a smaller number of places, principally *Kosmichesky Issledovatl* (its English language iteration was called *Cosmic Research*).

The National Academy of Sciences and its SSB continued in denial, not realizing the degree to which the door had closed on its role in deciding space science programs. The National Academy of Sciences, the parent body, continued its self-assigned task of defining the American program of space science in *Introduction to the United States space sciences program*, on 12th March 1959. The SSB continued to invite proposals through 1959 for a ten-year space science program, along with payloads and scientists responsible. The seven headings were atmospheres, the ionosphere, energetic particles, electric and magnetic fields, gravitational fields, astronomy and biosciences, each divided into immediate and long-range objectives. These priorities did inform both the Explorer program and others over time.

Not until 1960 did the SSB realize that it was not in the driving seat. The SSB closed its seven subcommittees, but assigned itself the new task of evaluating NASA's performance. An initial battleground was the early post-army Explorer missions developed by Goddard, before 37-1-1 came into effect. Goddard had selected its own in-house scientists and appeared to offer no opportunities to the rest of the scientific community. This was partly in the interests of speed, but also in the absence of any other procedures. The SSB described this as a 'vicious practice', but rather than complain directly to NASA the board worked around it, instead going – albeit unsuccessfully – to the President's Science Advisory Committee. The SSB lacked the stamina to continue this battle, not least because many of the leading space scientists actually joined NASA, especially the younger ones. Many of those involved in the IGY, V-2 and Vanguard programs went to Goddard, where the Explorer program developed.

Wernher von Braun leaves NASA

The relationship between the SSB and NASA evolved into one that was long and productive but punctuated by acrimonious incidents. Newell called it a 'love-hate' relationship. He also took the view that because the Academy had done so much to shape the priorities of the IGY and space science in the mid-1950s, it took an over-proprietary view of space science. Newell's view was that since NASA was, in law and in practice, responsible for space science outcomes, it could not reasonably hand over responsibility to an external agency.

In the battle between the scientific community and the new national institution, NASA quickly won the opening, decisive skirmish. NASA in effect inherited a space science program from the IGY, V-2 panel and Vanguard groups and did not need to invent one. NASA developed a set of working papers on a space science program early in 1959 and published *The United States National Space Sciences Program* which listed atmospheres, the ionosphere, energetic particles, electric and magnetic fields and astronomy as priority areas of research within the NASA space science program. According to Newell, although the program was recast and re-titled from time to time, its substance remained remarkably stable. Biosciences were later added in 1959 and aeronomy (a Sydney Chapman phrase) in 1960. None of this was significantly different from what the Academy and SSB had argued for.

The final policy statement of the Eisenhower administration, *U.S. policy in outer space* (26th January 1960), made the matching of Soviet efforts the principal pre-occupation, identified key areas of scientific research as the atmosphere, ionosphere, energetic particles, electric and magnetic fields, gravitational fields, astronomy, bio-sciences and geodesy and attempted to re-assure the scientific community of the continuing importance of space science.

Enter President Kennedy, Administrator James Webb

These priorities were to be quickly tested. President-elect John F. Kennedy soon received advice that the space science priority of the Eisenhower period should be maintained. Advisor Richard Neustadt wrote him a memo entitled *Problems of space programs* on 20th December 1960, warning him that the program was being dragged into achieving 'Sputnik-type firsts' to beat the Soviet Union, absorbing resources at the expense of everything else. The chairperson of Kennedy's transition team on space, Jerome Wiesner, sent him the *Report to the President-elect of the ad hoc committee on space* on 10th January 1961, which re-iterated the importance of space science. Wiesner recited American achievements in space science, such as the radiation belts and astronomy, where the country now held leadership. America's best scientists had participated in the IGY and the U.S. had advantages in instrumentation that offset disadvantages in propulsion. Wiesner appealed for a higher priority for space science and made the observation that 'we must not delude ourselves into thinking that it will be easy for the USA to maintain in the future a prominent position in space science. The USSR has a number of competent scientists. It will be easier for them to catch up with us in instrument development than for our engineers to catch up with the Russians in the technique of propulsion. Thus, we must push forward in space science as effectively and as forcefully as we can.'

The conventional narrative is that American space resources were irrevocably committed to the race to the Moon at the expense of all else – such as space science – from the time of the President's address to the Congress on 25th May 1961. President Kennedy's memorandum for the Vice-President, dated 20th April 1961, had asked the question of what space program promised 'dramatic results which we could win' and considered such alternatives as space stations. The impression created is of an abandonment of a 'rational' model of a space program prioritizing scientific knowledge, in favour of the political imperative of defeating the Soviet Union in space. It is important, though, to remember that Kennedy sought budget allocations for the Moon program 'above and beyond' the space spending already approved by the Congress, which would have included the science program.

President Kennedy's first decision on the space program was an important one: Who would be the new administrator for NASA? James Webb's arrival on 1st February 1961 was greeted with apprehension, for he was a southern Democrat with no known interest in space science and certainly did not come from the north-eastern elite. Webb (1906-1992) came from Tally Ho, North Carolina and had worked small-time, small-town jobs. His father was a schools' superintendent and his mother shared in the ideals of education.[27] After a spell in the Marine Corps, Webb worked in the Congress. He was an admirer of the New Deal and became a skilful administrator for government and the private sector with an interest in technology.

James Webb

Webb surprised the space scientists in November 1961, however, by replacing the Office of Space Flight Programs with the Office of Space Science, reporting directly to the Administrator. He put Newell in charge, later giving him the direction of both Goddard and JPL. This arrangement has remained in place ever since, except for the years 1974 to the early 1980s, though it was retitled as the OSSA, or Office for Space Science and Applications, in 1967. Reared in the tradition of the New Deal activist role of government, Webb began a program of engagement with the universities, providing funds for grants, fellowships and laboratory equipment to test instruments and prepare proposals. In the course of his field tours, he would often to call in to universities unannounced, their scientists and administrators at first puzzled and unused to such high-level attention. Established universities and field centres were suspicious. He went to out-of-the-way locations, urging these universities on to make ever more ambitious contributions to the space program, to the consternation and distrust of senior, traditional scientists. He also tried to get different disciplines together in the cause of space science, long before it was fashionable. The engagement with the universities helped build the NASA constituency. Webb's arrival led to a considerable expansion of the space science program. Indeed, when the President suggested that some money be transferred from space science into Apollo, Webb resolutely refused. He always argued for and would present a 'balanced' program – one which absolutely included space science – and would then argue the merit and demerits of each aspect with the Congress and other players.

Webb's program of engagement with the universities had an impact that is still evident to the present day. The Sustaining University Program (SUP) was very much his invention and was rooted in his own interest in education, as well as a desire to spread its benefits beyond the elite colleges of the north east and end the way in which people who had to settle for southern universities felt second class. Webb sneaked SUP through the NASA budget – with Kennedy's blessing – and by the mid-1960s had a budget of $25m. Although tiny compared to the overall NASA budget, 4,000 new graduates received $6,000 each in a system of awards, fellowships and grants distributed to all the universities, who were encouraged to go out and entice talented students and professors. He reversed the logic of students going to colleges with one of colleges seeking out students. There were also bricks-and-mortar grants to the universities to encourage them to found space science laboratories, so that they would not just do theoretical work but get their hands dirty by designing, making and building instruments, or even entire spacecraft. These universities were expected to build links with private technology companies in their neighbourhood or state and were also expected to be multi-disciplinary, 'getting the physicists and astronomers and chemists together,' as Webb once said.

This policy had several outcomes. First, Webb built up a big constituency of individuals and universities who had a vested interest in the space program (and made them less likely to criticize). Second, he paved the way for their direct involvement in conceiving,

President John F. Kennedy and James Webb

proposing and managing space projects, as became ultimately and especially evident in the Explorer program. The involvement of universities with the program to this very day may be ultimately attributed to Webb's interest five decades ago. Third, he decentralized excellence, attracting talented students into the space program from all over the country and ensuring that out-of-the-way places, including the south and places heretofore considered as educational backwaters, could become contributors to such missions (the location of participating scientists now is good evidence of this).

Although the SSB had been put back in its box, James Webb engaged with the board to encourage it to fulfil the role of providing long-term strategic advice. The SSB organized a 'summer study' on space research in Iowa in summer 1962, involving a hundred participants in a broad range of disciplines, together with NASA staff. This began a process of annual reviews whereby the SSB provided useful advice and suggested priority missions and others of lesser importance. By the end of 1962, this pattern of decision making for space science had begun to settle down and normalize.[28]

James Webb wrote to the President on 30th November 1962, stating: 'Space science includes the following distinct areas: geophysical, solar physics, lunar and planetary science, interplanetary science, astronomy, and space biosciences. At present, by comparison with the published information from the Soviet Union, the United States clearly leads in geophysics, solar physics, and interplanetary science. Even here, however, it must be recognized that the

President John F. Kennedy and Wernher von Braun

Russians have within the past year launched a major series of geophysical satellites, the results of which could materially alter the balance.' Webb justified an expansion of the space science program, on the grounds that 'a broad-based space science program provides necessary support to the achievement of manned space flight leading to lunar landing,' especially the radiation belts through which astronauts would have to travel to reach the Moon.[29]

Few Americans made comparisons to the Soviet program of space science, so Webb's letter was unusual in doing so. As far back as 1958, in *A national space program for the United States* presented on 26th April, Hugh Dryden, then director of NACA, had commented approvingly on Soviet aims in geophysics, meteorology, physics and astronomy, Earth's atmosphere and cosmic rays, noting their similarity to American objectives.[30] Although early Soviet satellites had scientific objectives (Sputnik 3 of May 1958 had an impressive list), the USSR did not formally articulate a comprehensive space science program until the introduction of the Cosmos program on 16th March 1962. This listed such objectives as the exploration of the upper and lower atmosphere (including cloud formation); measurement of charged particles and their concentration; detection and measurement of cosmic rays and corpuscular fluxes; study of the Earth's magnetic field; measurement of short-wave emissions from celestial bodies; propagation of radio waves in the atmosphere; the measurement of Earth's radiation belts; solar radiation; meteorites; and the study of the ionosphere.

The first round of the Cosmos series followed a program of activities similar to the early Explorer: meteorites; the magnetosphere and radiation; ultraviolet and gamma radiation; and the magnetic field. Both programs had similar starting points and similar states of knowledge, so the close comparison between the two is unsurprising. The principal difference was that although the USSR used solar cells from early in its space program, most early Cosmos missions were battery-powered and tended to transmit intensively for short periods, in contrast to Explorer where solar-cell-based operations lasted much longer. This led some to undervalue the information that was returned. NASA itself was initially dismissive of Soviet space science. The report *Evaluation of USSR vs U.S. output in space science*, which had been requested by the Congress, concluded that nearly all the original work in space research had come from the U.S. program, that the U.S. had published many more papers and that the USSR had 'done very little in space science;' the start of a paper of withering criticism.[31] This was uncharacteristically unfair, for Soviet achievements in spaceflight, both to date and to come, were impressive. However, they suffered, in the Cold War and beyond, from the increasing distance between the two superpowers, in which the circulation of information on Soviet space outcomes in the west diminished almost to the inconsequential.[32]

Goddard: From Agricultural Research to Space Research

One of the most important places in the story of Explorer is the Goddard Space Flight Center (GSFC), which became the home of the Explorer program.[33] It was named after the pioneer of modern rocketry, Robert Goddard (1882-1945), who flew the first modern small rocket on 16th March 1926. Physically it was not so far from NASA Headquarters in Washington DC, at the time temporarily located in a private building coincidentally but fittingly called the 'Cosmos Club'.

When NASA was formed on 1st October 1958, it took over several NACA facilities: Wallops Station, Virginia; Langley Research Center, Hampton, Virginia; Lewis Research Center, Cleveland, Ohio; Ames Research Center, Moffett Field, California; and the Flight Research Center at Edwards in California. As William Pickering, Director of JPL, pointed out to Keith Glennan on 24th March 1959, NASA was 'an amalgamation of a number of pre-existing programs and facilities.' Some of them had existed for some time (Langley dated to 1920) and all guarded their independence jealously. JPL and NRL were not NACA, but belonged to the army and navy respectively. Space science was sub-divided between lunar and interplanetary (which went to JPL) and the rest (which went to Goddard). Goddard built on a number of institutional features of the old NACA, specifically that it attracted competent personnel, had good research facilities and believed in strong in-house capacity.[34] Harry Goett from NASA Ames became the centre's first director in September 1959.[35]

Goddard was also chosen as the location for manned spaceflight and was originally the location of the Space Task Group (STG) which did the groundwork for the first U.S. manned space flight. In 1960, Goddard sketched the original design for a project APOLLO (upper case letters used) to take over after Mercury and bring crews to orbiting space stations, long before Apollo became the Moon program. Because Goddard did not yet have the room for the STG, it temporarily worked from Langley with a view to moving later to Goddard in Maryland. Mission control for the Apollo missions was originally planned to be in Goddard, with building 8 even intended for that purpose.

In August 1961, NASA Headquarters decided to establish a dedicated, permanent site for manned spaceflight, with 20 possibilities considered. In effect, this meant that it would move out of Goddard. Although a number of candidate sites came from California and Louisiana, most of those involved in the selection came from Texas, so it was no surprise that it ended up in that state (Houston). Some personnel did not want to move, so they either stayed in Langley or transferred to Goddard. In the meantime, mission control for the Mercury program remained in a blockhouse at Cape Canaveral (ignominiously bulldozed in 2016) before Houston took over from the Gemini program onward.[36]

Nevertheless, Goddard did play a key role in the manned program which is often forgotten. As the Manned Space Flight Network (MSFN), it hosted the communications from the Mercury program to the end of the Shuttle era. This was because Goddard inherited the Minitrack network built for the Naval Research Laboratory for Vanguard and the IGY, which evolved into the Satellite Tracking and Data Acquisition Network (STADAN) and the MSFN with stations worldwide.[37] All communications from Mercury to Apollo were actually routed through Goddard-based systems, although this was not well known.[38] Although Langley devised the global network of 18 stations for the Mercury program, the tracking, computer measurement systems and signals were all routed through Goddard, which was equipped with the two of the most powerful computers of the day, the legendary IBM 7090. Goddard was proud of the key role it played in tracking American astronauts in orbit, from John Glenn onward. Goddard was always swift to point out that the landing on the Moon was not broadcast direct to Houston, Texas, but relayed first of all to Goddard's three prime sites (Goldstone, Madrid and Honeysuckle Creek), then onward to Houston and the rest of the world: 'We got the signals from Apollo 11 first. Before Houston, before everyone else.' Goddard also handled the early Shuttle missions, until signal transmissions moved over to the on-orbit Tracking and Data Relay Satellite (TDRS) system. As the TDRS system neared completion, Goddard began to close ground stations, keeping some open for Explorer satellites such as IUE, IMP 8 and COBE. TDRS had its own command centre in Las Cruces, New Mexico, opened in September 1987, which is still within Goddard's jurisdiction. Goddard's Space Network continued to run the TDRS beyond the last Shuttle mission (21st July 2011) and was the main link for the International Space Station to the ground thereafter.

Goddard was NASA's first all-new field centre. NASA's first deputy administrator Hugh Dryden was aware of surplus government land near to the Beltsville Agricultural Research Center at Greenbelt in Maryland and, believing that many of the Vanguard NRL personnel lived in Maryland, thought it might suit them (in fact, most of them lived in neighbouring Virginia). Greenbelt was an unusual town, one of three model communities formed by the Resettlement Administration in 1935; its sister towns being Greenhills, Ohio and Greendale, Wisconsin. They provided low-cost housing, gardens and cooperatively-owned community facilities for unemployed people and families on modest means, protected against encroachment by a surrounding greenbelt on which further building was prohibited. Inspired by Eleanor Roosevelt, these green towns were crescent-shaped and surrounded by forests and fields, with homes focused on parks and playgrounds, forsaking the normal, squared grid street pattern. Around 450 Goddard employees settled in there straightway. In 1964, though, three highways were driven through Greenbelt, dividing the town into five sections. The population grew to 21,000 and traffic congestion became a significant problem. Some people continue to live happily in the original homes.[39]

Greenbelt under construction [credit: Farm Security Administration]

In no time, Senator Glenn Beall of Maryland proudly announced that the new outer space agency would establish a laboratory there, bringing 650 new jobs to the state. NASA convened a meeting to design what was then called the Beltsville Space Center on 12th February 1959 and gave it five designated functions: planning, management, research, fabrication and operations. It was charged with developing an active space science program, communications and meteorological and geodetic satellites. The 223 hectares allocated – wooded, swampy and crawling with snakes – were formally taken in charge and construction began of 'building 1', which opened in September 1960. Eight buildings were to be built for $27m, with a centre complement of 2,000 people (forty years later, its numbers had climbed to 11,800 in 30 buildings).

On 1st May 1959, NASA Administrator Keith Glennan announced that henceforth it would be known as the 'Goddard' Space Flight Center. The site was formally dedicated on 16th March 1961, in cold, sleety, windy weather which failed to put off the crowds, with Robert Goddard posthumously awarded the Congressional Medal that was presented to his widow Esther. It was a big day for the centre, with speeches and a press tour (taking in the assembly process of the S-3 energetic particles satellite), with the music provided by the University of Maryland band.

The early days of Goddard were not for the faint-hearted. NASA's entry pay rates were low compared to elsewhere, the prospects were uncertain and the early office conditions were primitive, while engineers were unused to working with scientists. Despite that, early working starts and long weeks quickly became the norm, such was the dedication. One early photograph shows where the solar panels on a satellite were calibrated – in the car park! Another shows technicians assembling satellites in ordinary work clothes and not in clean rooms, but no satellites seem to have come to harm as a result.

The early space scientists were typically between 35 and 45, had fought in the war, had been educated under the GI Bill and were young, energetic and patriotic. There was also a culture clash, because Goddard, like other NASA centres, was run like a military institution with a chain of command and orders that were expected to be followed. The scientists had come from universities where people were expected to do their own thinking, work odd hours and not necessarily regard obedience as a virtue. Ed Weiler, a later director of Goddard and a scientist with a military background, once described managing scientists as like 'trying to herd cats – the only thing to do was to leave food (money) out for them and then they would follow.'[40] Typical of an early 1960s workforce, it was highly gender segregated, the only women being the 150 secretaries and typists and the librarian, with a 'wives' club' launched in 1961. First to break through the glass ceiling was Eleanor Pressly, who became head of the vehicles section and in 1963 received the third annual Federal Women's Award. Many years later, Goddard acknowledged the contribution of African American women and men, singling out tracking expert and later assistant chief of research Melba Roy Moulton, who joined Goddard as far back as 1959 and Emmett Chappelle, an astrochemist who joined Goddard in 1966. Dorothy Hoover was another prominent African American woman at Goddard, identified by the highly acclaimed book and movie *Hidden Figures* which did so much to begin to right the historical record.[41]

Most of the original building work was completed over 1962–3. Goddard took over the IGY Minitrack network, with its stations in Antigua, West Indies; Quito, Ecuador; Lima, Peru; Santiago, Chile; Woomera, Australia; and Esselen, South Africa. Stations were later added at Blossom Point, Maryland; Fort Myers, Florida; Goldstone, California; St John's Newfoundland; East Grand Forks, Minnesota; Fairbanks, Alaska; Rosman, North Carolina and Winkfield, England. This network was later brought under what was called the GSFC Space Operations Control Center.

Goddard developed two 26m dishes, with the first in Fairbanks, Alaska and the second at Rosman, a $5m facility located in Pisgah National Forest in North Carolina. The new Rosman dish was almost 37m high and weighed 300 tonnes. It was situated some 600km from Goddard, but was chosen due to the lack of radio interference from being located in a forest, so that it could hear the 5w signals from distant satellites 100,000km away. It was also clear of air routes, in a natural depression and, being national forest, was gifted by the state. All data collected there went directly to Goddard by microwave and there was a permanent staff of 65. As for the overseas stations, although locally operated it was normal for NASA engineers to work there; there were 12 in the later Zanzibar station, for example.

An early consideration was the balance between work done in-house and that contracted out. There was a single program manager for the Explorer and sounding rocket program: Marcel Aucremanne from 1963 and John Holz from 1966. Most of the early Explorers were built in-house, but Explorers 30 and 37 were jointly managed with the NRL; 25 and 40 with University of Iowa; and 31 with the Canadian Defence Research Board. The in-house work required considerable floor space for fabrication, testing and simulation. The early rule-of-thumb was that at least one satellite project should be done in-house at any one time and Goddard soon had to learn how to procure and tender work with the giants of American industry, like Hughes and Grumman, as well as how to recruit its own staff of scientists and engineers. For example, Explorer 6, Goddard's first satellite, involved two universities (Chicago and Minnesota) and one private company, Space Technology Laboratories.

The new management staff to be assigned to the Beltsville Space Center first met in conference to discuss its organization and functions on 16th February 1959. They were told that the anticipated program requirements were for a science program of 30 satellites, 15 probes and 155 sounding rockets, as well as application satellites, tracking and project Mercury. Over time this grew and in a news release on 14th March 1961, NASA stated that 'Goddard Center programs embrace unmanned scientific research and exploration of space; study of the Earth's upper atmosphere; study of the Earth itself from the space viewpoint; unmanned technological utilization of space for practical purposes, such as weather forecasting and global telephone, radio and television communications; and near-space tracking and data handling. During the present decade, in the execution of this program, the center plans to launch at least 96 scientific satellites and 28 applications satellites.'

The role of Homer Newell in the development of the science program has already been mentioned. Two other key personalities were John Naugle and Frank McDonald, both from the University of Minnesota, who arrived at Goddard on 1st July 1959. One of McDonald's first decisions was that each mission should have a project manager *and* a project scientist, for he recognized that not all scientists were necessarily good managers. This became a feature of the Explorer program. McDonald eventually became NASA's chief scientist in 1982.

Finally, an important early decision was to allow access to the data coming from the science program in general and Explorer in particular. Traditionally, Principal Investigators

(PIs) owned all their own data for all time, but Homer Newell told Abe Silverstein, head of the NASA Office of Space Flight Programs, that NASA should release the data from its missions to the broader scientific community and archive the data for future users. Newell developed the policy for handling data from NASA missions, giving first access to investigators who had conceived space experiments or who had developed instruments to gather data in space before releasing it to the rest of the world. In 1966, Newell established the National Space Science Data Center (NSSDC) at Goddard and this became both the central archive and the mechanism to distribute the data to any scientist upon request. Such a policy of openness has been a priceless gift to the world's scientific community. Moreover, to develop ties with universities and the research community, the Goddard Institute for Space Studies was established in New York City in 1961 close to Columbia University, tasked with theoretical studies and the analysis and interpretation of data. Dr Robert Jastrow was made its first director and began with a staff of only three. Its purpose was to enlist the interest of university scientists in the space program, starting in the country's largest metropolitan centre. It welcomed scientists on two-year research terms, as well as foreign scientists. Jastrow ran the centre until his retirement in 1981 and he died 2008. He was succeeded by James Hansen (1981-2014) and then Gavin Schmidt.

Another early decision at Goddard was to set up a library and appoint a librarian, beginning the process of accumulating and conserving documentation. For space scientists and enthusiasts though, perhaps the most important event was in September 1995 when Goddard joined the world-wide web and began to put up its information on line.

Conclusions

The space age in general and the Explorer program in particular had their roots in the rebuilding of post-war science. Whilst originally conceived in the resumption of polar exploration, this reconstruction took the form of the International Geophysical Year (IGY). The IGY came at the juncture of a number of both connected and unconnected developments: the building by Germany of the modern rocket, the A-4 and the arrival of German rocketeers in the United States; interest in the idea of an Earth satellite; and the development of new missiles in the shadow of the Korean war. The V-2 Upper Atmosphere Panel brought together the rocket designers and upper atmosphere scientists for the first time in one of the key building blocks of the Explorer program.

The stories of *Vanguard vs von Braun*, set against the onrushing Soviet interest in orbiting an Earth satellite, the drama of the first Sputniks and the Vanguard failure, are well known and have been well told elsewhere. With the hindsight of history, two elements stand out. First, neither side had a name for their first orbiting satellite, clearly demonstrating the absence of modern public relations teams. The Soviet side quickly acquired the title 'Sputnik'. The American title 'Explorer' – a hasty, impulsive, overnight decision handed to General Eisenhower – proved to be an inspired one. The other remarkable feature was the lack of a coherent government view as to how press and publicity should be handled. Contrary to popular perceptions that the American space program was open from the start, we learn instead of rockets brought out under cover of tarpaulins in darkness, decoy strategies to pretend that no launch was contemplated, strict penalties for leaks and a policy of covering up any possible failure. Arguably, the success of Explorer 1, revealed at a crowded, enthusiastic, middle-of-the-night press conference, showed how the new world of spaceflight had hit a raw nerve of popular enthusiasm that made a hidden program thereafter unthinkable. That the appetite for the new endeavour was unstoppable became ever clearer when the discovery of radiation belts became front cover news.

The Explorer team was indeed fortunate to have the benefit of James Van Allen's talents, not to mention his foresight and opportunism in designing his radiation counter so that it could be fitted to Explorer at short notice. Coupled with the introduction of the humble but vital tape recorder, the inspirational interpretation of the Explorer 3 readings made possible the finding of the radiation belts.

After such a spectacular start, the future of the Explorer program would ultimately depend on the manner in which the new space program developed and the last years of the Eisenhower administration saw the rapid establishment of the architecture that endures to the present time. With that, the key issues became: *How important will space science be? What form will it take?* and *Who will decide?* Although they may not have realized it at the time, the new NASA essentially embraced and extended the IGY program, making it the core of its new mission. All the mid-1950s hard theoretical work by scientists in designing a space science program was rewarded, for it had a favoured position in the hierarchy of NASA priorities at the very start.

Or so it seemed. The first week of NASA saw the arrival not of a space scientist, but a champion of manned spaceflight and the adoption of a man-in-space program. This set the stage for a confrontation with the National Academy of Sciences and its Space Science Board, which had seen itself as determining the space science program. NASA moved quickly to make it clear that it would be making these decisions, a choice which the shell-shocked scientists took a long time to absorb. A hasty reading of these developments suggested that space science – and with it programs like Explorer – had been quickly degraded to give way to spectacular 'beat-the-Soviets' man-in-space programs.

In reality, this was not the case. The early NASA was still sufficiently flexible to reinstate the space science program as a priority, with a budget to match. Non-scientists like Silverstein and Glennan had the imagination and sense to make room for and co-opt prominent space scientists like Dryden and Newell, giving them scope to do their job. Space science got a purpose-built field centre in the form of Goddard, perhaps the most substantial infrastructural investment of the agency's first years, providing space science with the strongest possible institutional base. NASA also set down procedures and rules which delineated the roles of NASA as a state agency and the Academy as a scientific community. The exclusion of the SSB did not have the negative consequences initially feared and over time, the scientific establishment was to exercise a distinct but distant role in guiding priorities. The issue of who should determine space science – a government agency or the scientific community – can be argued to-and-fro and there is no perfect answer, but the early NASA actions had the merit of setting down clear lines of engagement from the start and of recognizing the importance of avoiding conflicts of interest.

The arrival of President John F. Kennedy and Administrator James Webb, set against a backdrop of continued Soviet achievement, re-opened the question of the priority to be given to space science. The decision to go to the Moon in May 1961, a program subsequently pursued relentlessly and ultimately successfully by James Webb, at first sight suggested a diminished role for space science as priorities and resources shifted to the resource-hungry Apollo program. In the event, the budget for the Apollo program was 'new money', not taken from space science. James Webb proved to be a supporter and defender of space science, not least because he had the intelligence to realize that gains in scientific knowledge in one part of the space program benefitted all the others. His enterprising, democratizing spirit brought space science to the universities, not just the well-established ones but a wide range across the nation. It is fitting that the new space telescope should be named after him.

References

1. Saegesser, Lee D: *U.S. Satellite proposals, 1945-9.* Spaceflight, vol. 19, no. 4, April 1977; see also *Prelude to the space age.* Spaceflight, vol. 24, no. 11, November 1982.
2. Parkinson, Bob: *A prehistory of outer space.* Spaceflight, vol. 55, no. 10, October 2013.
3. Ruley, John: *Homer Newell and the development of U.S. space science.* Poster, undated.
4. LePage, Andrew: *A Redstone in New Hampshire.* www.drewexmachina.com, 30th October 2015; for an account of the origins and development of the Redstone, see LePage, Andrew: *Old reliable - the story of the Redstone*, 2nd May 2011.
5. The name of the facility has since been through numerous evolutions. In 1964, all the facilities were renamed the John F. Kennedy Space Center and 'Cape Kennedy'. Cape Kennedy reverted to Cape Canaveral in 1974, with the NASA facilities retaining the name Kennedy Space Center (KSC). The area still includes military facilities, such as Cape Canaveral Air Force Station.
6. Histories give different dates for the decision: 19th July, 29th July and 3rd August. The overall time period, though, is clear. See LePage, Andrew: *Vanguard - America's answer to Sputnik.* Space View, December 1997.
7. Wilson, Andrew: *Jupiter C/Juno 1 - America's first satellite launcher.* Spaceflight, vol. 23, no. 1, January 1981.
8. LePage, Andrew: *Project Orbiter - prelude to America's first satellite.* Space View, January 1998.
9. Morgan, George D: *Rocket girl - the story of Mary Sherman Morgan, America's first female rocket scientist.* Prometheus, 2013.
10. The technically correct name for the 4-stage launcher was Juno 1, as the Jupiter C was a 3-stage nose cone test rocket. Contemporary accounts and subsequent NASA histories principally used the term Jupiter C and this is the rocket popularly associated with the first launch. Both terms will be used here.
11. Mudgway, Douglas: *William H. Pickering - America's deep space pioneer.* Washington DC, NASA, 2008.
12. LePage, Andrew: *Explorer - America's first satellite.* Space View, February 1998.
13. McLaughlin, William: *Explorer 1 at 25.* Spaceflight, vol. 25, no. 2, February 1983.
14. Goddard employee obtained first Explorer 1 photo. *Goddard News*, 15th March 1983.
15. LePage, Andrew: *Vintage micro - the original standardized microsatellite.* www.drewexachina.com, accessed 5th July 2014.
16. LePage, Andrew: *Vanguard and its legacy.* Space View, 1st February 1999.

17. Ludwig, George H: *The first Explorer satellites*, text of remarks delivered at James Van Allen's 90th birthday celebration, University of Iowa, 9th October 2004.
18. *Explorer's legacy* at http://explorer.lib.iowa.edu/.
19. Massey, Harrie *et al: Scientific research in space.* London, Elek, 1964.
20. Wilmot Hess (ed): *Space science.* London & Glasgow, Blackie, 1965.
21. *Memorandum from Presidential Science Advisor James Killian to the President*, 23rd August 1958, from *Eisenhower archives.*
22. O'Donnell, Franklin: *Explorer 1.* Caltech, 2007.
23. These and subsequent extracts are taken from John M Logsdon (ed): *Exploring the unknown - selected documents in the history of the U.S. civil space program, NASA History series, 1995: vol. 1: Organizing for exploration; vol. 2: External relationships; vol. 3: using science; vol. 4: Accessing space; vol. 5: Exploring the cosmos; vol. 6: Space and Earth science.*
24. *Space unlimited - satellite launchings by the USA in connection with the IGY.* U.S. Information Bureau, undated.
25. Newell, Homer: *Beyond the atmosphere - early days of space science.* Washington DC, NASA, SP-421. For the history of this early period, see Naugle, John: *First among equals - the selection of NASA space science experiments.* Washington DC, NASA, 1991; and NASA historical reference collection and document series: *Exploring space*, vol. 5; *Exploring the unknown*, vol. VI.
26. Portree, David: *NASA's origins and the dawn of the space age.* NASA, Monographs, §10.
27. Bizony, Piers: *The man who ran the Moon - James Webb, JFK and the secret history of project Apollo.* London, Icon Books, 2006.
28. The following are the summer study reports: *Review of space research* (1962); *Space research - directions for the future* (1965); *Planetary exploration, 1968-75* (1968); *The outer solar system - a program for exploration* (1969); *Lunar exploration - a strategy for research* (1969); *Venus strategy for exploration* (1970); *Priorities for space research 1971-80* (1970); *Outer planet exploration* (1971); *Scientific uses of the Space Shuttle* (1973); *Future exploration of Mars* (1974); *Infrared and submillimetre astronomy* (1975); *Solar physics* (1975) and *Institutional arrangements for the space telescope* (1976).
29. NASA Historical reference collection.
30. Gorn, Michael: *Hugh Dryden's career in aviation and space.* Washington DC, NASA, 1996.
31. NASA, published 27th February 1961, from John M Logsdon (ed): *Exploring the unknown*, vol. 5: *Space science: Origins, evaluation and organization*, 157-162.
32. See this author with Zakutnyaya, Olga: *Russian space probes.* New York, Praxis-Springer, 2011.
33. Descriptions of the early Goddard are taken from Naugle: Rosenthal, Alfred: *Venture into space - early years of the Goddard Space Flight Center.* NASA, U.S. Government Printing Office, 1968; Wallace, Lane: *Dreams, hopes, realities - NASA's Goddard Space Flight Center, the first forty years.* Washington DC, NASA, 1999; Rosholt, Robert: *An administrative history of NASA, 1958-1963.* Washington DC, NASA, 1966, SP-4101.

Foundations

34. *Birth of NASA - the diary of T. Keith Glennan.* NASA special publication 4105, 1993.
35. Subsequent directors were John Clark (1965-76), Robert Cooper (1976-9), Thomas Young (1979-82), Noel Hinners (1982-7), John Townsend (1987-90), John Klineberg (1990-5), Joseph Rothenberg (1995-8), Alphonso Diaz (1998-2004), Edward Weiler (2004-8), Rob Strain (2008-12) and Christopher Scolese (2012-).
36. Powell, Joel: *A vanishing era.* Spaceflight, vol. 58, June 2016.
37. Cape Canaveral, Florida; Corpus Christi, Texas; Eglin Air Force Base, Florida; Kauai, Hawaii; Guyamas, Mexico; Point Arguello, California; White Sands, New Mexico; Muchea, Woomera, Australia; Kano, Nigeria; Zanzibar, Kenya; Bermuda; Grand Bahamas and Grand Turk, British West Indies; Grand Canary, Spain; Canton, Kiribati; and ships *Rose Knot Victor* and *Coastal Sentry Quebec*.
38. Von Ehrenfeld, Manfred 'Dutch': *The birth of NASA - the work of the Space Task Group, America's first true space pioneers.* Springer, Heidelberg, 2016.
39. *Goddard News*, vol. 3, §8; *Goddard News*, May 1987.
40. Wright, Rebecca; Johnson, Sandra; Dick, Steven: *NASA at 50 - interviews with NASA's senior leadership.* Washington DC, NASA, 2008.
41. Shetterly, Margot Lee: *Hidden Figures. The untold story of African American women who helped to win the space race.* London, Harper Collins, 2017.

2

Early Explorers

This chapter looks at the early period of the Explorer program once it was taken from the army by NASA, from Explorer 6, the first 'NASA Explorer', up to Explorer 55, the last officially numbered craft. The first NASA Explorers benefitted from a more powerful rocket, an upgraded Juno I. The new Juno II had greater thrust on the first stage (up from 37,640kg to 86,025kg), was 92cm longer and had 20 seconds more burn time. The upper stages, with their small, powerful but imprecise rockets, remained the same.[1] Critically, the weight of payload that it could orbit rose from 11kg to 43kg, with a possible orbital high point (apogee) of 480km. The new Juno was highly improvised, as its imperfect reliability was to illustrate.

Greater payload capability meant that the next Explorer satellites could be larger. Van Allen wanted satellites to fly higher – out to 1,000km – through the auroral zones to look down on them. These new Explorer satellites were called the 'S' series ('S' for scientific). The first of these was S-1, which weighed 40kg, was shaped like a spinning top and was spin-stabilized. It had 3,000 solar cells for power with 15 nickel-cadmium batteries and was 76cm high. S-1 carried instruments for studying cosmic rays, solar ultraviolet and x-rays, meteorites and Earth's heat balance. It was to work for a year, after which it would be turned off by a timer.[2]

Under its new management, NASA took the opportunity to define the Explorer program:

'Geophysical satellites and probes of various configurations developed to study the space environment and upper atmosphere surrounding the Earth, including such phenomena as radiation fields, cosmic rays, micrometeoroids, temperature, magnetic field, solar radiation, ionospheric studies, air density, solar plasma and gamma rays'.[3]

Explorer S-1, which should have been Explorer 6, was launched on 16th July 1959. It was the last of three failures in a row. Control was lost after 5.5 seconds due to an electrical failure, at which point it went off course, turned through a horizontal 90° and was destroyed by the range safety officer, crashing 75m from the pad and 90m from a crowded bunker of spectators. It was one of the most spectacular early failures at Cape Canaveral, one that is frequently re-shown in video, although rarely identified as Explorer S-1. Under the new policy of *no-flight, no-name*, it was never formally identified, but it is included in the table (see Appendix I). This was not a good start for the new NASA management.

Explorer 6: NASA's First Explorer

Honour was restored the following month with what subsequently became Explorer 6. This was not the S-1 replacement but a totally different spacecraft, which received the designator S-2. It is something of an anomaly in the Explorer program, since it was a prototype Venus probe.[4]

Explorer 6 was the creation of Space Technology Laboratories (STL), a spin-off of the Ramo-Wooldridge company which had attracted USAF funding to develop a rocket able to send a small (38kg) spacecraft to the Moon, the idea being to beat the USSR to this next target. The first American Moon probes, called Pioneer and launched in 1958, were unsuccessful but created great excitement as the nation followed the launches from Cape Canaveral.[5] When a Soviet probe reached the Moon in January 1959, the U.S. Air Force and STL moved the goalposts to target Venus and determined to beat the USSR there instead (in Moscow, consideration was also given to a Venus probe in 1959, but the first such mission did not take place until 1961). STL was also motivated by the desire to impress the new NASA, in an attempt to win contracts for future launches. STL decided to launch two spacecraft to Venus, on 3rd and 7th June 1959, with arrival at Venus that November, a mission called Able-4. To test the technologies, the precursor test, Able-3, would fly in February 1959 and the backup that April.

At this stage, the respective roles of the military and the civilian space agency NASA had not fully been resolved. Military programs were in effect given a civilian identity. On the Pioneer 2 mission, for example, the USAF colours had been painted over with the new NASA logo. It was decided to locate Able-3 within the Explorer program – and thus subject to NASA authority – as S-2 or Explorer 6. This was a new design, whose most striking feature was its paddlewheel solar panels, necessary for the provision of electrical power during the long transit to Venus. There were other important innovations. Thermal design was crucial, because the probe would get ever hotter as it travelled sunward, so a new system of expanding and contracting wire coils was fitted. Because communications with the probe would be required at a distance of 40 million km, as it neared Venus, digital communications replaced analogue for the first time, which would reduce noise and enable data to be stored and compressed, though the term 'telebit' was used then rather than 'digital'. The probe also transmitted at a much higher frequency, at 378MHz rather than the normal 108MHz, though this was not advertised at the time because it was also used on air force projects. The instruments were based on those carried by the first Moon-bound Pioneers: a Geiger-Muller tube (University of Minnesota), a radiation counter in the form of a scintillator (STL), a proportional counter (University of Chicago), a magnetometer (search-coil and fluxgate, also STL), a micrometeorite detector (USAF) and a camera, the package costing $5 million. The transmitters would be used to measure electron density and very low frequency transmissions, such as whistlers from Earth's lightning and possibly extra-terrestrial sources as well. In the interests of speed, cost and reliability, as much equipment as possible was bought from local electrical and hardware stores.

Explorer 6 preparations

In the event, getting the spacecraft off the ground took much longer than expected. Now that the U.S. had actually put some satellites into orbit, NASA management insisted on a policy shift from 'getting it launched' to 'getting it right'. Able-3 was delayed several times, to the point that the prospect of getting a Venus probe (Able-4) away that June quickly vanished. The two Venus probes were then re-designated as Moon probes with the aim of getting the first pictures of the far side of the Moon – unless the Russians did so first – and a deep space mission. Able-4, now a Moon probe, was prepared for launch that autumn. On 18th September 1959, the Atlas rocket carried out a 24-second flight readiness firing, but a propellant line ruptured and the entire rocket was destroyed in a monstrous explosion, the only consolation being that the payload had not yet been installed. This was more than unfortunate, however, for the USSR's Automatic Interplanetary Station (later called Luna 3) got the much-prized pictures of the Moon's far side the following month. NASA then took an Atlas rocket from the Mercury program, designed to put America's first astronaut into orbit, fitted the Able-4 on top and launched it on 26th November 1959. At 45 seconds, the shroud was ripped away, destroying the third stage and the payload, although the lower stage continued to fire. In response to this, small holes were drilled in the shroud for the subsequent mission to prevent the build-up of pressure and resulting explosion.

Pioneer 5, the deep space explorer, was a great success, being launched inward to the Sun on 11th March 1960. Buoyed by this, NASA launched two more spacecraft based on the S-2 Explorer 6 design. The first of these, called P-30, was launched toward the Moon on 23rd September 1960, but the second stage lost thrust and started tumbling and the probe crashed after 17 minutes. The last, P-31, was launched on 15th December 1960, but this time the second stage ignited prematurely at an altitude of 12km, blowing up the whole stack. That was the ignominious end of the first stage of the American lunar program.

Even though the original purpose of Able-3 had been overtaken by events, work continued on it, now with the intention of launching it as a scientific satellite in its own right. STL was still keen to compete as a contractor within the new NASA, so the company worked hard to get Able-3 flight ready throughout the spring and summer. It would use the new Thor missile, built by the air force in the mid-1950s as a rival to von Braun's Jupiter, continuing the army-air force rivalry. Thor arose from a decision by Goddard and missile contractor McDonnell to put the three Vanguard upper stages on the top of the Thor missile, thereby removing the most troublesome part of the Vanguard program, its first stage. The version used here was called the Thor Able. As an IRBM it was tested intensively, with no less than 450 launched over its lifetime. Thor was well known in Britain, where the Royal Air Force equipped four squadrons with them as Britain's first missile deterrent. The rocket later became the British Blue Streak launcher and Thor also formed the basis for the first Japanese rocket, the N-1. The Thor first stage was a 21.5m tall, 2.4m diameter liquid-fuel rocket, using RP-1 kerosene (paraffin) and liquid oxygen for 68,000kg thrust. Being the first version of the Thor used for spaceflight, it received the title A, or A for Able, under the army-navy phonetic alphabet (e.g. 'c' for Charlie, 'd' for dog, 'e' for easy, etc.), though the second stage was also called 'Able' and the third 'Altair'. It was the only Thor Able in the Explorer series. From Explorer 10 onward they were launched on the next version, the Thor Delta, later more simply called the Delta.

Explorer 6 was a small, 64kg spheroidal paddlewheel satellite with four 50cm square solar panels to recharge the batteries of the 66cm diameter satellite. It was the first to use large solar panels (or 'paddles' as they were often called). The mission's stated objectives were to study trapped radiation of various energies, galactic cosmic rays, geomagnetism, radio propagation in the upper atmosphere and the flux of micrometeorites, as well as testing a scanning device to photograph the Earth's cloud cover. The satellite was spin-stabilized. Each experiment apart from the television scanner had two outputs, digital and analogue. Two VHF transmitters were used to transmit the analogue signal and were operated continuously, while the UHF transmitter used for the digital telemetry and the TV signal was operated for only a few hours each day. The program manager and scientist was John Lindsay.

The Thor Able functioned perfectly on 7th August 1959, heading out 47° over the sea and dropping the first stage into the ocean off Bermuda, while the Smithsonian observatory later spotted the third stage 8,000km high. The satellite's signals were picked up by the British radio telescope at Jodrell Bank near Manchester 12 minutes later, as soon as it crossed the Atlantic. It was also picked up in Singapore 40 minutes after that. That was the end of the good news, however, because one of the solar paddles snagged in the cord holding it down under the fairing, resulting in only 63% of planned solar power, a level that decreased over time. The apogee reached was 42,400km, much higher than the 36,600km intended, but with a low perigee of 245km, giving an orbital period of 765 minutes at 47°.

Explorer 6: NASA's First Explorer 55

Explorer 6 launch

The first task for Explorer 6 was to send a picture to Earth, so it began scanning our planet in strips on 14th August from an altitude of over 30,000km. The strips were then assembled in Los Angeles and eventually released, with gaps, at a press conference at the end of September. That the Earth was a curved object was about the most that could be said of it. Had its earlier version taken a picture of the Moon as intended, that image would probably have revealed nothing other than that it was round too, which we also already knew. Further pictures were taken, but perhaps wisely not distributed. It was, nevertheless, the first picture of the Earth taken from orbit and, as one commentary put it, 'at that stage, every victory mattered.' There was no reference in the publicity about how this was supposed to be practice for getting pictures of the far side of the Moon before the Russians.

Scientific data were more convincing. Explorer 6 passed through the radiation belts twice a day and made 113 traverses, transmitting 18 hours of data a day for 40 days from its eccentric orbit and mapping Earth's magnetic field out to 42,450km. Its instruments picked up the radiation belts at a lower level of intensity than previous probes – until 16th August when Earth was rocked by a magnetic storm for two days, prompting impressive aurorae. This was the first of several storms, with others on 22nd August and 3rd and 20th September. Explorer 6 was able to follow the way in which the Earth's magnetic field contracted, recovered and expanded back to its normal size. The magnetometer found strong fluctuations in an unstable region between R5 and R7, while Explorer 6 also found

56 Early Explorers

a depression at R5 that might indicate the existence of a ring current around the Earth, something that had been speculated since 1830.[6] It further found that the outer radiation zone was actually two regions separated at 15,000km. Explorer 6 was the first satellite to experience resonance – its orbit (period and inclination) being perturbed due to the movement of the Earth-Moon system and mutated to 6,740–48,800km – and it was able to make a contour map of radiation (iso intensity), finding that the magnetic field was perturbed and deformed. Eventually, its equipment began to fail, with the reduction in power causing a lower signal-to-noise ratio affecting most of the data, especially near apogee. One VHF transmitter failed on 11th September 1959 and the last contact with the payload was made on 6th October 1959, at which time the solar cell charging current had fallen below that required to maintain the satellite equipment. Explorer 6 managed to return 827 hours of analogue and 23 hours of digital data.

Explorer 6 first image of Earth

More detailed results were assembled later.[7] The University of Chicago's triple coincidence telescope reported that there appeared to be high energy radiation of 10–20MeV on the inner side of the Van Allen radiation belt, 531km thick, at about 1,995km from the Earth. The total counting rate maximum was about 1,400 counts cm^2/sec, but with no protons with energies greater than 75MeV nor electrons more than 13MeV. The University of Minnesota measurements of the belt showed the same region of hard radiation, but with wide variations in intensity. Between them, they showed that the radiation belt had a 'complicated and variable' structure. The recording rate for micrometeorites was low.

Its mission over, Explorer 6 was the target of the world's first anti-satellite test. Called Bold Orion, it was a nuclear-tipped missile developed by the USAF, with eight single-stage tests launched from a B-47 bomber between May 1958 and June 1959, reaching 100km. A double-stage version was tested between December 1958 and October 1959. The last of four tests, this time launched from a B-52 at 10,700m on 13th October 1959, was aimed at Explorer 6 at an altitude of 251km. The missile came within 6.4km of its target and was declared a successful intercept (had the warhead been detonated, it would undoubtedly have destroyed the satellite).

Even with a low perigee of 230km, Explorer 6 should have remained in orbit for twenty years, but the ability of the Moon to perturb satellite orbits, dragging down the perigee, had not been fully appreciated and Explorer 6 burned up on 1st July 1961 after less than two years in orbit. This of itself was an important learning point. Explorer 6 was a substantial scientific success. It had developed new technologies and paved the way for Pioneer 5. Despite its improbable background, the mission showed what quality engineering and science could achieve under pressure.

Explorer 7 (S-1A): Earth's Radiation Balance

Explorer 7 was the re-build of the 16th July 1959 failure, now designated S-1A. It began a series of missions that assessed Earth's radiation budget and was launched 13th October 1959. The 42kg magnetospheric, spin-stabilized satellite's structure comprised two truncated conical fiberglass shells joined by a cylindrical aluminium centre section. The spacecraft was 75cm tall and 75cm wide at its equator and was powered by 3,000 solar cells on both the upper and lower shells. There were 15 nickel-cadmium batteries on its equator which also helped to keep up the spin rate. A pair of 1w, 20MHz antennae projected outward from the middle, while a 108MHz tracking antenna was placed on the bottom of the lower shell. Explorer 7's telebit telemetry was able to interrogate 24 functions and then miniaturize and dump the information.[8]

Explorer 7 was also known as 'the kitchen sink' or 'heavy Explorer' because of the large number of instruments aboard (six), weighing 31.7kg, three-quarters of its mass. The experiments addressed solar x-rays, Lyman α rays (NRL), micrometeoroids (Goddard, this failed), cosmic rays (University of Iowa), heavy nuclei (disrupted by interference) and Earth radiation balance. The program manager was H.E. LaGow. Explorer 7 orbited from 523–857km, 50°, 99 minutes, which put it above the F_2 layer in daytime and above the F layer at night-time. It made about 4,000 radiation observations daily, or 432,000 monthly, with the outcomes being transmitted to an IBM 704 computer. The tracking beacon became inoperative on 12th December 1959, but data continued intermittently until 24th August 1961 when it went off air. This was a good record at the time.

Explorer 7's prime purpose was to measure the Earth's radiation balance and the changing nature of its climate. Dr Verner Suomi, a meteorologist with the University of Wisconsin, found that there were large-scale outgoing radiation fluxes and losses which appeared to be closely connected to cloud levels.[9] Dr Suomi had written his PhD in 1953 on how cornfields absorbed energy and radiated it back into space, a miniature example of the planet's energy balance. A key instrument was the radiometer which he developed with his colleague at Wisconsin, electrical engineer Robert Parent.[10] They were apprehensive

Explorer 7 preparations

about its success, as they had been at Cape Canaveral that July to see the Explorer S-1 failure. In the event, their instrument was able to measure the global heat budget and quantify its seasonal changes. They went on from Explorer 7 to devise instruments on TIROS and subsequent weather satellites that analysed Earth's energy balance with increasing sophistication and subsequently provided the first observations of climate change. Once the principal investigators had finished with the data, NASA released tapes from the thermal radiation experiment in 1967.

Explorer 7 revealed slow changes in the inner radiation belt up to a factor of two, but rapid changes in the outer, up to 100-fold. Explorer 7 recorded the inner parts of the radiation belts, plotting radiation densities during a magnetic storm on 28th November 1959 as well as reporting changes in the flux of its inner zone. Intense radiation was recorded on one orbit as Explorer 7 passed through an aurora. The principal outcome of this encounter was a revision of the model of how the radiation belts powered the aurorae: not by the leaking of the radiation belts (they did not have the energy to do so) but by solar storms. It not only detected the main solar storms but also milder ones and showed how common they were. Explorer 7 clearly showed the correlation between solar activity levels and the intensity of radiation in the belts. The Goddard instrumentation recorded the first penetration of a sensor in flight by a micrometeoroid.

Explorer 7 the night before launch

S-46, the next Explorer early the following year, failed on 23rd March 1960. Radio contact broke down and the third stage failed to ignite. It was a magnetospheric satellite designed to provide coverage of the radiation belts for a year. This was a return to the pencil-shaped design of Explorer 1, at 18cm diameter, 53cm long and 16kg in weight, with four sets of solar cells on a rectangular box. Under the *no-fly, no-name* rule, it did not receive a number.

Explorer 8 (S-30): Finding High Helium

Explorer 8 was one of the first in-house Goddard spacecraft, which used the S-1 type design but with a 45-day battery rather than solar cells due to a concern that solar cells would produce their own mini electric fields. Weighing in at 41kg, it comprised two truncated cones with the bases attached to a cylindrical equator. It was intended to be an ionospheric satellite and received the designator S-30. The outer shell of aluminium had a diameter of 76cm and a height of 76cm. The 108.00MHz transmitter had an average power of 100mw and functioned for the life of the battery pack. S-30 was spin-stabilized at 30rpm. The data system comprised both continuous operation and real-time transmission, but computerized interpretation of the data proved impossible and it had to be sorted by hand. The satellite's purpose was to obtain measurements of electron density, electron temperature, ion concentration, ion mass, micrometeorite distribution and micrometeorite mass in the ionosphere between 400–1,600 km. It would also study the temporal and spatial distribution of the ionosphere and its variation from full sunlight to full shadow or night-time. Explorer 8 carried an RF impedance probe, an ion current monitor, a retarding potential probe, a two-element and a three-element electron temperature probe, an electron current monitor, a photomultiplier and microphone micrometeorite detector, an electric field meter, a solar horizon sensor and thermistor temperature probes.

Explorer 8 preparations

Returning to the controversy of *Who should decide?* in space science, the scientists and experiments for Explorer 8, 10, 11 and 12 were decided upon by Goddard, without open competition, but were drawn from those who had made themselves known to the SSB following the 4th July telegram. Even though the SSB had been formally shut out of decision-making, its hard work in sifting scientific priorities by no means went to waste. The program manager was Robert Bourdeau, who was both manager and project scientist, selecting the experiments.

Explorer 9 (S-56): Introducing the Scout and Wallops Island

Explorer 8 launch

Explorer 8 was launched on a Juno II on 3rd November 1960. Its ion trap mapped nitrogen and oxygen in the Earth's environment under 120km, oxygen up to 1,000km, a helium layer – an important discovery – and then hydrogen from 2,000km out.[11] Explorer 8 measured diurnal electron temperatures between 392–2,200km, finding daytime temperature to be about 1,800K and night-time about 1,000K. The ion probe reported the mean mass of the ions at 16 atomic mass units, indicating the predominance of atomic oxygen. Above 800km, helium ion was found to be an important constituent of the ionosphere, while at 2,200km there was a heavy predominance of helium ions over hydrogen ions. The electrical measurements found that the spacecraft was charged, but the negative charge changed to positive with altitude. Explorer 8 sampled the ionosphere for its electrical and chemical make-up and to get a better picture of the distribution of cosmic dust, as well as testing the ionized cloud that forms around a spacecraft due to its interaction with the ionosphere. The satellite transmitted 500 bps every second during the 1,300 hours when it was broadcasting and signals were received until 27th December when the battery failed.

Explorer 9 (S-56): Introducing the Scout and Wallops Island

The next launch, on 4th December 1960, marked two important milestones: it was the first satellite launched from Wallops Island, Virginia, America's second launch base and the first launched on the Scout rocket, a solid-fuel rocket specifically tailored for putting up small satellites. Unfortunately, it did not go well due to a second stage malfunction. The S-56 atmospheric satellite – in effect a 4m diameter balloon with instruments – failed due to what was euphemistically called a 'procedural deficiency' in the checkout of the ignition system. It would have been Explorer 9, but under the *no-fly, no-name* policy, it received no number.

Wallops Island was named after John Wallop, the deputy surveyor of Virginia from 1672. It became the Chincoteague Naval Air Station until taken over by NACA on 27th June 1945, when it was renamed the Wallops Auxiliary Flight Station and, from 1974, the Wallops Flight Center. Wallops was mainly a sounding rocket base, with no less than 13,000 launchings between 1945 and 1988, the first of which was a 5m long research rocket, Tiamat, fired 4th July 1945. Far up the coast from Cape Canaveral, it did have similar features: a sand spit sea front, with a bay, inland waterways and the Atlantic lapping its shores to make it a picturesque location. Scout launches finished in the 1990s, but Wallops Island gained a new lease of life in the 21st century by launching the Russian-engined Antares rocket to send the Cygnus freighter to the International Space Station. Wallops was the only exclusively civilian launch site, since the original Cape Canaveral was an air force station while Vandenberg was and remains an air force base.

Early Wallops Island ramjet test

Defined by NASA Langley Research Center in 1958 and built by the Vought Corporation of Dallas, Texas, Scout was a four-stage solid-fuel rocket, intended to be larger than a sounding rocket but smaller than the more powerful rockets now coming on line. With Explorer missions in mind, it was called 'Scout' to go with the name 'Explorer', although

Explorer 9 (S-56): Introducing the Scout and Wallops Island

the unofficial text cited the need to replace the stopgap Juno II with something purpose-built and more reliable. Scout was to be a simple, low-cost launcher for small payloads, taking advantage of improved knowledge of solid-fuels in the 1950s. The $1.8m contract went to Chance Vought over 12 other candidates.[12] The original specification was for a launcher able to lift 60kg into 500km orbit and there was also a 90% reliability requirement, considered unrealistically demanding at the time. Costs were kept down by the time-honoured practice of taking stages from existing rockets – the first stage from the Polaris missile, the second from the Sergeant and the third and fourth from Vanguard. Scout was managed by Langley but was handed over to Goddard at the very end, in January 1991. It was the first solid-fuel-only rocket to put a satellite into orbit and the only early American launcher to be checked out horizontally prior to launch (this was the Soviet system).

Scout proved to be a long-lasting launcher, with the payload capability of the final version increased to over 300kg. It proved ideal for small payloads such as Explorer and became the principal American light launcher until its retirement with SAMPEX in 1992 and its replacement by Pegasus. The original nomenclature was X-1, X-2, X-3 and X-4, followed by A, B, D, F and finishing with G1 (neither C nor E flew). There were 118 launches over 1960–94, the last being a military satellite on 9th May 1994. The first two suborbital tests in 1960 went well, but the first orbital attempt failed. A feature of its trajectory was that after third stage burnout (171 seconds), there was a long wait until 591 seconds before the fourth stage was ignited at the top of the climb, burning until 624 seconds. The end of Scout launches was nearly fatal for the Wallops station, which was at risk of closure until the arrival of the Antares rocket for the International Space Station in the new century.

Explorer 9 launch on Scout

Scout was 21.9m tall, 1m in diameter and weighed 36.6 tonnes. Its four stages had a thrust of 52,155kg; 22,675kg; 6,170kg; and 1,360kg respectively. It was improved over time, becoming longer (23m), wider (1.14m) and heavier (47 tonnes), with thrust of 49,115kg; 28,660kg; 12,925kg and 2,675kg respectively for the D version in 1972. Although it used solid rockets already in existence in 1958 (called Algol, Castor, Antares and Altair), subsequent improvements led to a 250% better performance over the following decade.

The next launch attempt, another 7kg S-56 atmosphere satellite that was designated Explorer 9, was successful on 16th February 1961. Earlier attempts had been made to get 4m balloons into orbit, called Beacon 1 (Juno I) and 2 (Juno II), on 23rd October 1958 and 14th August 1959, but both failed. On Beacon 1, the Juno I began to vibrate at 90 seconds, with the payload breaking away at 149 seconds and the rocket exploding a minute later. It was the last Juno I, so nothing could be done for this rocket, but the decision was taken to fit an aerodynamic shield to Juno II. At the time, Beacon was not formally considered part of the Explorer program, but was integrated later.

Before the space age, it was thought that the atmosphere ceased to exist at around 30km (indeed, 99% of it is below this height), but an early discovery from space probes was that elements of the atmosphere reached, however tenuously, to 60,000km. The limited atmosphere at orbital height fluctuated between day and night and according to the 27-day solar rotation and 11-year solar cycles, when it heated and cooled becoming more and less dense. A nitrogen-filled balloon, being extremely sensitive to air density, was an ideal means of testing this fluctuation. In August 1960, the large Echo balloon had made a promising start to balloon-based orbital measurements.

Explorer 9 was a 3.66m inflatable sphere, constructed from alternating layers of Mylar polyester film and aluminium foil, with painted polka dots 5.1cm in diameter to distribute the Sun's heat evenly. The folded balloon was packed into a 21.6cm diameter, 48.3cm long tube and mounted in the nose of the Scout's fourth stage. On separation of the third and fourth stages, a nitrogen gas bottle with bellows inflated the sphere and then a separation spring ejected it out into its own orbit. The satellite had a tracking signal powered by a small bank of solar cells, the intention being to compare the balloon's orbit with a theoretical orbit in a complete vacuum. The two hemispheres of aluminium foil were separated with a gap of Mylar at the spacecraft's equator and served as the antenna, with a 136MHz, 15mw beacon for tracking purposes. Power for the beacon was supplied by solar cells and rechargeable batteries.

Explorer 9 was therefore a double first: the first launch from Wallops Island and the first launch into orbit by an all-solid-fuel rocket. Explorer 9's balloon remained in orbit until 9th April 1964 by which time the next in the series, Explorer 19, had been launched (later balloon missions were Explorers 24 and 39). On this first launch though, the beacon failed and tracking relied on the optical Baker-Nunn camera network. Despite this, the density of the atmosphere was calculated at an altitude of 700km and set against data from the big Echo I balloon two times further up.

A week later (24th February), the attempt to launch a 34kg S-45 solar astronomy satellite on a Jupiter Juno II from Cape Canaveral – what would have been Explorer 10 – failed when the third stage did not ignite. The S-45 used the same design as Explorer 7 and 8 and its purpose was to transmit low-power signals on six frequencies between 20 and 960MHz to test how they were affected by the ionosphere.

Explorer 10 (P-14): Journey into Plasma

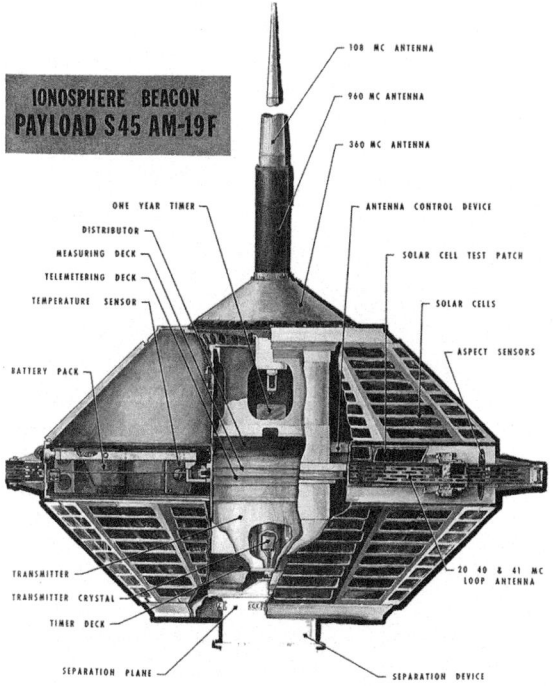

Explorer S-45

Explorer 10 (P-14): Journey into Plasma

Just as Explorer 9 marked the introduction of a new launcher, so too did Explorer 10, in this case the Thor Delta (over time, the term 'Thor' was gradually dropped and 'Delta' used on its own). Explorer 6 had already used the first iteration of Thor, the Thor Able and the change of name was accounted for by the second stage. This was derived from the Thor Able's second stage, originally called Able and now called the 'Delta'. It retained the Altair third stage.

The Delta became the classic American medium-lift rocket. NASA commissioned Thor Delta as a civilian launch vehicle on 29th April 1959.[13] It was originally intended as an interim launcher until a more suitable one was found, with NASA making it clear at the time that its first order for 12 would also be the last order. That seemed a wise decision when the initial launch dumped the first Echo orbiting balloon into the Atlantic Ocean in May 1960, but that was the end of the early bad luck as the next 15 missions were all successful. By early 1963, NASA had ordered 40 more, its success being attributed to the Thor being well tested as a military missile. Its second and third stages were taken from the Vanguard program, so its bugs had been ironed out earlier. The Delta was continuously upgraded – 14 times in the next thirty years – eventually becoming one of the most reliable launch vehicles ever.[14]

66 Early Explorers

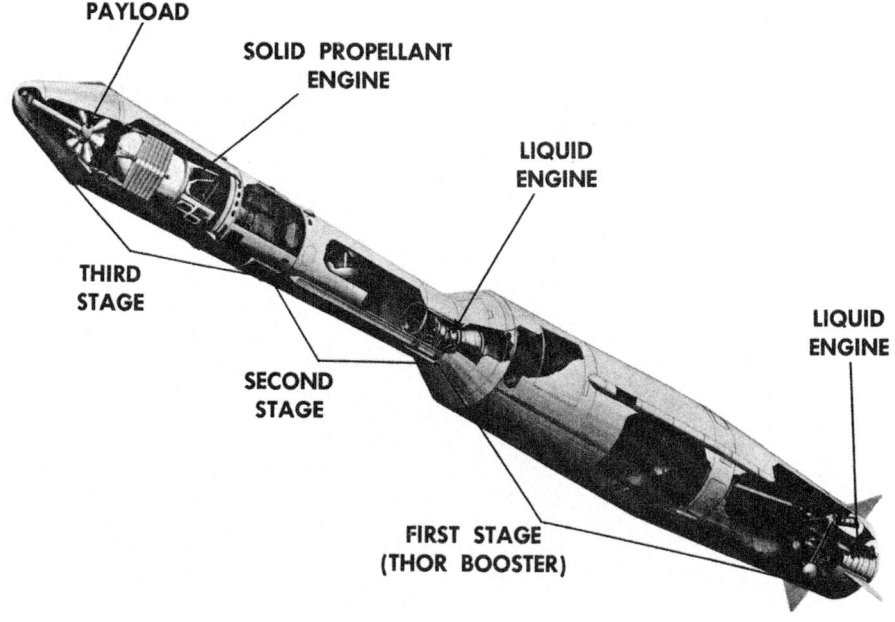

Thor Delta

Delta developed one of the most complex numbering systems of all American launchers. It began simply enough with A, B, C, D, E designators (for example, Delta A launched Explorer 14 and 15 and Delta J launched Explorer 38), but in 1972, these gave way to a four-digit numerical system based on the type of first, second, third and fourth stages (1000, 2000, 3914, 3920: for example, Delta 1604 launched Explorer 47, Delta 1913 launched Explorer 49, Delta 1600 for Explorer 50 and Delta 1900 for Explorer 51). Delta production was closed down in 1984 when all payloads were moved to the Shuttle, but after the *Challenger* accident in 1986 it was re-opened and old stock even taken out of retirement (one launched the COBE Explorer). The length of the Delta grew from 27m (Delta B) to 38m (Delta II). The Delta II (always using the Roman 'II') was introduced in 1989 with designators such as 6925, 7326, 7420, 7425, 7920 and 7925. In 1997, the Delta (including Thor Delta) had its last failure, having at that point flown 241 times with only 13 failures, mostly in the earlier stages of the program. The Delta II was still flying into the new century and was used for the WISE Explorer in 2009. The Delta comprised a first stage fuelled by RP-1 (paraffin) with liquid oxygen; small, thin, solid-fuel rockets on the side that peeled off as the rocket accelerated; a second stage using storable nitrogen-based fuels; and, when required, a final solid-fuel third stage. On a typical mission, its first stage would burn for 150 seconds and its second, small two-tonne stage for 115 seconds. This was followed by a 360-second coast and then an even smaller 500kg solid propellant ABL248 third stage would burn for 40 seconds. The first Thor Delta had a first stage thrust of 216,800kg, a diameter of 2.4 and height of 35m.

Explorer 10 (P-14): Journey into Plasma

Delta II

Explorer 10 was the first satellite to be wholly designed, assembled, built, tested and flown from Goddard. The program manager was 30-year-old James Heppner, who was also project scientist and PI. He had come from the NRL, where he had fired sounding rockets in Alaska and then made instrumentation for Vanguard 3. His original plan was to measure the magnetic field of the Moon, which he felt could best be done by hitting the Moon and measuring the field just before impact. NASA turned down this approach, on the basis that it was at edge of its launcher capacity at the time and it did not want yet another lunar failure. Heppner was encouraged to redesign the mission, called P-14, as a non-lunar mission and it was launched on 25th March 1961 (it is the only early Explorer with a 'P' designator). The satellite reached R42.3 in an extremely elliptical orbit more than half way out to the Moon, 130° from the Sun.

The mission objective of Explorer 10 was to investigate magnetic fields and plasma as it passed through the Earth's magnetosphere and beyond. The extremely elliptical orbit was chosen to reach the point where the effects of the magnetic field should be negligible. Explorer 10 was, in effect, the first plasma probe. The batteries had only a limited life of 52 hours, although they functioned for 60, so it was possible to get data only on the ascent of the first orbit. Explorer 10 was cylindrical, battery-powered and weighed 35kg. Its payload consisted of a rubidium-vapour magnetometer, two magnetometers and a multi-grid plasma probe to determine the flux, energy spectrum and directionality of very low energy protons in the plasma.

68   Early Explorers

Explorer 10

Despite its limitations, Explorer 10 duly obtained useful information on the upper atmosphere and magnetic fields.[15] Magnetic conditions were quiet at the start of its journey until it encountered distortions at R8. On the second day, there was a solar flare which led to disturbed magnetic conditions. Radiation levelled off at 30γ at R15 and was negligible at R19, but the interplanetary field began to build up again at R21.5 when solar plasma was detected. There were large fluctuations in magnitude and direction for the next five hours, probably due to shock waves. Explorer 10 crossed the boundary from R22 to R47 six times, finding radiation levels of 20–30γ inside and 10–15γ outside, suggesting that the boundary wavered in solar wind gusts. At R37, there was an increase in intensity, coinciding with the arrival of disturbances on Earth itself. Plasma was detected out to R38.5. Data suggested a field boundary around 140,000km out when solar plasma affected Earth's magnetic field like a stream around a rock. This was important in identifying the outer extreme of the radiation belt, the boundary of the magnetosphere. Explorer 10 measured the solar wind at 300 km/sec, confirming measurements made by the Soviet Moon probes and suggested a geomagnetic cavity in solar wind and the existence of solar proton streams. Explorer 10 is credited with discovering the magnetotail behind the Earth, now known to be 1.29 million km long, as it flew on the night side of the Earth.

Explorer 11 (S-15): Gamma Rays

Explorer 11 was launched on a Jupiter Juno II from Cape Canaveral on 27th April 1961, the last successful Juno II. It was a 37kg S-15 astronomy satellite, probably the first real astronomy satellite and the first use of a gamma ray telescope, so it was a historic mission. The objective was to map the distribution of high energy gamma rays above 50 MeV and the two scientists behind the satellite were William Kraushaar and George Clark of the Massachusetts Institute of Technology (MIT). Gamma rays penetrate the atmosphere only down to 36km, while gamma rays from cosmic rays are broken up, so it is necessary to go above the atmosphere to observe them.

The satellite was a spin-stabilized, octagonal aluminium box 30.5cm by 30.5cm by 58.5cm, on a cylinder 15.2cm in diameter and 52.2cm long. Its key instrument was a gamma ray detector 51cm high, 25.4cm in diameter, weighing 13kg and comprising a sandwich crystal scintillator and a Cerenkov counter surrounded by a scintillator. Its most important feature was a system to filter out non-gamma rays and cancel their detection. Explorer 11 was also designed to map the direction of gamma rays relative to the galactic plane, the galactic centre, the Sun and other known radio noise sources; to relate the measurements to the cosmic ray flux density and the density of interstellar matter; and to measure the high energy gamma ray albedo of Earth's atmosphere.

The spacecraft achieved an orbit of 480–1,858km, a period of 104 minutes and an inclination of 28.9°. It was a poor start because the orbit was high enough to enter the lower reaches of the radiation belts, saturating the electronics and damaging the solar cells. This high point also obscured gamma ray counts for a large part of the mission, meaning they were only available for 5% of the time. The tape recorder also failed early on, so it could only transmit real-time data. Explorer 11 operated until early September when the power began to fail and then data ceased after 224 days.

The spacecraft could not be pointed, but it was deliberately deployed in a tumble calculated to get a broad scan of the sky in the range above 50MeV to an accuracy of 0.1 seconds. The satellite was to tumble in such a way as to scan the celestial sphere along the galactic plane, also measuring Earth's reflectivity to gamma rays. It was left attached to its Juno II fourth stage in orbit, which was a clever way to reduce its spin rate to 5rpm. Data were transmitted to NASA Goddard, then onward to MIT's IBM 7090, the top of the range computer of the day with a 32kb memory and speed of 32kflops/sec. The data supplied filled 32km of microfilm.

Explorer 11 detected the first gamma ray sources outside the solar system and is credited with finding that cosmic gamma radiation comes from all directions. Altogether, Explorer 11 picked up 22,000 cosmic ray events and 127 gamma ray events, of which 105 came from the Earth while the 22 of cosmic origin were not clustered in any particular region. The outcomes of the mission were first extensively publicized in *Scientific American* the following year by William Kraushaar and George Clark in *Gamma ray astronomy*. Later, in *Explorer 11 experiment on cosmic gamma rays*, Kraushaar was able to make an assessment of the average intensity of cosmic gamma rays based on 141 hours of observations (for the record, $3 \times 10^{-4} cm^{-2} sec^{-1} sterad^{-1}$).[16] The energy spectrum of positive particles was found to peak at 500eV and the number density of the plasma protons found to range from 6 to 20/cm^3. Perhaps the most significant outcome was that Explorer 11 did not find the steady rate of gamma ray signatures which would match the steady state theory, thereby giving a lift to the then insurgent theory of the big bang.

The next attempt, a 33kg S-45A solar astronomy satellite launched by a Jupiter Juno II, failed at Cape Canaveral on 24th May 1961, with the guidance system losing power and the upper stage failing to ignite. This was intended to replace the failed S-45 launch from 24th February, so this was the second failed S-45 launch. It was also the tenth and last Juno II, which was retired ingloriously after six failures, one Moon probe (Pioneer 4) and three Explorers (7, 8, 11). Henceforth, larger payloads would take the Thor and smaller ones the Scout.

The next attempt, an 84kg S-55 micrometeoroid satellite on a Scout X-1 from Wallops on 30th June 1961, was also doomed. This time, the third stage Antares motor failed to fire and the rocket dived into the Atlantic 100km downrange. The S-55 was a micrometeoroid Explorer, but it would later be followed by successes in the form of Explorers 13 (S-55A), 16 (S-55B) and 23 (S-55C).

Explorer 12 (EPE-A, S-3): The Magnetopause

Starting with Explorer 12, the program began to use a new subset of designators, now being divided into such categories as Energetic Particle Explorers, Atmosphere Explorers and so on, even though the 'S' system continued in use and these were still 'S' satellites (S-3 in this case).

Explorer 12 was the first Energetic Particle Explorer (EPE – 14 was the second, 15 the third and 26 the fourth) intended to study radiation belts, cosmic rays, solar wind and magnetic fields. Explorer 12 was launched from Cape Canaveral on a Thor Delta on 16th August 1961. This 38kg astronomy satellite to collect data on radiation and the solar wind was the first project of Frank McDonald, a State University of Iowa cosmic ray expert, as project scientist. The objective of this series was to measure cosmic ray particles, trapped particles, solar wind protons and magnetospheric and interplanetary magnetic fields. Accordingly, it was sent into an eccentric orbit out to nearly 78,000km. For 90% of its time, the satellite was to be within the Van Allen belt region, where it was to describe the protons and electrons trapped therein; to study all particles coming from the Sun, but especially occasional very intense bursts of high energy protons; to study the cosmic radiation from outside the solar system; and to correlate particle phenomena with the observed magnetic field in space about the Earth.

Explorer 12 was octagonal, 48cm from side to side and 68cm tall, including a magnetometer boom. It had four solar panels and was spun at 28rpm, but due to the influence of the Sun on the solar panels, this increased over time to 34rpm. Its instruments were a proton analyser to measure the proton flux and distribution of energies beyond R6; a three-core fluxgate magnetometer to measure Earth's magnetic field from R3 to R10; a trapped-radiation experiment (four Geiger counters); a cosmic-ray experiment (double telescope for cosmic rays, single crystal for energetic particles, Geiger-Mueller telescope); an ion-electron detector to measure particle fluxes, type and energies in and above the Van Allen belt for protons below 1 MeV and electrons below 100 KeV; an experiment to determine the deterioration of solar cells resulting from direct exposure to radiation, to be compared with the effectiveness of glass filters in preventing degradation of solar cells; and an optical aspect experiment to determine its orientation in space against time.

Explorer 12

Flying out to over 76,000km, Explorer 12 traversed the radiation belts every 26 hours. It went to the edge of the magnetosphere on the sunlit side, making nine passes through the magnetosphere between R9 and R10. Explorer 12 provided 2,568 hours of data over three months until its transmitter failed due to a short circuit on 6th December.[17] Data were received for 90% of its active lifetime.

Explorer 12 achieved important outcomes and was credited with identifying Earth's magnetopause, the boundary with interplanetary space, as well as for finding that the radiation of the belts was not too strong to prevent manned spaceflight. As it traversed the radiation belts, Explorer 12 noted a sudden decrease in radiation at R8.2. In the direction of the Sun, it found a very sharp outer limit, a distinct outer edge to the geomagnetic field 96km across, which was later called the magnetopause. The ion-electron detector found that protons were present in the outer Van Allen belt, a region once thought to contain only electrons. Electrons were detected out to R8, with the maximum flux at R6 to R7. The edge was actually at R10.3, measuring 125γ on the inside and 10.3γ on the outside. Beyond that, magnetic fields were variable in direction and intensity. Explorer 12 found radiation far outside what had been considered the boundary of the radiation belts, also reporting wide variations in the intensities of electrons and finding protons trapped in the outer belt. It also found intense fluxes of low energy protons and its scintillation counter found more protons in the outer belt than the inner belt. Explorer 12 confirmed the existence of a low energy proton current ringing the Earth along an east-west direction, perpendicular to the north-south spiralling motion along the geomagnetic field lines.

Some of its most important observations, starting 28th September, were of high-speed particle streams which were later attributed to solar Coronal Mass Ejections (CME). Their recording coincided with a large geomagnetic storm at Earth, suggesting their close association. These connections were so intriguing that the backup spacecraft was later launched as Explorer 14 on 2nd October 1961, to continue where 12 left off.

Overall, Explorer 12 was responsible for a substantial revision of the understanding of the radiation belts. Earlier Explorers had indicated two radiation belts, one of high energy protons at 3,200km and a second of high energy electrons at 16,000km, but Explorer 12 found low energy protons in the upper belt and electrons in the lower belt, suggesting it was essentially one region of varying intensity and greater complexity. The belts were one big trapping region, a storage reservoir for solar particles. The belt turned out to be egg-shaped, 64,000km out on the sunny side, but could be compressed to 40,000km, with the tail being 160,000km long. Finally, the passive solar cell experiment had a clear outcome – the solar cells suffered badly when passing through a concentration of protons in the radiation belt, but there was no noticeable effect on those protected by glass.

Explorer 13 (S-55a): Unlucky with Micrometeorites

On 25th August, Explorer 13 reached orbit on a Scout X-1 from Wallops Island. The purpose of the mission was to determine both the volume of meteoroids and the risk they posed to spacecraft and to determine the degree to which spacecraft could be protected. Micrometeorites are made of stone, iron and nickel, travelling through space at terrific velocity and constantly raining on the Earth. The degree to which meteorites, micrometeorites and cosmic dust constituted a hazard to spacecraft, manned or not, was unknown at this stage – and nobody foresaw that the real danger would later be man-made debris. Earth was known to be struck by meteor showers from time to time throughout the year, possibly by leftovers from disintegrated comets or asteroids, or the elements that formed the solar system, with velocities ranging from 35,000km/hr to over 200,000km/hr. Early American deep space missions (Pioneer 1, Mariner 2) had indicated that the volume of meteors decreased 10,000:1 the further one travelled from Earth; indeed, the USSR's Mars 1 probe had reported deep space to be 'virtually empty', but it was important to know for certain.

Explorer 13 was the first micrometeoroid satellite of the S-55A series. The 86kg satellite used a body converted from the top stage of the Scout launcher itself and was a cylinder 1.93m long and 61cm in diameter. The instrumentation comprised cadmium sulphide, wire-grid, piezoelectric, pressurized and foil micrometeoroid detectors, designed to measure both the punctures and the degree of their penetration. There were, in effect, five sub-experiments: one was a battery of pressurized cells which released pressure on puncture. The second, foil gauges, measured impact by a change in resistance. The third marked a change in resistance when impact broke a wire grid. The fourth, photoelectric cells, detected light transmitted through aluminized Mylar sheets. The fifth recorded impacts on piezoelectric crystals.

The final stage of the launching did not go as expected, with the fourth stage igniting while the third stage was not only firing but still attached below. 'Hot staging', as it is called, is a legitimate procedure and is sometimes used intentionally, but there must be a clean

Explorer 13 installed on a Scout

separation which did not happen this time. The orbit was too low – some say not much more than 100km – and it decayed in three days, by which time Explorer 13 had transmitted only 13 minutes of data but had not recorded any meteorite hits. The engineers redesigned the system for the technically more complex, but ultimately safer cold ignition.

Explorer 14, 15 (EPE-B, -C; S-3a, b): Companions, *Starfish*

Explorer 14 and 15 were companion craft designed to study space weather. Both were launched in the same month, though several weeks apart, on Thor Deltas from Cape Canaveral more than a year after the previous Explorer. The program scientists were Frank McDonald (for 14) and Wilmot Hess (for 15).

The aim of Explorer 14 was to provide a more complete picture of solar decline and its effects on the artificial radiation belt by determining its particle population. The principal scientists were Frank McDonald (cosmic rays), Leo Davis (ion electron detector), Michael Bader (proton analyser), Brian O'Brien (trapped particle radiation), Lawrence Cahill (magnetometer) and Gerry Longanecker (solar cells). They aimed to find the top of the artificial radiation belt, which explained the mission's high-altitude target of 100,000km. After

Explorer 12 had found a cavity at the interface between the Earth's magnetosphere and the solar plasma at R13, Explorer 14 aimed to explore it further by going as far out as R16.

While the primary objective of Explorer 4 had been the *Argus* tests, a less advertised but similar objective for Explorer 14 and 15 was to measure *Starfish*.[18] This, the most powerful atomic blast ever in the atmosphere at 1.4 MT, had taken place during the summer, on 9th July 1962. The explosion was the most powerful, longest duration and most visible, 400km above Earth and had lit up the night sky over the Pacific for hours. Parties were held along the U.S. west coast to watch the blast and subsequent aurorae, which stretched from San Francisco to Samoa to New Zealand and were so bright that one could read a newspaper in the middle of the night. The dangers and environmental consequences of these experiments, which nowadays would cause world-wide protests, were played down at the time, but the first hints that they were underestimated came when radio communications were disrupted across the Pacific. Three satellites were badly damaged (Ariel, Transit IVB and TRAAC) and a fourth, the popular Telstar communications satellite, put out of action. Only later was it admitted that, 'it had not been expected that the artificial radiation would be as intense as it proved.'[19]

Explorer 14 launched from Cape Canaveral on 2nd October 1962. Originally, it was a 40kg S-3A solar astronomy satellite for magnetospheric studies, the second in the S-3 series following Explorer 12 (with 15 and 26 to come). Its highly eccentric 36-hour orbit moved in and out of the magnetic field and meant that it was visible for 23 hours a day on the apogee side at 100,000km, sufficiently far out to mark the boundary between Earth's magnetic field and interplanetary space. Explorer 14 carried an experiment to measure the rate at which radiation degraded solar power cells, while its own solar cells were covered with 60mm glass for protection. Its instruments comprised a cosmic ray detector, an ion electron detector, a solar cell measurer, a proton analyser and a trapped particle radiation experiment. Explorer 14 was spun at 10rpm, which gradually decreased to 1rpm by 8th July 1963.

Following on from Explorers 6 and 12, Explorer 14 duly found the magnetic cavity at R10.5 and noticed many fluctuations in radiation and changes in direction, indicating the distortions and compressions of the magnetic field. Explorer 14 probed the tail to R16, finding it in the range 40 to 50γ and what appeared to be an electron tail on the night side of the Earth. The boundaries of the magnetosphere were mapped with more precision than before, showing the way that it flared away from the Earth in a definite shape. There was also possible confirmation of Explorer 6's suggestion of a ring current on the night side of the Earth.

The spacecraft worked well, except for the period from 10th–24th January 1963 and after 11th August 1963, when the encoder malfunctioned and then ended the transmission of useable data. Despite that, there had been good data for 85% of its active lifetime, 6,500 hours altogether. Explorer 14 developed what was called a 'seasick motion' or 'precessional effect', also termed 'coning' or 'torqueing', with its spin rate rising from over 9 revolutions a minute to as many as 12. This was possibly due to the ion-electron density experiment causing a magnetic reaction.

Explorer 15 was launched from Cape Canaveral on 27th October 1962 and was also a 45kg solar astronomy satellite. Its orbit was a much less eccentric 5 hours 15 minutes.

Explorer 14, 15

Although Explorer 15 used an identical bus, its instrumentation was different, with two instruments to measure radiation flux and the pitch angle of electrons (both University of California San Diego, with Bell Telephone), a magnetometer (University of New Hampshire), an ion electron experiment (Goddard), an optical aspect sensor (Goddard) and a solar cell damage experiment (Bell). The de-spin weights did not deploy which meant that the spin did not slow down and this degraded two of its instruments. Explorer 15's timing brought an unexpected bonus because the Soviet Union retaliated to *Starfish* by detonating three explosions of its own over the Caspian Sea, on 22nd October, 28th October and 1st November 1962. This allowed Explorer 15 to observe one recent explosion and a further two immediately, although they were less intense than *Starfish* and decayed more quickly. Explorer 15 had detectors for 1MeV to 5MeV and they were quickly able to detect the new belt of electrons and their sharp inner edge.

Between them, Explorer 14 and 15 uncovered how *Starfish* radiation at low altitude dispersed into the atmosphere and at high altitudes into space, but decay was slowest at intermediate altitudes of 1,600km to 3,200km and was expected to last ten to 20 years. Low- and high-altitude electrons deteriorated quickest; the lower ones because of their interaction with the atmosphere and the higher ones due to the influence of magnetic storms. Explorer 15 was able to compile a map of the *Starfish* and Soviet radiation belts and their intensities and was able to compare *Starfish* against the Soviet explosions, showing differential rates of radiation decay at different latitudes.[20] Explorer 15 found that the

Soviet explosions of 28th October and 1st November 1962 had little effect on high energy proton flux at high altitudes and both Explorers detected the arrival of a limited, new belt after the 1st November explosion. Generally, however, the new electrons disappeared after a few days, with the low energy particles dissipating fastest. Explorer 15 lasted until February 1963, having transmitted for 2,067 hours.

The two Explorers were themselves damaged by *Starfish*, with both suffering intermittent electronic failure and Explorer 15 experiencing under-voltage to the electrical supply, but they suffered much less than others, like Ariel and Telstar. A similar electromagnetic pulse nowadays, in a much more electronic world, would disable cars and planes and fry the internet.[21]

At this point, the superpowers had created several artificial radiation belts (*Teak, Orange*, three *Argus*, one *Starfish*, three Soviet), with each country having proved its prowess. Then wiser counsels prevailed and both countries signed a nuclear test ban treaty in summer 1963, one of the last but most commended achievements of the Kennedy presidency. Some new nuclear countries persisted with testing in the atmosphere, but worldwide disapproval (especially a wine boycott) ended the last round of French testing in the 1990s.

Explorer 16 (S-55B): First Meteorites

Explorer 16 was a follow up to the failed Explorer 13 and was a 100kg S-55 micrometeoroid spacecraft that was launched on a Scout X-3 from Wallops Island on 16th December 1962. It was a small cylindrical satellite, 70cm in diameter, 2m long and was sent into an orbit out to 1,000km. It was built around the fourth stage of the Scout launch vehicle to which it remained attached. Not long after entering orbit, it signalled that its 5mm metal sheets had been perforated by micrometeorites, the first such observations.

Explorer 16 comprised helium-filled pressurized cells; two foil gauges and copper wire grids, designed to trigger electric circuits whenever struck; cadmium sulphide cells to emit light when penetrated, indicating the size of the object; impact detectors with three levels of sensitivity to indicate momentum; and test groups of silicon cells to compare the effectiveness of different thicknesses from 1mm to 5mm with an unshielded group. Each time there was an impact, the currents travelling through both would be broken and an impact recorded. The cadmium sulphide detector had first been tested on Vanguard 3 and both experiments were intended to be light, uncomplicated and undemanding on telemetry. The Goddard physicist, Luc Secretan from Switzerland, expected five to ten hits a month.

In fact, there were 16 penetrations in first 29 days: ten copper penetrations of 4mm and one of 8mm, plus two stainless steel foil gauge penetrations and three cadmium sulphide cell penetrations. All of them seemingly came from very small objects, possibly the size of a grain of sand. By the end of six months, Explorer 16 had registered 41 punctures up to 1mm and 11 up to 2mm, but none above 5mm.[22] The spacecraft operated for seven months to July 1963 and was considered a success. These impacts prompted further investigations, leading later to the large Pegasus micrometeorite satellite launched by the Saturn 1.

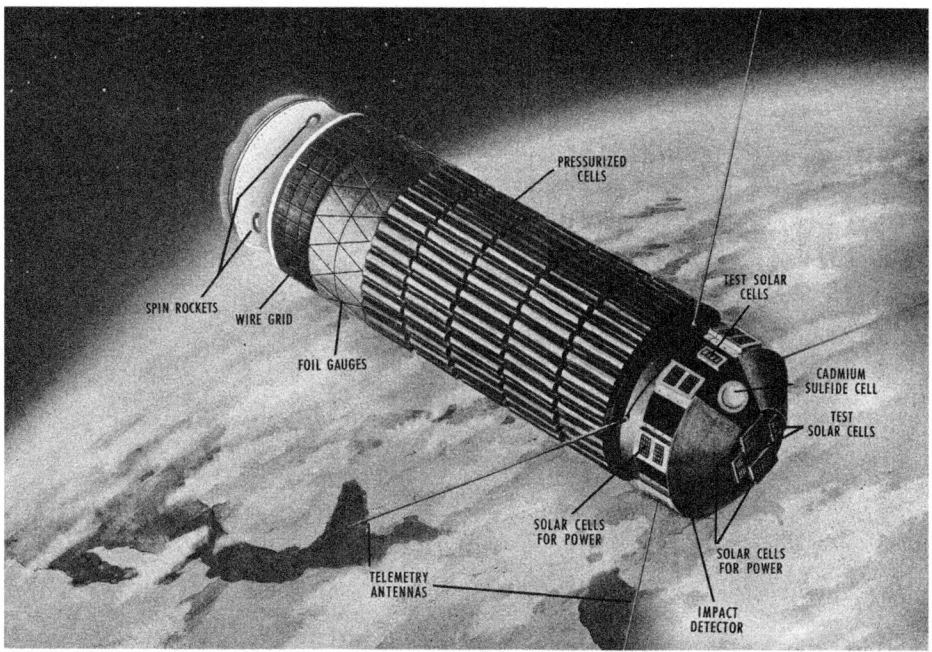

Explorer 16

Explorer 17 (AE-A, S-6): Finding the Helium Belt

Explorer 17, launched in darkness on 3rd April 1963, was a 185kg S-6 globe-shaped spacecraft on a Thor Delta B from Cape Canaveral. It reached a perfect 255–920km orbit after a two-day countdown and was Delta's 16th straight success. It was the first Explorer for atmospheric research, designated Atmosphere Explorer-A (AE-A) and would be followed later by Explorer 32 (AE-B), 51 (AE-C), 54 (AE-D) and 55 (AE-E), so this became a long series. Explorer 17 was a spin-stabilized, 95cm diameter ball made of stainless steel, its shell being only 25/100th of an inch thick and was vacuum sealed to prevent contamination. Its purpose was to study density, pressure and composition, making direct samplings of such atmospheric constituents as helium, nitrogen and oxygen and flying over geographical parts of Earth never covered before, greatly extending earlier sounding rocket launches. It carried four pressure gauges for the measurement of neutral particle density, two mass spectrometers to measure neutral particle concentrations and two electrostatic probes for ion concentration and electron temperature measurements. As an indicator of the growing capability of Explorer satellites, its 500mw power was capable of supplying 40 separate digital information channels. Explorer 17 made two daily passes over Blossom Point tracking station in Maryland, executing four-minute data dumps each time. It had an orientation system designed to focus on the Earth, Moon or Sun, but began with the Moon, which is why it was set for a night-time launch.

Explorer 17

The mission ended when its battery failed on 10th July 1963, but Explorer 17 tripled the volume of measurements of the constituents and parameters of the upper atmosphere, making them worldwide. One aim was to follow the indication of helium by Explorer 8 and its findings indeed suggested that Earth was surrounded by neutral helium in a belt at 250–900km (other figures are given).[23] The mission suggested the need for more data from different perigees, which was done with Explorer 51, 54 and 55. Given that the upper thermosphere was governed by the lower thermosphere, there was a need to dip lower into it to find out more.

Explorer 18 (IMP A, S-74): Discovering the Bow Shock, the Tail to the Moon

The IMP (Interplanetary Monitoring 'Probe', later 'Platform') series was proposed by Frank McDonald in May 1961 as consideration began for a manned lunar landing program. He argued that a proper understanding of the Earth-Moon environment was essential to anticipate the level of protection needed for astronauts and this would require a program of monitoring by an overlapping series of satellites to measure the radiation

Explorer 18 (IMP A, S-74): Discovering the Bow Shock, the Tail to the Moon

environment in Earth's magnetosphere and near-lunar space. Subsidiary goals were to develop the capability to predict solar activity and flares, to study the relationship between the Sun and Earth and to determine the quiescent properties of interplanetary magnetic fields and how they were related to solar particle fluxes. The idea was to fly a probe into an extreme elliptical orbit three-quarters of the way out to the Moon to reach multiple sampling points, typically R15 to R30, whizzing around the Earth at its closest at 201km. It was designed to explore the teardrop shape of Earth's magnetosphere. The first three IMPs were octagonal and they were small spacecraft, only 71cm across, 20cm high and 62kg in weight, with four solar panels 66cm long and 46cm wide incorporating 2,880 solar cells able to supply 38w and with a boom extending 2m from the top for the magnetometer. It was not much bigger than a human and that was with instruments extended.

Frank McDonald was responsible for the design of the scientific payload for the first three IMP missions, selecting experiments to investigate magnetic fields, cosmic rays, the solar wind and the flow of particles away from the Sun's corona. This suite was the last in-house selection by Goddard before Homer Newell set down new procedures whereby instrument selection would be made by NASA headquarters, with peer review by the scientific community and taking account of NASA HQ advisory committees (see chapter 1). At the autumn 1961 Particles and Fields Subcommittee meeting, the committee accepted McDonald's recommendations but agreed that the new procedures would be followed thereafter. Procedures aside, the series proved to be one of the most productive in the Explorer program. The instruments were a rubidium-vapour magnetometer, two fluxgate magnetometers, an orthogonal Geiger counter telescope, an ion chamber and Geiger counter, a low energy proton analyser, a plasma probe and a thermal ion electron sensor. The three 2m-long magnetometers (two fluxgate, one rubidium-vapour) were designed to measure the magnitude and direction of the magnetic field, but were mounted to enable them to avoid the spacecraft's own miniscule field. The solar panels had 11,520 cells providing power for 13 silver-cadmium batteries and there was a tiny 4w transmitter, weighing only 750g, sending on 136.145 mc/s using four 66cm antennae. A typical session was 82 seconds long and comprised 795 data bits. The satellites were spun at 22rpm, but the elongated, 93-hour orbit meant some lengthy periods in Earth's shadow without power. Paul Butler was the project's manager.

The program was quite fast, taking only two years from drawing board to orbit. IMP was a development of the Energetic Particle Explorers (12, 14, 15) but had a number of innovations: modular design, integrated circuits, on-board computer data handling, spin-and-despin control systems and biological decontamination. IMP also had its own numbering system, with letter codes before launch (A, B, C to J) and number codes after (1, 2, 3 to 8 etc) in addition to the Explorer names (18 etc). The two intended for lunar orbit (achieved with Explorer 35 but not with 33) were also called AIMPs, for 'Anchored' IMPs (which were AIMP 1 and 2). To further confuse things, they did not fly in alphabetical order, with F going before E and I before H. The literature of the time often referred to an individual spacecraft by any one of these designators, so for clarity (and in keeping with the practice elsewhere in this text) the Explorer designator is most commonly used. Table 2.1 shows the IMP series of missions.

80 Early Explorers

Table 2.1 IMP series

IMP Designator	Explorer Designator	Type
A, B, C	18, 21, 28	First series
D, E, or AI (Anchored IMP)	33, 35	Lunar orbit
F, G	34, 41	
I	43	New drum design
H, J	47, 50	Same design as I, near-circular orbit

After a launch delay, Explorer 18 was launched on 26th November 1963. The Delta had a new third stage motor, the X-258 with 2,585kg thrust, designed to achieve the required 153-hour orbit 197,000km out. But it had to be precise, for a little too much thrust and it would leave Earth orbit altogether. However, all went well and the scientific outcomes came quickly and were substantial, an indicator of more to come in the Explorer series. They were first presented by Norman Ness in a seminar at Goddard on 15th December 1964. Explorer 18 was credited with being the first to observe the bow shock at R30 and how it moved back and forth up to one radius at a speed of between 10km and 200km/sec, with ions and electrons heated in the shock. Behind the shock wave was a turbulent region of highly energetic plasma, compressed on the sunny side and trailing away from the Earth in ever-widening bands of weakening intensity up to 300,000km. The magnetometers found a shock front at R10 and a trailing teardrop tail far out.[24]

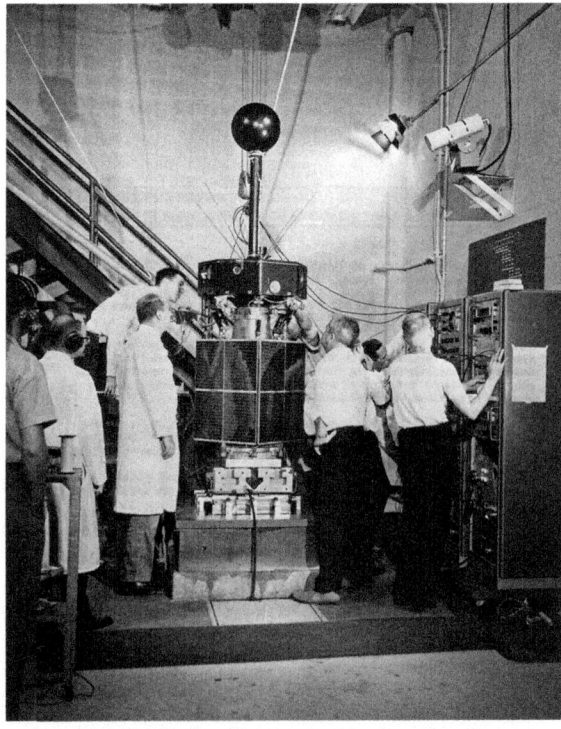

Fitting out an IMP

Explorer 18 headed out in the direction of the Sun, initially encountering a steady field before meeting the magnetopause at R13.6, with a sharp change and much turbulence, later considered to be a transitional region. Then it met a second change at R20 with radiation steady at 4-5γ and with some changes in direction. The upstream was quiet and steady, but there was a disturbed and turbulent downstream.[25] Norman Ness described a neutral zone – which had been hypothesized before but never detected – as a thin sheet, a permanent part of Earth's environment and almost void of magnetic activity. Although its role was unknown, it was possibly responsible for speeding particles into Earth's polar region, either directly or via the Van Allen belt to cause aurorae.

More was to come. On 14th–15th December, Explorer 18 noted that the normally steady interplanetary magnetic field had become quite disturbed, rising by 10γ. This coincided with the Moon being in a line with the Earth and suggested the possibility that Earth's magnetic tail was very long and stretched out to the Moon, far beyond the R50 line originally calculated. Further investigation of the tail found that it had a sheet-like, magnetically neutral surface populated with energetic electrons which could cause auroral displays. Explorer 18 made the first detailed night-time mapping of the magnetosphere, taking measurements out to R30. The probe found that the tail was a permanent extension of the geomagnetosphere, with an average magnitude of 8γ. Explorer 18 found isolated bursts of electrons to 30keV in the tail and great temporal and distance variations in magnetic storms. It also found a gusty wind, almost radial from the Sun, generally in the range 300 to 500km/sec, with the presence of protons and helium nuclei. Explorer 18 provided the first accurate direct measurement of the interplanetary magnetic field and measured interplanetary plasma travelling at 250 to 440km/sec.

Explorer 18 survived eight hours in the Earth's shadow, a predicted event, on 7th May 1964. This was the longest time that a satellite had been in such cold, so there was great relief when Santiago station in Chile picked up its signals again. Those signals became less reliable and too weak to pick up thereafter and it was formally abandoned on 10th May 1965.

Explorer 19 (ADE-A): First from the West Coast

Explorer 19 launched on a Scout X-4 from California on the west coast on 19th December 1963. It was a 7kg, 4m balloon identical to Explorer 9 for atmospheric density studies. Whereas Explorer 9 orbited over equatorial regions, Explorer 19 was destined for polar latitudes where air density might be more affected by the polar lights. Its course took it as far northward as Thule, Greenland and as far southward as Antarctica.

Explorer 19 was the first launch from the west coast, at what is now generally called the Vandenberg Air Force Base (VAFB), named after Chief of Staff Hoyt Vandenberg (it is also called the Western Test Range). Located at Lompoc between Los Angeles (201km distant) and San Francisco (451km), Point Arguello juts out into the Pacific and makes possible a flight path over the ocean to polar orbit, either up or down the coast, that avoids populated areas (polar launches are not made out of Cape Canaveral, because they would quickly overfly land). The air force took over the old army facility of Camp Cooke north of Point Arguello in 1956 to train crews to handle the missiles that would later be launched

82 Early Explorers

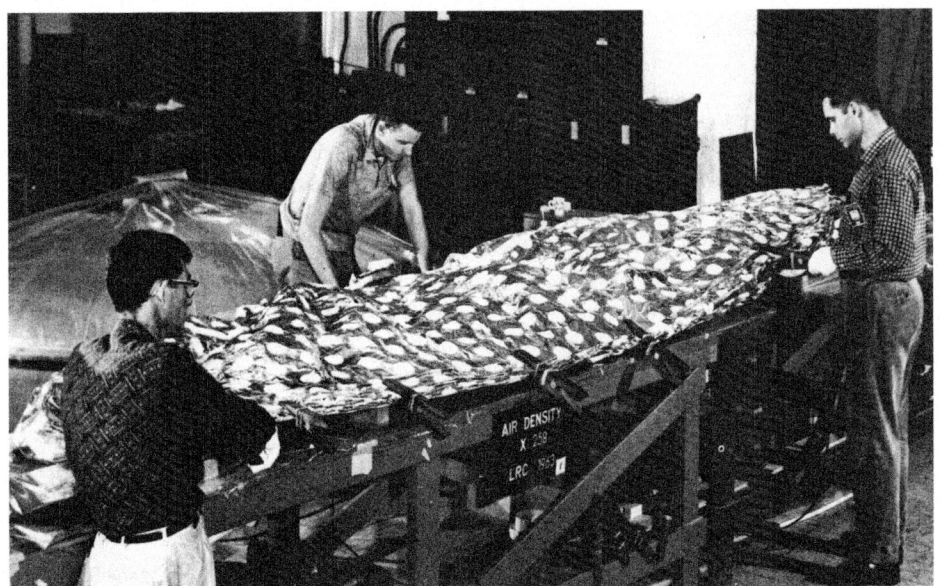

Explorer 19: rolling up the balloon

from Cape Canaveral, but it soon became a launch base in its own right. Over time, most military missions were launched from there, starting with the Discovery reconnaissance satellites, although Cape Canaveral was and is still used for some military launchings. Thor rockets were launched from Vandenberg from 1958 and a Scout pad was also installed, the principal purpose being to serve the U.S. Air Force version of the rocket, called Blue Scout. Scout's pad was called Space Launch Complex 5 (SLC-5, or 'slick 5' in the argot).[26] SLC-5 was built on one of the few coastal plains set in the rolling hills and canyons of the west Pacific coast, with an overview of the occasionally foggy sea. The adjoining pad, SLC-6 ('slick 6') became the west coast base for the air force Space Shuttle – a project that was abandoned after the *Challenger* accident. After the last Scout launch, SLC-5 was decommissioned and stripped.

Development of the Blue Scout proved far from simple. The air force hoped to adapt Scout as a military version to launch small military payloads but there were numerous failures of Blue Scout from SLC-5, depositing classified military payloads in the ocean that the USAF then had to retrieve before anyone else did. In the end, the air force abandoned the Blue Scout and instead used the ordinary civilian one; indeed, the last Scout launches were military. There were 69 Scout launches from SLC-5 between April 1962 and May 1994, of which nine failed, most at the early stage. The Explorers launched from there were 19, 20, 22, 24/25, 39/40, 52 and finally SAMPEX in 1992.

Explorer 19 launching on Scout from the west coast

Explorer 19's mission coincided with the International Year of the Quiet Sun, though it was actually two years 1964-5. The mission did not go as well as hoped, as its apogee was lower than planned and the beacon did not have enough power to be picked up by ground stations. As was the case with Explorer 9, Baker-Nunn cameras were used for optical tracking instead. In fact, Explorer 19 was launched while Explorer 9 was still being tracked, so that densities in two different portions of the atmosphere were sampled simultaneously. Together, they observed a tenfold increase in atmospheric density over the solar cycle.

The next mission was Beacon Explorer A, a 54kg S-66 magnetospheric satellite which launched on 19th March 1964 but failed when the Thor Delta B from Cape Canaveral failed to generate enough third stage thrust, though the cause was never determined. The satellite crashed into the South Atlantic where it was destroyed.

Explorer 20 (IE-A, S-48): Topsi

Explorer 20 was an ionospheric satellite, 'Ionosphere Explorer A', which weighed 44kg and was designated S-48. It was launched on 25th August 1964 on a Scout X-4 from Vandenberg – the second from there – and was the next step after the ionospheric satellite

Explorer 8, to be followed later by ionospheric physics Explorers 22 and 27. It was originally to have been launched on 13th March 1964, but the Scout launcher had to be sent back to the contractor due to problems with its electrical wiring.

The idea that the ionosphere bounced radio waves dated back to December 1901, when Guglielmo Marconi received a radio message across the Atlantic. This clearly could not have travelled in a straight line, as the Earth is curved. The following year, Oliver Heaviside and Arthur Kennelly suggested that these waves must have reflected off the atmosphere, or they would have escaped into space (although frequencies more than 15 mcs/sec do escape). Over time, it was learned that the ionosphere's efficiency as a reflector depended upon the number of free electrons and atmospheric density. During times of solar activity, the absorbing layer would broaden and lead to a blackout.

The purpose of Explorer 20 was to investigate irregularities in the Earth's ionosphere from a 990km orbit, in particular its structure and how it acted as an electrified mirror for long-range radio communications. It would do this by making topside radio soundings at six frequencies every tenth of a second, with the signals indicating the various electron densities. It was not the first topside satellite: that honour went to the Canadian satellite Alouette in 1962, which found that the ionosphere was complex and highly variable, with layers that wrinkle, shift, wax and wane.[27] Alouette had sent back 100,000 profiles but they were mainly vertical readings and did not give detailed horizontal profiles, which Explorer 20 now hoped to collect. Specifically, scientists hoped to gather information on patches and irregularities in the F_2 region, on the way in which radio energy was guided along Earth's magnetic fields, on the connections between density and magnetic disturbances and on the links between the F_2 region in the northern and southern hemisphere. In doing so, it would build on data collected by the British satellite, Ariel 1. At the same time, its ion mass spectrometer would collect information on ion concentrations and temperature, while it would also obtain readings of cosmic noise or radio noise coming from outside the solar system between 2MHz and 7MHz. The ion mass spectrometer was also made at University College London, which meant that there could be continuous data from Ariel 1 onward.

Explorer 20 was a cylinder tapering at both ends, with four telemetry antennae on one end and an ion probe on the other; in NASA's words, it looked like a chess pawn. It was a small satellite, 65cm diameter, 81cm high, with power coming from 2,400 solar cells. The antennae were long, two of them being 37m across and the other two being 19m across. Antenna deployment was expected to take 20 minutes, after which the satellite would be spun to 2.3rpm, slowing eventually to 0.45rpm. The orbit was set to be circular at 80° and the satellite would be ejected from the Scout's fourth stage at 55 minutes some 2,400km south east of the Cape of Good Hope, while moving northward over the Indian Ocean having already flown over Antarctica. The mission lifetime was set at between three months and a year. Explorer 20 was managed by Goddard, with the primary experimenter being the National Bureau of Standards Central Radio Propagation Laboratory in Boulder, Colorado.

Explorer 20 carried no tape recorder, so live data were sent to ten ground stations, calculated to provide 80% coverage. These were operated by the U.S. (East Grand Forks, Minnesota; Fort Myers, Florida; College, Alaska), Quito, Ecuador and Santiago, Chile; Canada (Resolute Bay, St John's) and Britain (South Atlantic, Singapore and Winkfield). Even though there were problems with telemetry and interference with the ion probe,

Explorer 21 (IMP B, S-74a): Bad Orbit Leads to Disappointment

Explorer 20

Explorer 20 operated for over a year, primarily because the turn-off command had been accidentally disconnected before launch and it continued transmitting long after it should have done so, into 1966. Explorer 20's observations were also matched against a Javelin probe launched on a sounding rocket on 23rd October 1964 out of Wallops.

Explorer 21 (IMP B, S-74a): Bad Orbit Leads to Disappointment

Explorer 21 was the second IMP, weighing 62kg and with identical instrumentation to the first. It was launched on a Thor Delta C on 4th October 1964. The intended orbit was 190km to 200,000km, but the third stage failed soon after ignition, leaving the spacecraft at an apogee of 95,000km and at a bad angle to the Sun, which left the battery underpowered and the spacecraft to overheat. The battery began to fail after two months. The spin rate and direction were also not what was planned. Data were transmitted to ground stations in Johannesburg, Woomera and Santiago and to range stations at Rosman, North Carolina and Carnarvon, Australia. They were useful for four months and intermittently for another two, but did not come from far enough out to observe the solar wind, so overall it was a disappointment. Its last transmission was on 13th October 1965.

Explorer 22 (BE-B, S-66a): Worldwide Geodetics

The first successful Beacon Explorer was Explorer 22, launched on a Scout X-4 from Vandenberg on 9th October 1964. The previous S-66 was originally manifested to launch in August 1963 and then failed to reach orbit in March 1964, so this was a program far behind schedule. The satellite was built under Goddard's direction at the Applied Physics Laboratory of the Johns Hopkins University in Maryland, which was later to become an active participant in the program. Its project scientist was Robert Bourdeau.

Explorer 22 was a 52kg windmill-shaped satellite designed to collect atmospheric, ionospheric and geodetic data. It was octagonal, 45cm diameter, 30cm high, made of fibreglass and honeycomb nylon, with four windmill solar panels 1.7m long. There were two 1.5m antennae and two dipole antennae for the beacon. In anticipation of radiation damage, twice as many solar cells were provided as initially necessary, intended to power the signals to 100mw, sent on 20Hz, 40Hz and 41Hz. This mission was the first time that lasers would be used to bounce beams off reflectors.

For the laser experiment, 360 silica glass reflectors were attached to the satellite as one-inch prisms. It also had a radio beacon, a passive laser tracking reflector and a Doppler navigation experiment. To orientate the satellite toward Earth and ensure it would be at the right attitude to reflect lasers, two bar magnets were combined with a zero spin rate. Explorer 22 was expected to make a comprehensive survey of Earth's atmosphere and ionosphere from 56km upward and to evaluate laser techniques for obtaining orbital and geodetic information from a circular 920km orbit at 80°, orbiting every 103 minutes. Two 25cm electrostatic probes would measure electron densities in the immediate vicinity of the spacecraft and total electron content between the spacecraft and the Earth. As signals passed through the ionosphere, ground stations could measure changes in the electron content and thereby create a vertical cross-section. Vacuum insulation was provided to protect the instruments from the risk of cold arising from the lack of spin, but if the temperature fell below 19°C, a trigger would command heat from the solar panels.

The purpose of optical and radio tracking was to determine the precise shape and size of the Earth. There was no tape recorder aboard, so the satellite performance data and electrostatic probe data could be observed only when the satellite was within range of a ground telemetry station receiving its signals. Lasers would also fire at the satellite at night-time from Goddard and Wallops Island to determine its position. The satellite was expected to show as a ninth magnitude star, so a 45cm optical telescope would be used in Wallops to locate it and then send flashes once per second. It carried mirrors so that it could be tracked by ground stations worldwide.

Explorer 22 was the most international American mission so far, sending signals to 80 ground stations for 50 scientific groups in 32 countries. There were ground stations in Antarctica, South America, western Europe, Australia, Greenland, Canada, Africa (Ghana, Kenya, Nigeria and Sudan), Hong Kong, India, Japan, Israel, New Zealand, Malaysia, Thailand and the U.S. (the largest number, including the Smithsonian network). Although NASA did not advertise this, it must have been one of the first missions to engage with African countries. The signals were designed to be so simple that they could be received by DIY stations costing less than $5,000 for an antenna, two radio receivers, a timing device and a recorder. In 1970, the French developed this model as the ISAGEX network.

Explorer 22

Data from Explorer 22 were discontinued in August 1968, with its role being taken over by Explorer 27. Geodetic maps, its main outcome, were handed over the European Space Research Organization (ESRO).

Explorer 23 (S-55c): Meteorites and Wind Shear

Explorer 23, a 134kg micrometeoroid spacecraft also called S-55c, flew on 6th November 1964 on a Scout X-4 from Wallops Island. It was the successor to S-55b, Explorer 16 of 1962 which, over seven months, had provided the first statistically significant sampling of meteoroids in near-Earth space, supplementary to other meteoroid data provided by Explorers 1, 7 and 8. Like them, the objective of S-55c was to provide data on the penetration of meteoroids and the ability of protective materials to resist them.

Explorer 23 was a cylinder of 60cm diameter and 2.3m long, built around the fourth stage of the Scout. Its weight, excluding the fourth stage and motor, was 96.4kg (134kg including fuel). There were two telemetry systems this time, 136.08mc/s and 136.86 mc/s. The first, a tracking beacon with a week's battery supply, would help the ground get a firm estimate of the satellite's orbit.

88 Early Explorers

Explorer 23 ready on its Scout launcher

There were three experiments: the first to determine the meteoroid penetration of stainless steel, the second to determine the penetration of high energy radiation and the third to measure air loads during the vehicle's ascent. There were also four types of detector: pressurized cells, cadmium sulphide cells, 24 impact detectors and capacity detectors to measure the degradation of solar cells and the effects of high energy electron radiation. The air load measurement sensors were designed to assess wind shear and turbulence between 7,600m and 12,000m in windy conditions of 110km/hr or more, measuring angle of attack, side slip and strain. The data were transmitted to Wallops Island, Blossom Point in Maryland and downrange stations in Bermuda, Antigua and Trinidad.

Explorer 23 worked for a year, to 7th November 1965, except for the cadmium sulphide cell detector which was damaged on lift-off and provided no data. During that year, the 25 micron cells were hit 50 times and there were 74 punctures on the thicker cells.[28]

Explorer 24/25 (ADE/Injun 4): First Double Mission

Explorer 24 and 25 were launched together – NASA's first double launch – on 21st November 1964 on the Scout X-4. NASA proclaimed this to be the first-ever dual satellite launch, but this was not the case as the USSR had already carried out two dual-launch missions earlier that year, the highly successful Elektrons 1/2 and 3/4 missions to map the radiation belts.[29] Explorer 24 was a 9kg Atmospheric Density Explorer (ADE), another

4m polka-dot balloon identical to the previous ones, Explorers 9 and 19. Explorer 25 was a 40kg magnetospheric satellite to collect radiation data and was also called Injun. The term 'Injun' was a personal invention of James Van Allen, coming from Mark Twain's 'Injun Joe' and all Injuns came from his University of Iowa. Injun 1 to 3 were not formally part of the Explorer program and failed anyway, but 4, 5 and 6 were (Explorers 25, 40 and 52).[30]

Because Explorers 24 and 25 were required in a near-polar orbit of 82°, Vandenberg was selected as the launch site, the orbit intended being every 115 minutes at 525–2,400km. Being launched in International Quiet Sun Year, 1964-5, their outcomes could be compared to earlier solar observations, while their dual orbits made it possible to compare air density directly with radiation data. The intention was to match air density measurements against atmospheric heat to assess the nature of their interactions, working in conjunction with Explorer 19 in a similar orbit. The perigee point was aimed for the north pole at the winter solstice, the south pole for the spring equinox, then the north pole for the summer solstice and back to the south pole for the autumnal equinox, thereby covering all the seasonal dimensions. ADE was built of four layers of 0.5mm thick aluminium foil and Mylar plastic (polyethylene terephthalate) which reflected both sunlight (for optical tracking) and radio waves (for radar tracking). To stop it from overheating and bursting, the surface was covered in 4,000 white spots. There was also a 136MHz radio tracking beacon inside, powered by four sets of solar cells.

The aim of the Injun Explorer was to measure the flux of corpuscular radiation in the atmosphere, sampling the concentration and energy distribution of charged particles. Injun was spherical and 60cm diameter with 50 flat surfaces, of which 30 had solar cells. It had 16 sensors to measure radiation, located on five omnidirectional detectors (three Geiger-Muller tubes to measure protons and electrons and two spherical retarding analysers on booms) and 11 directional detectors (including four Geiger-Muller tubes, two scintillation counters, three cadmium sulphide detectors and one junction detector). It also carried a permanent magnet to align the satellite with the Earth's magnetic force lines, with a magnetic damping rod to prevent it from spinning or tumbling, as well as a magnetometer on a boom to measure the alignment with the Earth's magnetic field. Injun carried a command receiver, tape recorder and two telemetry transmitters, one low power (72mw) transmitting continuously (136.29MHz) and one high power (3w) which transmitted five hours of tape-recorded data on command (136.86MHz). Also on board were two U.S. Air Force particle detectors to measure low energy particles. Injun digital data transmissions went directly to the University of Iowa.

The two satellites were launched southward, with orbital injection over Guatemala after nine minutes, 2,080km downrange and with first apogee over Madagascar. A timer activated valves to squirt compressed nitrogen gas into the balloon which separated when filled, a process completed 29 minutes after launch. Injun was then released 58 minutes later. Initial tracking was done by a Goddard trailer in San Diego and a range ship off the Guatemalan coast, as well as the NASA STADAN network (Johannesburg; College, Alaska) and the optical Baker-Nunn cameras of the Smithsonian Astrophysical Observatory. The principal scientist was Maurice Dubin.

Explorer 24 ADE re-entered the Earth's atmosphere on 18th October 1968. Explorer 25 Injun did not achieve magnetic alignment for several months, until late February 1965, but it returned radiation data until December 1966 and is expected to be in orbit for about 200 years.

Explorer 24

Explorer 26 (EPE-D, S-3c): After *Starfish*

Explorer 26 was a solar astronomy satellite weighing 46kg, launched on a Thor Delta C out of Cape Canaveral on 21st December 1964. It was Energetic Particles Explorer (EPE) D, also known as S-3c and followed on from Explorers 12, 14 and 15 in studying natural and artificial radiation belts, learning how energetic particles were injected, trapped and lost there. A second objective in mind was to assess the level of danger to the Apollo astronauts flying to the Moon. Even though nuclear testing in space had stopped, there was still interest in what had happened to *Starfish* radiation more than two years later. The Delta enabled the satellite to reach an oval orbit of 316km by 26,191km, period 456 minutes, inclination 20.1°. Compared to its predecessors, Explorer 26 had radiation hardening – probably in recognition of the *Starfish* damage – and improved telemetry which operated continuously. Its design lifetime was one year. Alois Schardt was program scientist for the EPE series and Leo Davis was project scientist for this mission.

Explorer 26 was constructed by Goddard and was an octagonal fibreglass honeycomb construction, 42cm high and 67cm diameter, with a thin aluminium skin. It had four solar panels of 1,536 cells each, to charge 134 silver-cadmium batteries. It was spun at 33rpm, eventually decreasing to 2rpm and then tumbling and coning at 1rpm. Telemetry was on a 2w transmitter with four 60cm antennae on 16 channels at 136.275mc/s sending to the STADAN network, which then comprised Blossom Point, Maryland; Fort Myers, Florida; Johannesburg, South Africa; Lima, Peru; Mojave, California; Quito, Ecuador; Woomera, Australia; College, Alaska; East Grand Forks, Minnesota; St John's, Canada; and Winkfield, England.

The five instruments were the same as Explorer 15, with instruments from the Bell Telephone Laboratories in New Jersey and two universities (San Diego, California and Durham, New Hampshire). The five experiments were:

- Electron Proton Angular Distribution and Energy Spectra solid state detectors to measure the spatial and angular distribution of electrons and protons;
- Electron Proton Directional-Omnidirectional Detectors to measure the intensity and angular distribution of protons and electrons;
- Magnetic Field Measurement Fluxgate Magnetometer, 85cm long, to measure the magnitude and direction of Earth's magnetic field;
- Ion Electron Scintillation Detector to measure the direction, time and position of low energy particles; and
- Solar cells damage detector, principally to measure the continuing effects of *Starfish*.

Apart from some deactivations due to insufficient power, the spacecraft systems functioned well until 26th May 1967, when telemetry failed.

Explorer 27 (BE-C)

Explorer 27 was the second successful Beacon Explorer, launched on a Scout X-4 from Wallops Island on 29th April 1965. The purpose of the 60kg follow-up to Explorer 22 was to use 360 fused silica reflectors to analyse the precise shape of the geoid by measuring Doppler shift, to measure electron densities and temperatures in the ionosphere with the help of 86 ground stations in 36 countries and to test laser techniques. The windmill-shaped spacecraft was to be launched into a near circular orbit of 1,000–1,080km at 41° on a Scout from Wallops Island, orbiting at 1 hour 45 minutes (quite different from Explorer 22's 80° angle from Vandenberg). It carried mirrors identical to those on Explorer 22 and had continuous wave transmitters on 162 and 324mc/s.

Responsibility for Explorer 27 rested with Goddard, with assistance from universities (Illinois, Pennsylvania State, Stanford), the Central Radio Propagation Laboratory and the Applied Physics Laboratory of Johns Hopkins University. Robert Bordeau was the project scientist. Otherwise identical to Explorer 22, the laser experiment was modified due to difficulties experienced by ground stations in confirming the return of the laser beam due to atmospheric interference. This time, each strike was not only reflected (assuming it happened), but recorded on the spacecraft itself (to confirm the event).

Explorers 22 and 27 found that the upper atmosphere comprised coexisting populations of ions, electrons, free radicals, neutral atoms and molecules, all stirred by the Sun.[31] Explorer 27 was turned off on 20th July 1973, because its frequency was needed by a new spacecraft.

92 Early Explorers

Watching an early Scout launch

Explorer 28 (IMP C, S-74b): Third into the Distant Radiation Belts

Explorer 28 was a 58kg IMP, the third of the series, which was launched on a Thor Delta C from Cape Canaveral on 29th May 1965, less than seven months after the disappointing Explorer 21. The spacecraft entered an orbit that looped through the radiation belts out to 262,000km, much further than the 220,000km planned. Additionally, its instrumentation gave problems, with the failure of the Massachusetts Institute of Technology (MIT) plasma probe, one of the magnetometers (both during launch) and the Ames proton analyser early on. Like its two predecessors, Explorer 28 had instruments for magnetic fields, carrying three magnetometers (one rubidium-vapour and two fluxgate). For cosmic rays, it carried a range energy loss telescope, two Geiger counters, a particle telescope, an ion chamber and two Geiger counter tubes. For the solar wind, it had a curved plate electrostatic analyser, a plasma probe and an ion electron sensor. As with the earlier IMP probes,

Explorer 28 was built by Goddard, with Frank McDonald as project scientist. Explorer 28 worked until late April 1967 when telemetry became intermittent, eventually ceasing on 12th May 1967.

Explorer 29 (GEOS 1): First GEOS

Explorer 29 was a 176kg satellite launched on 6th November 1965 on a Thor Delta E from Cape Canaveral. It was the first GEOS (Geodetic Earth Orbiting Satellite) and was also the first to use the Thrust Augmented Delta (TAD), designed to deliver 30% more power and which NASA hoped to use with the later IMPs or to lift a 650kg satellite to 400km. The extra power was achieved by adding three solid-fuel rockets at the side – the same rockets as the second stage of the Scout – each with a thrust of 27,000kg, giving a lift-off thrust of 158,000kg compared with 77,000kg for the original Delta. Each solid burned for 40 seconds. Goddard would later manage a further upgrade, widening the tanks to 1,371mm to increase burn time from 160 to 400 seconds.

Explorer 29 had its origins in the rival Vanguard program. Vanguard 1 scientists had found that its orbit was irregular when tracking it. The only possible explanation was that the mass of the Earth was itself uneven, even pear-shaped, being broader in the south than the north and with local mass variations (the Moon also has mass concentrations, or mascons, below the surface). Earth's shape had four important highs and five lows, such irregularities not being visible and most likely caused by the uplift of molten lava in Earth's core. They were sufficient to alter sea levels by several hundred metres and to make satellites move up and down from a perfect path by over 100m. The aim of the National Geodetic Satellite Program was to establish a single, coordinated triangulation network to map Earth's gravitational field, called the geoid. Several earlier satellites had geodetic objectives, such as Explorer 22 and 27, as well as the military ANNA (Army, Navy, NASA and Air Force), SECOR (Sequential Collation of Range) (army) and Transit (navy) navigational satellites. Accordingly, Explorer 29 was built by the Advanced Physics Laboratory at Johns Hopkins University in Maryland, the team that had developed the earlier Transit navigation satellites.

This became a new sub-set in the Explorer program, called the Geodetic Explorer, the Geodetic Earth Orbiting Satellite, or GEOS. The program's objectives were to determine the structure of the Earth's gravity field and to test a space-based geodetic system to an accuracy of 10m. The instruments were optical beacons (five flashing lights, each with four 670w bulbs, to be detected at pre-set intervals by at least three night-time telescopes at a time), two quartz prism laser reflectors, a radio range transponder, Doppler beacons and a range and range-rate transponder. Their purpose was to discover how Doppler radio tracking and changes in radio signal frequency could measure orbital variations. It was the first geodetic mission with radio beacons and a flashing light for tracking. The navy measured the Doppler shift on the three-frequency VHF radio beacon; the army measured the range of the signal; while the air force, NASA, the Smithsonian, and the Coastal and Geodetic Survey followed the optical beacon, between them offering almost worldwide coverage.

94 Early Explorers

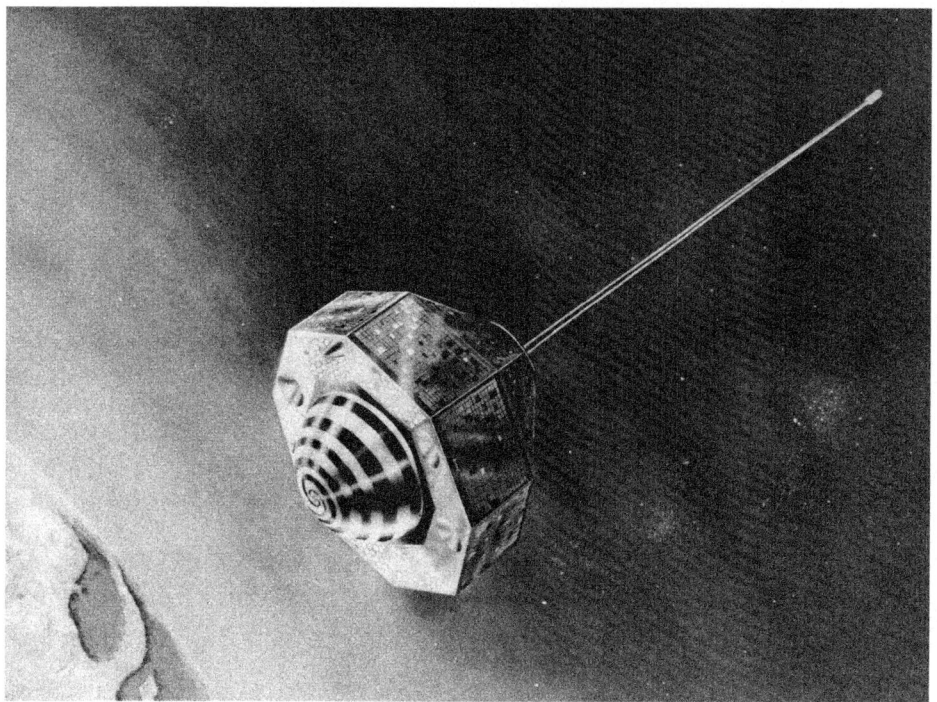

Explorer 29

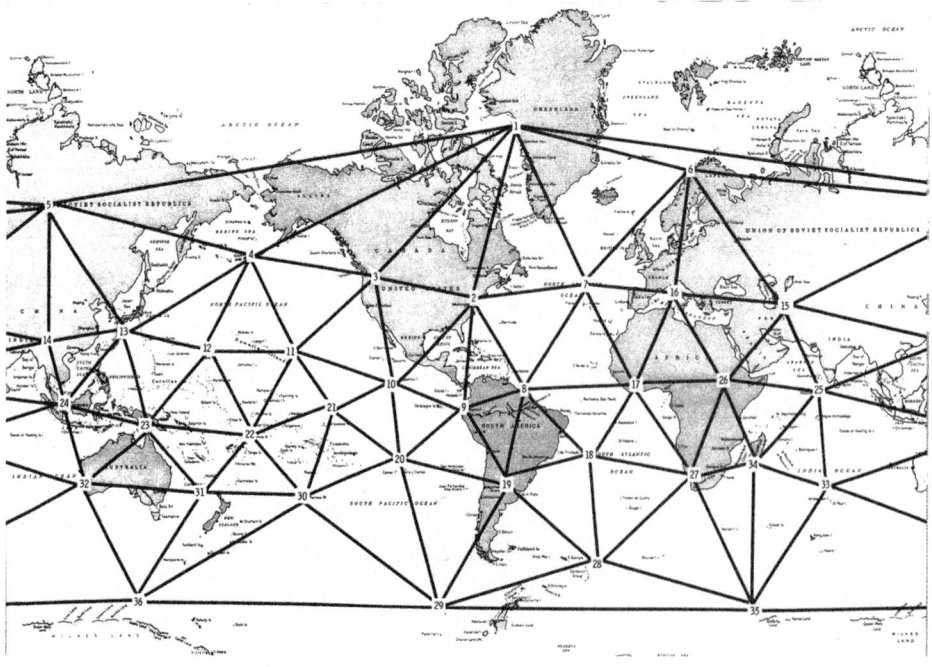

Explorer 29 geodetic control points

Explorer 29 was a hexagonal satellite with three solar cell systems. It carried a 20m boom (more formally a 'gravity-gradient stabilization system') intended to keep the flat side with the instruments facing downward toward Earth. The satellite was intended to orbit sufficiently high to avoid the atmospheric drag that would distort the results, but the Delta rocket over burned its second stage. Instead of cutting out at the scheduled time it burned to depletion, giving Explorer 29 an altitude of 1,113km by 2,275km – about 800km higher than intended – inclined at 59.4°. The satellite operated until January 1967.

The initial results, though, presented problems, because there were discrepancies of hundreds of metres between the measurements of the different agencies. These turned out to be inaccuracies in settling the precise geodetic location of the tracking stations, not in the satellite. This problem had still not been fully resolved by the time of Apollo 8, when inconsistent tracking values came in from the isolated island stations such as Hawaii, Guam, Ascension Island and the Canary Islands. Precision tracking could not be more important than in following astronauts out to the Moon and back, so data from GEOS 1 and 2 were applied in the months that followed to ensure that they were consistent and correct in time for Apollo 10 and the subsequent Moon landing.[32] This was another good, albeit unexpected, example of how Explorer complemented the Apollo program, just as James Webb had said it would.

Explorer 30 (Solrad 8): Solar X-ray Astronomy

Explorer 30 was a 57kg solar astronomy satellite, launched on a Scout X-4 from Wallops Island on 19th November 1965. The objectives of Explorer 30 were to monitor x-rays, measure solar emissions and provide real-time solar monitoring. It was the first of three to study solar x-ray and ultraviolet radiation and would be followed by Explorers 37 and 44. Solar flares are the most important aspects of the Sun's output that determine space weather, with their x-rays and enhanced ultra-violet light being potentially disruptive of communications since both are absorbed by our atmosphere. Explorer 30 continued the work begun by the NRL's Solar Radiation (Solrad) satellites, so this part of the Explorer program acquired the title Solrad as well. Solrads had their origins in navy missions with A-4s from 1949, Rockoons (rockets on balloons) from 1956 and sounding rockets from 1957. The first Solrad mission was in June 1960, detecting x-ray emissions. The navy launched a subsequent mission in June 1961, which tumbled, then successful ones in January 1964 and March 1965, the last providing 200 minutes per day of direct solar observations from 44-60Å.[33] NASA had also developed the Orbiting Solar Observatory (OSO), the first of eight planned missions being launched in March 1962 and the NRL flying instruments on OSO 2 in February 1965.

Explorer 30 was designed to complement OSO and it was timed for the final part of the International Quiet Sun Year (IQSY) 1964-5, giving it the additional title of 'IQSY Solar Explorer'. The international scientific community was invited to acquire real-time data directly. The mission also held out the potential for improving forecasting of space weather and how it might affect short-wave communications.

96 Early Explorers

Explorer 30

Explorer 30 was spherical in appearance, although in reality it comprised two hemispheres with 12 photometers installed along the middle and with solar cells mounted on both. There were four telemetry antennae mounted on the middle. Solar cells provided 6w to the nickel-cadmium batteries. Internal temperature was maintained between 10°C and 40°C, while the spin was controlled by two low-thrust vapour jets. It was the first solar radiation satellite with a recording system, intended to collect data from the four x-ray detectors for 24 hours and then transmit to the ground for five minutes, after which it would re-set itself. Its solar x-ray and ultraviolet photometers would point radially outward from its equatorial belt to view the Sun on each revolution. The mission was directed by NASA's Office of Space Science and Applications (OSSA), with Goddard responsible for tracking and data going to the NRL tracking station in Hybla Valley, Virginia. NASA's program scientist was J.M. Weldon; NRL's was R.W. Kreplin.

The mission encountered two technical problems. First, the spin rate fell from the intended 60rpm to 4rpm by September (data collection was difficult below 10rpm) and although ground control managed to spin it up again to 78rpm, this exhausted the gas supply and the spin rate fell back down to 10rpm by August 1967, so data quality became poor. Second, the recorder failed after a month, so only real-time data could be received thereafter until mission end in November 1967. One of Explorer 30's most important findings was that background solar x-ray activity increased in advance of a flare and it also gave the best definition to date of active x-ray regions on the Sun when it followed an eclipse over Greece in May 1966.

Explorer 31 (DME-A, S-30a): Red Arcs

Explorer 31 was the first and only Direct Measurement Explorer (DME) and was launched on a Thor Agena B from Vandenberg on 29th November 1965. It was a 99kg ionospheric probe, designed to work with the Canadian Alouette 2 launched on the same rocket so that they could collect atmospheric data simultaneously. Once in orbit, their apogees and perigees were within one kilometre of each other. Explorer 31 and Alouette 2 between them represented the most comprehensive attack on the ionosphere attempted to that point.

Explorer 31 carried seven experiments: a thermal ion experiment, a thermal electron experiment, an electrostatic probe, an electron temperature probe, a spherical mass spectrometer, an energetic electron current monitor and a magnetic ion-mass spectrometer. It did not have a tape recorder, so all data had to be received in real-time. Two experiments were provided by University College London's Mullard Space Laboratory. They immediately found electron temperatures of 5,000K, indicating heating in the higher ionosphere, possibly from solar wind. The inclusion of its ion probe meant that UCL had continuous data from Ariel 1 to Explorer 20 and now Explorer 31.[34] While Explorer 31 made direct measurements, the nearby Alouette 2 used radio sounding, with signals from both going to NASA stations, Canadian stations and Britain's Science Research Council stations in Singapore and Port Stanley. Explorer 31 flew further into the ionosphere than previous satellites, between 534km and 2,9932km.

A power failure reduced data levels by 50% in May 1966 and two of the seven experiments failed. From 1st July 1969, data were no longer collected regularly ('standby monitoring') but instruments were turned on again on 1st October 1969 to acquire data on a red arc event (nine minutes of data were received). Red arc events are extremely unusual and invisible to the human eye. They are the remnants of magnetic storms and happen when oxygen atoms in the ionosphere emit light after being excited by electrons higher in the magnetosphere. The monitoring station in Boulder, Colorado, could get no response to commands after that and the satellite was abandoned on 15th January 1971.

Explorer 32 (AE-B): Gravity Waves

Explorer 32 (AE-B) was launched on 25th May 1966 on a Thor Delta C1 from Cape Canaveral, the second Atmospheric Explorer (AE) satellite following Explorer 17, but 25kg heavier. It was the first Delta to use the FW-4 third stage motor, although it had been used on three Scouts earlier. Explorer 32 reached an orbit out to 2,700km, higher than intended due to an eight-second over burn of the second-stage. It was an aeronomy satellite which was designed to measure temperatures, composition, densities and pressures in the upper atmosphere directly on a global basis. Two specific objectives of the mission were to learn why change occurs in the upper atmosphere and to study short-term disturbances in the atmosphere caused by solar storms, for example in the polar regions or the South Atlantic Magnetic Anomaly.

Explorer 32 combined a hermetically sealed body, fuel cells and a spin-axis orientation system. It was an 87cm diameter sphere with a hermetic seal of stainless steel that was vacuum-tight, so that it could not measure its own discharges of gases or electrons. The satellite carried 2,064 solar cells, a turnstile omnidirectional antenna and two electrostatic probes coming from the middle. It was spun at 30rpm and utilized two long wires and weights as a de-spin mechanism. In contrast to Explorer 17, it relied on batteries and fuel cells, supplemented by its modest solar cell system to provide additional power up to 10w when commanded. Explorer 32 carried three small fuel cells (then called Gas Combination Cells) to absorb hydrogen and oxygen generated by the batteries and two heavy (80kg) silver-zinc batteries to generate 10,500 w/hours of electricity. It carried eight experiments:

- Two neutral particle mass spectrometers to determine the concentration and temperature of the neutral constituents of the atmosphere (atomic hydrogen, helium, molecular and atomic nitrogen and oxygen);
- An ion mass spectrometer to measure the concentration and scale of the ionic constituents of the atmosphere (hydrogen, helium, atomic nitrogen, atomic oxygen);
- Three magnetron density gauges to determine neutral particle density and height in the atmosphere;
- Two extending electrostatic probes to determine the concentration and temperature of thermal electrons in the upper atmosphere, with one always free of its plasma wake.

NASA's program scientist was R. Horowitz, with Larry Brace at Goddard. The satellite was equipped with an optical aspect system of four sensors, based on the Earth, Sun and Moon, to determine if it was operating in day or night and ensure correct interpretation of data. Explorer 32 aimed to quadruple data collection and used two identical solid state telemetry systems with a continuous loop recorder able to hold 2MB, transmitting for four minutes at 8.64 kb/sec either by command or pre-set recorder. The primary receiving stations were Rosman, North Carolina and Fairbanks, Alaska, which relayed data to Goddard by microwave. The secondary stations were in the STADAN network.

The two neutral-particle mass spectrometers failed six days after launch. Explorer 32 provided useful data for ten months (until March 1967), but then the sphere depressurized and the battery failed. One of its main outcomes concerned gravity waves. Writing in the *Journal of Geophysical Research*, researchers Dyson, Newton and Brace said that Explorer 32 had provided direct evidence that gravity waves in the thermosphere were associated with wave-like structures in the F region consistent with Travelling Ionospheric Disturbance (TIDs). The wave structures varied by ±10% to ±30% in electron and neutral density. The relationship between the electron density and temperature wave structure probably resulted from local cooling effects in the ionospheric plasma.[35]

Explorer 33 (IMP D): Further Away than Ever

Explorer 32

Explorer 33 (IMP D): Further Away than Ever

The next satellite, Explorer 33, was the first of three consecutive IMPs. As the IMP program got underway, Goddard's Norman Ness, now hailed as discoverer of the geomagnetic tail, came up with the idea of putting an IMP in orbit around the Moon. IMP was never part of the robotic Moon program (Ranger, Surveyor, Lunar Orbiter), but there was no law to say that it could not be. This began what was called the 'Anchored' IMP (AIMP) program, as in 'anchored' in the Earth-Moon gravitational environment. Of the two anchored IMPs, the first (Explorer 33) overshot the Moon, entering a wildly elliptical orbit beyond it, while the second (Explorer 35) achieved lunar orbit in July 1967. Both provided a high scientific return (another Explorer, 47, concerned with radio astronomy, also entered lunar orbit some time later). These probes are often forgotten in the story of the American Moon program and they still orbit the Moon even now. The formal objectives of the anchored IMPs were to:

- Investigate and obtain scientific data on the characteristics of the interplanetary plasma and the interplanetary magnetic field at lunar distances;
- Study the monthly interaction of the Earth's magnetic tail with the Moon;

- Measure interplanetary magnetic fields, solar plasma and energetic particles;
- Obtain data on dust distributions around the Moon; and
- Obtain information on the lunar ionosphere magnetic field (anticipated to be 40 times less than Earth's), lunar gravitational field and lunar radiation environment.

Although the Earth-Moon geomagnetic environment was the prime focus of interest, findings were expected to be helpful in paving the way for Apollo manned missions and improving solar flare predictions. The spacecraft were identical to the earlier IMPs, except that they had a retrorocket and their structure had to be reinforced to stand the load of its firing. The spacecraft took the form of a 70cm base, with retro motor on top, four solar panels at the side and the fluxgate magnetometer extending at both ends. The four panels could supply 43w to a silver-cadmium battery, sufficient for operations in shadow for up to three hours. The satellite carried a 6w transmitter operating on 136.020 mc/s and was tracked by Carnarvon; Santiago; Tananarive (Malagasy); and Rosman, North Carolina; with telemetry also sent to Kano, Nigeria; Johannesburg; Orroral, Australia; and Fairbanks, Alaska. The spacecraft was decontaminated on the orders of the Office of Planetary Quarantine, with swabs taken to ensure its effectiveness. These measures were rated at 97% effective, although the spacecraft still flew with an estimated 250,000 aerobic and anaerobic organisms on board (many of them spores) both within their parts and on their surfaces.

Explorer 33 had the opportunity to become the first American satellite to orbit the Moon (the Soviet Union had already circled the Moon with Luna 10, launched in March 1966). It had a three-minute launch window, with a margin of error of less than 1%, as there was no scope for a mid-course manoeuvre during the 72-hour coast to the Moon. The retrorocket could be burned only once, for lunar orbital insertion. The intended lunar orbit was 1,280–6,400km, ten hours, 175° where – critically – it would pass through Earth's magnetic tail 13 times a year. The chances of success were given at 80%, but if it did not, then an Earth orbit of 38,616km to 434,430km, period two weeks, was anticipated.

The spacecraft carried six scientific experiments, two passive and one engineering experiment (solar cell damage). One magnetometer was provided by NASA Ames, the other by Goddard. The Ames one was 2m long with a range of 0.2 to 200γ, while the Goddard one worked across 64γ. There were three radiation experiments, the first being an energetic particles experiment (University of California, Berkeley) to measure low energy solar electrons, energetic electron fluxes, low energy flares, low energy solar flare protons, solar protons over 12MeV and galactic cosmic ray intensity. The second was an electron and proton experiment (University of Iowa) to study electrons over 40KeV in the magnetospheric wake up to R60 and their spatial, temporal and angular distribution; to look at the incidence and intensity of low energy solar cosmic rays against protons and alpha particles, determining their energy spectra and angular distribution; and to study solar x-rays in the 0-14Å range. The Goddard thermal ion and electron experiment was to measure low energy electrons and ions, detect the presence of a lunar magnetosheath or shock front and observe the flow of solar wind. The solar wind experiment was a plasma probe from MIT to measure the angular and energy distribution of the proton flux in the range 100MeV to 5KeV.

Explorer 33 (IMP D): Further Away than Ever

Explorer 33 engine firing

The passive experiments were designed to analyse signals to determine how they were affected by lunar ionospheric radio waves and thereby determine the geodesy of the Moon (or, to be terminologically accurate, its selenodesy). The solar cell experiment was intended to measure solar cell degradation against various types of cover glass. The program scientist was A. Schardt, with Norman Ness at Goddard and the project manager was Paul Marcotte. The designed mission lifetime was six months.

Explorer 33 was launched using a TAD Delta from pad 17 at Cape Canaveral, the 39th Delta. The precise pathway required to reach the Moon necessitated a narrow launch window on 30th June, 1st July (when it got away), 2nd and 3rd July 1966. Lunar Orbit Insertion (LOI) was designed to be 4,800km ahead of the Moon, where the 414kg TE-M-458 thrust motor would fire for 22 seconds to enable capture and would be jettisoned two hours later.

In the event, all the three launch stages over-performed. The margins were small, each within limits and would not have presented a problem in isolation. Their cumulative effect, though, was to put lunar orbit insertion beyond reach of the small solid-fuel rocket. Goddard decided about four hours after launch that lunar orbit could not be achieved, but it would still fire the retrorocket to stop the spacecraft entering solar orbit. Accordingly, it was decided to use the fourth stage motor to achieve a highly eccentric Earth orbit out to 450,000km, the highest achieved at that time and one which still made approaches to within 35,000km of the Moon possible. Explorer 33 ended up in a highly elliptical orbit that carried the spacecraft beyond the distance of the Moon and NASA described it as orbiting both Earth and Moon.

The mission did suffer further problems, however. The magnetometer booms were stuck, but the firing of the retro-rocket jolted them open and they eventually operated for 27 months. The ion chamber experiment failed in August and then the two Geiger-Muller tubes. The battery began to fail on the 343rd day and the spacecraft's performance became erratic thereafter.

It is not known if anyone anticipated just how perturbed the orbit would become. Over the first three-year period, its perigee varied between R6 and R44, while its apogee varied between R70 and R135, with inclination from 7° to 60°. Jeremiah Madden's first flight report in December 1966 found that the orbital period was varying from 15 days to 30 days, but tending to become more circular. His second, the following May, pointed out how the orbit had been perturbed by the Moon to the point that completely new orbital predictions had to be made. Perigees ranged from as close as 20,000km to as far as 119,700km from the Earth, with apogees in the range 407,000 to 523,000km.[36] Twenty lunar passes were expected in the first two years. Explorer 33 came within 36,000km of the Moon on 8th July, the closest it ever came and coinciding with the Earth experiencing a shock wave from a solar flare. The spacecraft functioned until 21st September 1971.

There was a substantial scientific return from the mission and Goddard issued several progress reports on Explorer 33. In *Mapping of the Earth's bow shock and magnetic tail by Explorer 33*, Kenneth Behannon reported on how, during its first eight orbits of 16 days each, the triaxial fluxgate sensor mapped the Earth's magnetosheath and magnetic tail from the western flank of the bow shock to the eastern flank.[37] The bow shock was a well-defined surface far downstream extending to at least R75. He continued: 'Explorer 33 has also found that the Earth's bow shock is still a detectable boundary between the interplanetary magnetic field and the downstream magnetosheath at a geocentric distance of R75.7. This spacecraft has further revealed that the cross section of the geomagnetic tail is probably not cylindrical and that the magnetic field magnitude in the tail decreases with distance down the tail from the Earth. This magnitude decrease can be due both to a gradual expansion of the tail with distance and to a reconnection of magnetic field lines across the tail neutral sheet.' Multiple shocks were experienced on 4th October and Explorer met other storms on 15th July, 1st August and 14th September.

In *Observations of the Earth's magnetic tail and neutral sheet at 510,000km by Explorer 33*, Norman Ness, Keith Behannon, S.C. Cantarano of the University of Rome and C.S. Scearce finally resolved a conflict of data from the Soviet Luna 10 probe, where one scientist (Schmaia Dolginov) had suggested there was no magnetic field whereas others disagreed (e.g. Konstantin Gringauz). The explanation was that in the first case the Moon had moved into the neutral sheet, but then had moved into the tail. In summary, although Explorer 33 did not achieve the orbit intended, its wildly perturbed orbit brought in an unexpectedly large scientific haul. Norman Ness was able to report to a conference in Boston at the end of November that the tail of Earth's magnetosphere extended far beyond the Earth to 508,444km, some 120,675km beyond the Moon. Ness had joined Goddard in 1961 and became Head of its Extraterrestrial Physics branch. For his IMP and other work, he received the Arthur S. Flemming Award in 1969.

Explorer 34 (IMP F): From Deep Space to Polar Ice

Explorer 34 was a 104kg magnetospheric satellite launched on a Thor Delta E1 from Vandenberg on 24th May 1967. Although it was an 'F', it flew before IMP 'E'. This was the third set of the IMP program. The objectives of F and G were to study solar and galactic cosmic radiation, the solar plasma, energetic particles within the magnetosphere and its boundary layer and the interplanetary magnetic field. They were designed to build on the achievements of earlier IMPs in measuring the shock wave, the turbulent region of highly energetic plasma between the shock wave and the magnetosphere boundary and the transitional tail region. Whereas earlier IMP spacecraft were launched when solar activity was low, IMPs F and G were launched when solar activity was increasing.

NASA described IMP F as ten times more complex than previous IMPs. Its designed lifetime was one year, with a scheduled orbit of 224,000km by 258km, period four days, 66.5°. The spacecraft was designed, built, tested and managed by Goddard. It was octagonal, 71cm across, 28cm high and weighing 73kg, with two 1.8m magnetometer booms, four solar panels of 70cm by 51cm and with four times the data processing capacity at 460 bits/sec. Telemetry was transmitted continuously at 136.15MHz from four antennae extending at 45°. The four solar arrays carried 6,144 solar cells to generate 70w through silver-cadmium batteries. It was to spin once every 2.6 seconds.

Explorer 34 carried eleven experiments, totalling 23kg, to measure solar and galactic cosmic rays at the boundary of Earth's magnetosphere and interplanetary space. These included a three-axis fluxgate magnetometer provided by Goddard to measure the direction, magnitude and fluctuations of magnetic fields; and two radiation instruments. The first was a low energy telescope provided by Bell Telephone laboratories to measure electrons and protons, while the second was a low energy proton and electron differential analyser provided by the University of Iowa to measure the energy and direction of electrons and protons in the 100eV to 50keV range and to search for large-intensity low energy protons in the 10 to 100keV range between 12,800km and 33,600km. Six experiments were designed to measure galactic and solar cosmic radiation: a *range vs energy loss* cosmic ray detector for when the Sun was most active (University of Chicago), with a similar one from Goddard; an ion chamber from University of California; a counter for solar cosmic rays and solar electron events; a solar proton monitor (Goddard); a cosmic ray anistropy (Southwest Center for Advanced Studies) and a scintillation counter to measure the angular distribution of solar cosmic rays. There were also two plasma experiments: a spherical electrostatic analyser from TRW Systems to obtain a detailed energy spectrum of the solar wind, including its directional properties and search for particles arriving from sources other than the Sun; and a plasma experiment from Goddard and the University of Maryland to determine the composition and energy distribution of hydrogen and helium ions on the solar wind and in the transition zone between the wind and the Earth's magnetosphere. Finally, IMP F aimed to provide real-time solar radiation data for the forthcoming Apollo 8 and 9 missions.

A. Schardt was again the program scientist, while one of the scientists involved in Explorer 34 (and the subsequent 41) was the later leader of the Indian space program, Professor U.R. Rao. He subsequently published IMP outcomes on solar cosmic rays and their relationship with interplanetary magnetic fields, leading to the development of a convection-diffusion model of twisted interplanetary magnetic tubes in the form of an Archimedean spiral, as well as establishing the acceleration of energetic particles in shock fronts.

Once again, all did not go well. The spherical electrostatic analyser failed and the optical aspect system broke on 4th March 1969. Although the orbit went out to 214,000km, it was out of ecliptic at 67° and the lower than planned perigee of 242km meant that Explorer 34 would not have a long lifetime. It duly burned up some time between 28th April and 3rd May 1969.

Despite this, there was a good scientific return, which Goddard may not have been aware was used in the Soviet Union. Its radio wave instruments produced data on changes in solar protons and how they were absorbed by the ionosphere over the poles (what is called Polar Cap Absorption, or PCA). These data were combined with Explorer 10 to study the PCA event of 11th–18th April 1969, both from the Vostok station in Antarctica and the drifting north polar station NP-16.[38]

Explorer 35 (IMP E): Reaching the Moon

Explorer 35, launched on a Thor Delta E1 from Cape Canaveral on 19th July 1967, was the second attempt to enter lunar orbit. It was the 50th Delta launch and the 14th of the TAD version. The formal aims of the 104kg spacecraft were to study dust distribution, the lunar gravitational field and the radiation environment around the Moon, as well as interplanetary dust, solar and galactic cosmic rays and the magneto-dynamic wake of Earth in the interplanetary medium further afield. Explorer 35 also carried a micrometeorite flux experiment by Temple University, Philadelphia to measure the momentum, kinetic energy, velocity and source of dust particles, with two detectors (one directional, one omnidirectional) to calculate their distribution, source (Moon, deep space), mass and their connection to known or unknown meteor streams.

Although this was not stated explicitly, it appears that numerous remedial measures were put in place following Explorer 33, such as an improved trajectory and the addition of a new cold gas Freon-14 attitude control system to orientate the spacecraft both for Lunar Orbit Insertion (LOI) and while in lunar orbit. Lunar orbit was set for 480km to 45,600km, period between ten hours and 70 hours depending on the accuracy of LOI and inclination between 140° to 180°. Mission controllers were again prepared to settle for a highly inclined Earth orbit, 29,920km to 448,000km, 28° to 33°, if LOI failed. Instead of a preprogrammed LOI, Goddard would issue the command six hours after launch based on the most accurate measurement of the trajectory. Explorer 35 would transmit on 136.110MHz and its battery was able to store 11 amperes for 3.5 hours. Its spin rate was 25.6rpm.

This time the Delta behaved perfectly. On the fourth day, the 176kg solid-rocket motor fired for 21 seconds to achieve lunar orbit of 2,538–9,429km, 169°, 11.5 hours. The most significant finding of Explorers 33 and 35 was that the Moon had no large-scale magnetic field (or if it did, it was a thousand times weaker than Earth's), but there were small surface magnetic fields and a solar wind void existed behind the Moon. A map was drawn up of the small lunar surface magnetic fields. Explorer 35 also observed the way in which positive ions from the solar wind crashed directly onto the lunar surface. These data were later correlated with the packages left on the Moon by the Apollo astronauts. After six years, the spacecraft was turned off on 24th June 1973, one of the least well-known missions in the American Moon program.

Explorer 35 unfortunately came too late for the great storm of May 1967. This was a colossal radio burst disrupting Earthbound radio frequencies between 0.01 and 9GHz, the biggest for 80 years but only a tenth that of the 1859 Carrington event.

Explorer 36 (GEOS 2)

Explorer 36 was the second geodetic GEOS after Explorer 29 and was launched on a Delta E1 from Vandenberg on 11th January 1968. This 209kg satellite carried four optical beacons, two C-band radar transponders, a passive radar reflector, a radio range transponder, a range and range-rate transponder, laser reflectors, a laser detector, an interferometer beacon and Doppler beacons. It was designed to complement Explorer 29 from retrograde orbit by matching the distance measured by lasers against that of the laser altimeter. The mission did not go as well as hoped, due to problems with the power system, the timer clock and the optical beacon flash system. It was succeeded by GEOS 3 (9th April 1975), but that was not part of the Explorer program.

Explorer 37 (Solrad 9)

Explorer 37, a 198kg solar Explorer also called Solrad 9, was launched on 5th March 1968 on a Scout B from Wallops Island, but did not enter the intended orbit due to a failure in the fourth stage. Its aim was to measure and monitor solar x-ray and ultraviolet emissions at a time when the Sun was heading toward solar maximum, due late 1969. It was especially designed to follow solar flares, the most powerful emissions from the Sun, generally emitting over a few minutes and then decaying over hours, sending powerful streams of radiation across the solar system, including to Earth.

The spacecraft was 12-sided, 67cm high and 75cm across, with 24 solar cells providing 27w. The main part contained x-ray photometers, Geiger tubes, attitude control and spin nozzles. There were two ultraviolet photometers to measure radiation from 1,080Å to 1,350Å; a scintillation counter for 0.1Å to 0.5Å; and five x-ray photometers across 1Å-60Å. The NASA scientist was H. Glaser, with R.W. Kreplin for NRL. The spacecraft had two Sun sensors to orientate toward the Sun and was then stabilized at 1rpm with the help of small thrusters. Five photometers could relay data to a digital data storage system, taking in data from three solar radiation detectors over 14 hours for subsequent transmission to the ground, with the rest being transmitted in real-time. There were two analogue transmitters for continuous transmission. The 14 solar x-ray and ultraviolet photometers would point radially outward from their equatorial belt to view the Sun on each revolution. Transmission was on the 136MHz telemetry band and scientists and institutions were invited to tune in to receive and use them. Mission control was led by the NRL at Blossom Point, Maryland, assisted by Goddard, while data were handled by the Institute for Telecommunications and Aeronomy in Boulder, Colorado. Explorer 37's core memory failed on 11th June 1973 and the gas supply for its attitude control system was exhausted by 25th February 1974. This rendered it unusable and it was turned off.[39]

Explorer 38 (RAE-A): First Radio Explorer

Explorer 38 launched on a Thor Delta J from Vandenberg on 4th July 1968. Aside from being the first Radio Astronomy Satellite (RAS), the most remarkable feature of this 190kg Explorer was its 230m antennae, as long as a skyscraper, which extended like a giant X (half pointed skyward, half Earthward) and provided gravity-gradient stabilization.

This strange mission was designed to explore the radio sky. Until the 1930s, only emissions in infrared and ultraviolet close to the optical spectrum had been studied, but this range was extended in 1932 when Karl Jansky found that the Milky Way was a source of radio emissions across a much wider spectrum. Grote Reber (1911-2002), another American radio engineer, subsequently Australian resident, made a radio contour map of the Milky Way from 1940 to 1946 and the Sun was identified as a source of radio waves in 1942. The post-war period saw the construction of large radio telescopes in Green Bank, West Virginia; Jodrell Bank, England; and Puerto Rico. They found that whereas some stars were almost silent in the radio band, other stars, not to mention clouds of gas, were noisy. Over time, it became clear that many radio waves never even reached the ground: infrared, ultraviolet, x-rays, gamma rays and long radio waves, principally those below 10MHz.

The idea of a space-based radio telescope large enough to detect these waves, around 300m long, was first investigated in 1963 by Robert Stone and Joe Alexander, who wanted to find a way around the manner by which lower frequencies were more likely to be filtered out by Earth's ionosphere. By this stage, low frequency instruments had been installed on sounding rockets but there had been no system for pinpointing the direction of the signals. An x-shaped aerial would be able to address this. Work got under way on a Radio Astronomy Explorer (RAE) the following year, with Alexander and Stone appointed PIs. Their intention was to monitor the low frequency signals below 10MHz that normally bounced off the ionosphere back into space and did not reach the ground.

Accordingly, Explorer 38 was designed to measure the intensity of celestial radio sources, particularly the Sun, as a function of time, direction and frequency (0.2 to 20MHz); make a radio map of our galaxy; search for magnetic fields, electrons and ionized hydrogen between and beyond our galaxy; detect sporadic bursts from Jupiter (Jupiter was known to emit strong, sporadic bursts of radio signals, probably related to the passage of its 12 moons), solar bursts and radio emissions from Earth's own magnetosphere. Two spacecraft were built. The two RAEs were almost identical, with two 225m 'rabbit ear' antennae, an astronomical receiver for the range 33m to 667m and others to cover 38m to 1500m.

The cylindrical Explorer 38 was 90cm in diameter and 77cm high with four solar panels totalling 6,912 cells supplying 24w. The x-antennae were made of 5cm-wide copper and rolled out on a tape at 1.5cm/second, scanned by two slow-scan cameras at 13sec/frame. The process would take up to two weeks and these would be the longest booms extended in space to that point. Apart from the x-antennae, there was also a 40m omnidirectional dipole antenna, mainly to pick up Jovian signals, as well as a 200m boom to act as a stabilization damper and to prevent twisting or bending. The booms were coated silver on the outside to reflect the light and heat of the Sun and black on the inside to retain heat on the dark side of the Earth. The instruments carried were four-frequency Ryle-Vonberg radiometers operating from 0.45 to 9.18MHz, two multichannel total power radiometers operating from 0.2 to 5.4MHz, one step frequency V-antenna impedance probe operating from 0.24 to 7.86MHz and one dipole antenna capacitance probe operating from 0.25 to 2.2MHz. There was one high-power and one low-power transmitter for 137 and 138MHz frequencies at 10kbps and 0.4 kbps respectively. Explorer used Goddard's STADAN networks, with stations in Fairbanks, Alaska and Rosman, North Carolina taking data from the tape recorder. Other stations involved were Grand Canary, Spain; Carnarvon, Australia; South Point, Hawaii; and Clark Lake Radio Observatory, Borrego Springs, California. NASA's program scientist was Nancy Roman, with Robert Stone at Goddard.

Explorer 38 reached its orbit in stages: first, an elliptical orbit of 640km–5,860km; then, after six days, the 80kg motor circularized the orbit at 5,851–5,861km, 120°, 224 minutes. In the event, while Explorer 38 did observe the planets and ionosphere, Earth's radiation interfered with its galactic observations. Explorer 38 proved the concept but there was such a high level of interference from terrestrial sources – the magnetosphere, lightning and human signals such as radio and television – that its receivers were saturated for 25% to 40% of the time. Earth was unexpectedly noisy due to all these signal emissions, 40dB over the cosmic noise background. This prompted the idea that the successor mission, Explorer 49, should be sent to lunar orbit where the Moon could shield it from Earth's radiation, taking advantage of the experience of Explorer 35.

Despite this disappointment, 20 scientific papers were published. Explorer 38 made the first complete low frequency spectrum from 0.4MHz to 6MHz; the first radio noise scan of the galaxy; continuous observations of solar bursts; detected continuous radio emission from Earth's magnetosphere in low frequency, with sporadic bursts; and made the first survey of terrestrial radio noise. Its booms found that Earth emitted radio waves similar to Jupiter and the mission was notable for picking up sporadic low frequency radio bursts from the gas giant planet. The performance of Explorer 38's tape recorder began to deteriorate after two months in orbit and the spacecraft was retired by ground command on 31st December 1972.

Explorer 39/40 (AD-E/Injun 5): Second Double Launch and Whistler Finder

Explorer 39 was an Atmospheric Density Explorer (AD-E), a 9kg balloon launched on a Scout B from Vandenberg on 8th August 1968. Launched with it was Explorer 40, also known as Injun E or Injun 5, a radiation particle data collection mission. This was NASA's second double launch, with the balloon following Explorers 9, 19 and 24. The two satellites were targeted for similar orbit, 700km by 2,480km, 118.2 minutes, 82°.

The balloon was another 4m polka dot sphere, scheduled for the most active part of the solar cycle. Part of its task was to extend upper air research into the polar regions – hence the 82° orbit – comparing density and temperature variations there with lower latitudes. It was built at Langley Research Center from four layers of 0.5mm thick aluminium foil Mylar (polyethylene terephthalate), with 4,000 spots to distribute the heat. Tracking, conducted both visually and by a radio tracking beacon on 136.620MHz powered by four solar cells, was by the STADAN network and the optical Baker-Nunn cameras of the Smithsonian Astrophysical Observatory. The balloon was inflated by nitrogen about 25 minutes after launch and, once inflated, was released at 28 minutes. The chief scientists were Gerald Keating (Langley) and Luigi Jacchia (Smithsonian). Explorer 39 duly performed two sets of density experiments – systematic density variation and non-systematic density changes – based on sequential observations of the sphere by its radio tracking beacon and by optical tracking. The radio beacon went off air in June 1971, so cameras were used subsequently.

The other Explorer broke new ground. Explorer 40 was the heavier spacecraft, a six-sided 70.6kg cylinder, 72cm tall, 75cm diameter, with 12 detectors to measure the bombardment of the atmosphere by energetic particles from space and the intensity of very low frequency radio emissions. Like the rest of the Injun series, it was built by the University of Iowa. Injun 5 had six objectives: the comprehensive study of the downward flux of charged particles; Very Low Frequency (VLF) radio emissions in the ionosphere associated with the downward flux; geomagnetically trapped protons, alpha particles and electrons; solar cosmic rays; the continuing decay of the *Starfish* artificial radiation belt; and the temperature and density of electrons and positive ions of thermal and near thermal energy. VLF emissions were associated with energetic particles penetrating to mid-latitudes and causing a faint glow in the night sky. Explorer 40 had five booms – three with antennae and two with particle analysers – and carried 12 experiments in four groups:

- Low energy Proton Electron Differential Energy Analyser, able to measure particles of very low energy, 5eV, with three detectors and three tubes;
- Solar State Detector Experiment: proton electron telescope and alpha particle helium nuclei detector;
- VLF experiment, a magnetic loop antenna on a 3m folding boom aligned with Earth's magnetic field, to pick up radio noise between 30Hz and 10KHz; and
- Spherical Retarding Potential Analyzer experiment – one analyser to record electrons, the other positive ions.

Injun 5/Explorer 40 had a two-channel tape recorder for eight hours of storage, which it could replay in 22 minutes. It was released about 58 minutes after launch and its instrument deployment was completed at 70 minutes. Tracking was done by the University of Iowa, which had its own tracking station in Iowa City, supplemented by stations in Johannesburg and College, Alaska. The chief scientist was James Van Allen. Injun 5 made the first global survey of convection electric fields using its double-probe electric field experiment at low altitude (677km to 2,528km). It made dawn-dusk passes over the northern polar region and found abrupt reversals of the convection velocity between 70° and 80°, generally consistent with sunward flow on the equatorial side and anti-sunward flow on the poleward side. It also found inverted V electron precipitation bands in local afternoon or evening regions.[40]

Above all, Injun was notable for measuring whistlers, the bursts of very low frequency radio waves produced by lightning that are familiar to short wave radio listeners as gurgling and trilling sounds. The first Injun was a University of Iowa satellite designed by Van Allen and whistler expert Dan Gurnett, although the first VLF detector was not carried until Injun 3. On Injun 5, the VLF receiver detected not just the traditional whistler, but a chorus of them, in effect the singing of the electrons in the radiation belt. Like the radiation instrument on Explorer 1 ten years earlier, Injun scientists did not know how to interpret the signals and again suspected an instrument malfunction. In the event, this mission marked the beginning of tracking of the audio signature of plasma flow, solar storms, the radiation fields and their boundaries. The spacecraft systems performed well except for the solar cell power dump device, which delivered a low level of power to the experiments and affected the tape recorder. The spacecraft was off-air from 31st May 1970 to 18th February 1971, again in early June 1971 and then became inoperable.

Explorer 41 (IMP G): Longer Life?

Explorer 41 was a companion to 34, but after the highly elliptical orbits of the earlier IMPs, Norman Ness suggested using the retro-rocket motor to achieve a higher perigee, thereby lengthening mission lifetime. In the end, its nine experiments transmitted data for three and a half years.

Explorer 41 was a 174kg IMP to collect radiation data between Earth and the Moon and was launched on a Thor Delta E1 from Vandenberg on 21st June 1969. It had 12 instruments weighing 27kg to study solar plasma, magnetic fields and cosmic rays from an orbit of 344km–216,000km, period four days, 85°, for one year. One of its functions was to provide real-time solar radiation data for Apollo 11, which was about to start its count down at Cape Canaveral. Explorer 41 comprised a 25cm high octagonal platform of 70cm diameter, which carried the experiments, with the fluxgate magnetometer on a 2m fibreglass boom. Data went to Goddard and STADAN on 136.080MHz, through four 40cm antennae. Four solar panels with 6,144 cells provided 70w power – which declined to 47w after a year – with 13 silver-cadmium batteries. It was spun at 27.5rpm. The chief scientist at Goddard was Frank McDonald and the five principal investigators included James Van Allen. The experiments were:

- Three-axis fluxgate magnetometer to measure direction, magnitude and fluctuations of radiation;
- Eight galactic and solar cosmic ray experiments: solid state cosmic ray detector, solid state detectors, low energy proton and alpha detector, ion chamber, low energy solar flare electron detector; solar proton monitoring experiment, scintillation counter for cosmic rays, low energy telescope;
- Three solar plasma experiments: plasma detector, two low energy proton and electron differential energy analysers to cover a broad range.

The spacecraft began to experience drag at perigee on 20th December 1972 and did not re-appear above the horizon three days later, so was presumed lost. Its data were combined with Explorer 47's to measure beryllium 9 and 10 in cosmic ray particles, to estimate their age at 20 million years.

Explorer 42 (*Uhuru*): From Kenya to the X-ray Universe

Explorer 42 was the first Small Astronomy Satellite (SAS), a 143kg probe launched on a Scout B from off the coast of Kenya in east Africa on 12th December 1970. That is independence day in Kenya and was where the satellite got its name (*Uhuru* is Swahili for 'freedom'). It was the first of three small astronomy Explorers, with the other ones to operate in the ultra violet and infrared bands. This was a famous Explorer, for several reasons.

110 Early Explorers

Uhuru was the first American spacecraft launched from outside the United States and San Marco was an unlikely launch base, a classic case of what would now be called lateral or out-of-the-box thinking. The idea originated in 1961 from Professor Luigi Broglio, founder of the Italian space program, in conjunction with the Italian Space Commission and the University of Rome. He responded to President Kennedy's call for international cooperation in space with the idea of a joint project with the United States to launch small satellite payloads into equatorial orbit from Kenya using the American Scout launcher. Italian engineers duly travelled to the United States to learn how to operate the Scout rocket. The Kenyan coast offered the advantage of a direct ascent to an equatorial orbit, which avoided overflying the damaging South Atlantic Magnetic Anomaly. It also provided a downrange path over the ocean, away from inhabited areas. It would have been technically possible, though needlessly expensive of energy for such a small satellite, to get Explorer 42 into such an orbit using the Thor rocket out of Cape Canaveral by making a dogleg manoeuvre southward to the equator.

San Marco launch site

San Marco was also the first off-shore launch site. For this, Italy constructed a launch platform (San Marco) from an old oil rig towed from Charleston, South Carolina and converted an Italian oil rig into a control centre (Santa Rita). Legally speaking, at 5km from shore they were outside Kenya's territorial limits, but the project did have the blessing of the Kenyan government and was linked to tracking facilities there, in Nairobi and Mombasa. The Scout rocket was brought along on a barge and erected onto a triangular-shaped railing on the platform, then pointed in the launch direction and fired. The first San Marco launch of a Scout took place in April 1967, putting up an Italian satellite of that name. By way of historical footnotes, the San Marco platform was last used in 1988 and the Scout rocket is no longer in production, the last launch from there being San Marco 5 on 25th March 1988. Italy later abandoned the site in favour of its Vega rocket for the European Space Agency, launched from Kourou in French Guyana. The San Marco concept was later adapted for the Sea Launch project whereby the Russian-Ukrainian Zenit 3 rocket used a seaside base in Los Angeles to launch large communications satellites into orbit from the equatorial Pacific from 1999.[41] Kenya reappeared in space affairs when its Malindi tracking station, near the San Marco site, became a key point in the Chinese manned space program.

Uhuru was the first x-ray Explorer and was designed to investigate the x-ray universe. Its objectives were to survey the celestial sphere and search for sources in the x-ray, gamma ray, ultra-violet and other spectral regions. Its primary task was to develop a catalogue of celestial x-ray sources by systematic scanning of the celestial sphere in the energy range from 2keV to 20keV. The satellite rotated at 5 rpm in order to best sweep the sky. The experimental section had two identical halves to ensure redundancy, each housing an x-ray detector with a 1x10° and 10x10° field of view, the location to be matched against star and Sun sensors. *Uhuru* was aligned with Earth's magnetic field though an electrified magnet so that the spacecraft could be pointed anywhere, swivelling around to any newly-located source. Accuracy was expected to be in the 1 to 15 arc minute range.

The satellite was expected to collect more information in a day than sounding rockets had done in the previous eight years. X-ray sources had already been found by an American sounding rocket in 1962 and 40 had now been detected, all believed to be in our galaxy and

Santa Rita

some coinciding with visible objects. It was hoped that the number of known sources would rise from 40 to over hundred. They were powerful sources, up to a thousand times more powerful than the Sun. The strongest was x-1 Scorpius, a faint star but with 100 times more x-ray power than visible light, while one in Virgo had 70 times more power. Some were linked to exploded stars (supernovae) or quasars.

Explorer 42 *Uhuru* was a 63kg experimental cylinder mounted on an 80kg control cylinder. This was another new idea: a standard bus with scientific instruments carried separately. The control section was 61cm diameter by 1m long, housing four solar paddles with 27w power, 3.9m tip to tip. It had three antennae. The four solar paddles charged a six-amp, eight-cell, nickel-cadmium battery. For stabilization, there was an internal wheel with a magnetically-torqued control system to point the spin axis of the spacecraft at any point in the sky using a star and Sun sensor. There was an emphasis on redundancy, so that the system could withstand several simultaneous major failures without compromising its scientific objectives. The launch was controlled from the Santa Rita platform and its shore-based site, which would hand over control to NASA Goddard as soon as it entered orbit. The primary tracking station was in Quito, Ecuador, on the equator, to which it would dump 96 minutes of data at 30 times recorded speed every orbit. Data were stored on a single orbit storage tape recorder downlinked each time over 3.4 minutes at 1kbps. In the event of electrical storms over South America, the French Centre National des Études Spatiales (CNES) station was on standby in Brazzaville, Congo and additional stations could be utilized in Kourou, French Guyana and British stations at Ascension Island and Singapore, all close to the equator line. The NASA program scientist was Nancy Roman, with Marjorie Townsend at Goddard and the principal investigator was Riccardo Giaconni. The prime contractor for the SAS satellites was the Applied Physics Laboratory of Johns Hopkins University in Maryland, where they were built under Goddard's direction. Communications then were still quite rudimentary, with Goddard able to follow the launch only by telephone.

Uhuru was one of the most memorable success stories of the Explorer program, greatly exceeding the expectations placed on it. Four months after launch, *Uhuru* had duly surveyed 95% of the galactic plane and celestial sphere. The tape recorder had failed, but the planned backup system permitted acquisition of 60% of the data. First results were published in *Science* on 28th January 1972, by Carl Fichtel and his SAS colleagues. By then, *Uhuru* had already found 125 x-ray sources. They were different from common radio pulsars; some came from Seyfert galaxies and others from quasars; some were associated with binary star systems; and they supported the then-new theory of black holes. Most x-ray sources were from the Milky Way or nearby. The pulsar universe was a very strange one: one pulsar flashed in Centaurus every five seconds and would then go dormant for 30 minutes each two days. Crab nebula x-rays were picked up at 30 pulses a second, a remnant of the 1051 CE supernova. Explorer 42's most spectacular observations were of Cygnus x-1, the second known x-ray pulsar 1000 parsecs away, its period being calculated at 0.073 seconds. Indeed, *Uhuru* became famous as the satellite that identified the Cygnus x-1 black hole. The star had a dark companion so massive – judging by its orbital motion and x-ray energy transfer – that it could not be a neutron star but must be a black hole.

Marjorie Townsend with Explorer 42

Explorer 42 scientist Nancy Roman

114   **Early Explorers**

Uhuru found 339 objects, of which 161 were x-ray sources, created the first x-ray sky catalogue and led to more scientific papers that year than any other subject. Most of these sources were previously unreported. It found x-ray pulsars around binary companions, which were later determined to be neutron stars generating power by accreting matter; as well as extended x-ray emissions from clusters of galaxies, the first evidence of hot gas between galaxies.[42] The catalogue of 339 objects ranged from common x-ray sources to binary sources where material from one source impacted on a companion (called *brennstrallung* or braking radiation) to new x-ray pulsars, all quite different from the Crab. Seven new sources were supernova remnants, while six were binary stars. The strongest sources had x-ray luminosities at least a thousand times the luminosity of the Sun. Even the weakest x-ray source was 20 times more intense than radiation from the strongest radio source. Thirty-four sources had been identified with known objects, but many were new. Outside the galaxy were sources associated with ordinary galaxies, giant radio galaxies, Seyfert galaxies, galactic clusters, quasars and diffuse background x-radiation. Few can have imagined such a large haul from such a small satellite.

Explorer 43 (IMP I): New Drum Design

IMP I was a new IMP design and only one was flown. Built at Goddard, it was the largest and most advanced in the IMP series, with a new digital encoder data processing system. Explorer 43 was a 288kg IMP launched on the Thor Delta M6 (the first with six strap-on boosters) on 13th March 1971 to study the magnetosphere. It provided the first observations of the quiet-time interplanetary electron component in the 20 to 2MeV range.

This IMP looked different, a large drum shape of 16 sides, 1.82m high by 1.26m diameter. There were two 3.5m experiment booms, a 4.48m loop antenna, six electrostatic field measurement antennae (four were extended as far as 61m) and eight telemetry antennae. Power came from 48 solar cells. There were four other booms: two of them 4m long with sensors and the other two 1.5m long for attitude control. Attitude was controlled by Freon thrusters. Telemetry was on 136.170MHz and 137.170MHz and the satellite spun at 5rpm. A new computer was carried, the SDP, able to store 256 words and 4,000 bits of information, transmitting at 1.6kbps. The NASA program scientist was E. Schmerling, with Frank McDonald the chief scientist at Goddard.

Explorer 43 was designed to work with Explorers 33 and 35. By this stage, IMPs had generated 125 scientific papers, having measured and mapped the magnetic field, the shock front boundary and the turbulent transition region behind it (the magnetopause), the magnetospheric tail and the neutral sheet. One of the challenges of this mission was to try find an explanation as to how energetic particles trapped in the magnetosphere got there; what the acceleration mechanisms were and how they were lost; and why the level of galactic cosmic rays tended to increase during periods of *minimum* solar activity. Its primary purpose was to investigate the nature of the interplanetary medium and the interplanetary-magnetospheric interaction during a period of decreasing solar activity, including characteristics of the solar wind and interplanetary fields and the modulating effects of cosmic rays. Subsidiary objectives were to study the radiation environment of

Earth-Moon space; the quiescent properties of the interplanetary magnetic field and its dynamic relationship with particle fluxes from the Sun; solar flares; the properties of low frequency radio waves from the terrestrial magnetosphere, the solar corona and the Milky Way and determine their relationship to magneto-ionic properties of the solar system and the galaxy; and the electric field in the interplanetary medium.

The 12 experiments included new ones for electric fields and radio astronomy. Experiments covered solar and galactic cosmic rays (6); solar plasma (2); magnetic and electrical fields (3) and radio astronomy (1). Tracking of the mission was by STADAN, in particular at Fairbanks, Alaska; Carnarvon, Australia; Rosman, North Carolina; Tananarive, Madagascar; Santiago, Chile; Fort Myers, Florida; Quito, Ecuador; and Ororral Valley, Australia.

The mission required a very precise launch window of 10 minutes to achieve the correct angle of the spacecraft to the Sun. Like its predecessors, it was aimed at highly elliptical orbit, of 232km by 194,000km, 29°, period four days, though this was perturbed over time. An orbit like this meant that it slowed greatly, to only 1,380km/hr, as it reached its high point, but at its low point it was zipping by Earth at almost 36,000km/hr. The spacecraft re-entered on 2nd October 1974, having worked for three years. Six experiments were still working perfectly and four others in a degraded state, with only two having failed.

Explorer 44 (Solrad 10): Mapping the Celestial Sphere

Explorer 44 was Solar Explorer C, also called NRL Solrad (Solrad 10) and was launched on a Scout B from Wallops Island on 8th July 1971. It was companion to Solar Explorers A and B of 1965 and 1968. The 12-sided, 76cm diameter, 58cm tall satellite weighed 118kg and was designed to point continually at the Sun. It had four 18cm by 53cm windmill panels and spun at 60rpm, adjusted by firings of a low-thrust ammonia and hydrazine gas engine. It had two transmitters, on 137.710MHz (continuously) and 136.380MHz, with a 54kbit memory store dumping for five minutes to the NRL tracking station at Blossom Point, Maryland.

Explorer 44 carried 15 experiments addressing solar electromagnetic radiation (x-ray and ultraviolet), x-ray stellar radiation and solar flares and other solar activity, for better prediction of space weather. There were eight solar x-ray monitors; one solar electron temperature monitor; two solar Lyman α monitors; a solar ultraviolet monitor; a solar ultraviolet continuum flash monitor; two background x-ray level monitors; two solar hard x-ray monitors; a solar excitation F layer photometer; an anti-solar temperature thermistor; and a stellar x-ray variation large area proportional counter.

A particular role for Explorer 44 was to provide solar warnings for the upcoming Apollo 15 mission, launched later that month, although Explorer 44 coincided with a period of declining solar activity with the next maximum due 1979–80. The program scientist at NASA was J.M. Weldon and the scientific program manager at NRL was R.W. Kreplin. Explorer 44 mapped the celestial sphere using its high-sensitivity x-ray detector. The last data were received on 25th February 1974 and it decayed on 15th December 1975.

Explorer 45 (S³): New Small Scientific Satellite

Explorer 45 was the second launched from San Marco, taking a Scout B on 15th November 1971 to study the magnetosphere and energetic particles. It was sent into an egg-shaped orbit 26,000km out over the equator – the most elliptical for a Scout to date – to investigate the causes of world-wide disturbances due to magnetic flares. It was hoped to achieve three 'firsts': the first detailed measurements of the particles of the inner magnetosphere; the first measurement of particles in the ring current; and the first simultaneous investigations of particles, magnetic fields and electrical fields. It had the good timing to be working during the historic magnetic storm of August 1972.

The launch was observed by the Italian minister for research and ten members of the Italian parliament. Goddard's team, accompanied by Apollo astronaut Jim McDivitt, was located in a hotel in Malindi beside a national park, 30km away. The most challenging part of the assignment was getting aboard the San Marco platform, which involved a sea journey and then being lifted aboard by a crane in a Billy Pugh net, the type used by the Sea King helicopters to hoist returning Apollo astronauts out of the sea.

Explorer 45 was called Small Scientific Satellite, sometimes S3 or S³, with seven scientific and three engineering experiments from NASA and two universities to investigate the environment of the Earth's inner magnetosphere. Specifically, it was to investigate the cause of world-wide disturbances associated with large solar flares, when low energy protons became suddenly trapped in the magnetosphere but were later lost. In more detail, it was to study the development of magnetic forms and the ring current; investigate the relationship between aurorae, magnetic storms and the acceleration of charged particles in the magnetosphere; study time variations in charged particles; examine electrostatic waves; follow the movement and intensities of protons and electrons; and look at magnetic field strengths, fluctuations and electric fields in the inner magnetosphere. Its scientific experiments were:

- Channel electron multipliers experiment, to make a census of electrons and protons from 700eV to 25keV with charged particle counters;
- Solid state proton detector with two telescopes to measure protons, alpha particles (helium atoms) and heavier ions from 25keV to 2.35MeV;
- Solid state electron detector to measure electron intensities from 35keV to 400KeV;
- Fluxgate and search-coil magnetometers for 0.6 to 3,000γ and 1 to 3,000Hz;
- Electric field experiment using two long booms and four spectrometer channels to measure low frequency variations from 0.3Hz to 30Hz; and
- Electric field experiment to measure high frequency changes from 20Hz to 300kHz.

The three engineering experiments were to measure aerodynamic heating on the satellite at the point of shroud jettison, 104km; measure radiation damage on small integrated circuits in the most intense parts of the radiation belt; and test a star-sensing system.

S³ weighed 52kg and was proclaimed at the time as the most innovative and compact satellite to date, reprogrammable by ground control to respond to sudden events. It was made with riveted, rather than machined, thin-sheet aluminium, which reduced costs and weight. The eight-sided polyhedron, 68.6cm across and with solar cells for its 18-cell battery, incorporated two 2.74m booms (electric field sphere detectors), two 60.9cm booms (search coil magnetometers) and four 60.9cm antennae. The spin rate was 7rpm. Explorer 45 was intended for equatorial orbit at 3.5°, 222–28,876km, 8 hours 37 minutes, to operate for a year. Data were transmitted at 446bps and it was tracked by STADAN stations at Johannesburg, South Africa; Orroral, Australia; Quito, Ecuador; and Rosman, North Carolina; with scientific data sent on to NASA Goddard. Transmissions were on 136.830MHz and wide-band analogue on 139.95MHz. The program scientist at NASA was Lawrence Kavanagh, the project scientist at Goddard was Robert Hoffman and the program manager was Gerald Longanecker, who went on to manage the International Ultraviolet Explorer.

The mission got off to a bad start, as it began coning in its 233–26,895km orbit and was 24° off angle by orbit 20. After 3,000 attitude control commands and 20 complete reprogramming instructions were sent to the spacecraft, the coning was reduced to 1.8° and then stabilized.

The highlight of the mission was the great solar storm of 4th–6th August 1972 when Earthbound instruments went off-scale. On orbits 819 and 820, Explorer 45 observed the contraction of the magnetopause, its reversal of direction in a few seconds, huge fluxes of streaming protons, an increase of radiation by 500γ and intense broadband noise and hiss. It was the largest disturbance ever recorded in space and Explorer 45 captured the six key events in the process.[43] The storm occurred at the mid-point of the final two Apollo missions (16 in April 1972, 17 in December) and could well have been a danger to astronauts travelling on the Earth-Moon route at that time.

After two years of operation, project scientist Robert Hoffman announced that S^3 had made the first measurement of stimulated electromagnetic emissions in the magnetosphere and of alpha particles in the equatorial region, as well as the interaction of electromagnetic waves with charged particles and the closest approach to Earth of the boundary of the magnetosphere. In the first two years, the data processing system on S^3 had been reprogrammed 541 times.

S^3 was turned off on 30th September 1974, by which time its reprogrammings totalled 789. Seventeen scientific papers had been presented and 12 submitted, with a further 68 conference presentations. S^3 had made detailed studies of the Earth's ring current during the magnetic storm main phase, had found that alpha particles in the equatorial region were 100 times more numerous than expected and had measured the boundary of the magnetosphere at 25,000km at its closest approach.

Explorer 46 (MTS): Double Protection

Explorer 46 was a 136kg Meteoroid Technology Satellite (MTS), launched on a Scout D-1 from Wallops Island on 13th August 1972. It followed in the footsteps of Explorer 16 and 23 and the three large Pegasus satellites. It had a double wall, called a multi-sheet bumper, to protect against penetration. This was two thin sheets of metal 12.7mm apart, an approach that turned out to be six times more effective than a single wall of the same width.

The objectives of Explorer 46 were to measure the meteoroid penetration rates in the protected target while obtaining data on meteoroid velocity and flux distribution. The satellite was developed by NASA Langley under the guidance of the NASA Office of

Aeronautics and Space Technology. It was a 3.2m long cylinder with 12 panels of 96 detectors in a windmill shape. There were two telemetry systems, one for all the instruments and one reserved for the bumper experiment. NASA Goddard was responsible for tracking and telemetry and William Kinnard was the project scientist at Langley Research Center. The orbit intended was 491–8,815km, 98 minutes, 37.7°. The instruments comprised:

- Multi-sheet bumper, 7m across, its detectors filled with gas, to register and telemeter loss of pressure;
- Twelve box-shaped velocity detectors at various locations along the spacecraft;
- Impact flux detectors, with 64 detectors to assess the population of very small particles.

Explorer 47 (IMP H): Circular Orbit

IMP H was the last part of a sub-series with IMP J, but using the IMP I drum design. Explorer 47 departed on a Delta 1604 (six strap-on boosters) from Cape Canaveral on 23rd September 1972. At the time, data were still being received from IMP E, F and G. By this stage, IMPs had provided accurate measurements of the interplanetary magnetic field, the magnetosphere boundary and the shock wave and had also detected the extended geomagnetic tail plasma sheet, as well as finding that the permanent magnetic tail did not co-rotate with Earth. The primary objective of IMPs H and J was to obtain a more detailed understanding of these regions. This time, a nearly circular 13-day orbit (202,000–236,000km, R31 to R37) was used to monitor the interplanetary medium and the geomagnetic tail plasma sheet for an entire year. Such an orbit required a powerful 124kg kick motor. The secondary purpose was to perform detailed, near-continuous studies of the interplanetary environment during several rotations of active solar regions; and to study particle and field interactions in the distant magnetotail, including cross-sectional mapping of the tail and neutral sheet. The mission came at a time of declining solar activity and its orbit would put it in the geomagnetic tail plasma sheet for 6.5 days a year. IMP H was the largest and most complex of the IMP series.[44]

Explorer 47 followed the design of 43, being a 16-sided drum-shaped spacecraft weighing 376kg and measuring 157cm high by 135cm diameter. The spacecraft was powered by solar cells and a chemical battery and scientific data were telemetered at 1.6kbps. It was spun every 1.3 seconds (a spin rate of 46rpm), adjusted by a Freon-14 cold gas monopropellant. There were three rings of solar arrays, two 3m booms for magnetic fields experiments, two 1.2m booms for attitude control and eight antennae. Explorer 47 carried thirteen instruments designed to measure energetic particles, plasma and electric and magnetic fields. There were eight energetic particles experiments, two fields experiments and three plasma experiments. The energetic particles experiments comprised two for cosmic rays, one for energetic particles, one for charged particles, one for electrons and isotopes, one for ions and electrons, one for solar electrons and one for low energy particles. The fields experiments comprised one for magnetic fields and one for plasma waves, while the plasma experiments comprised two for solar wind and one for ion composition. The instruments came from Goddard, the National Oceanic and Atmospheric

Administration (NOAA), Los Alamos Scientific Laboratory, the Atomic Energy Commission and TRW systems, while the universities involved were Chicago, Johns Hopkins, CalTech, Maryland, Iowa and MIT. There were also three engineering experiments, for thermal coatings with 12 samples, data processing and glass solar cells. The program scientist was L.D. Kavanagh, with Norman Ness at Goddard. The two telemetry systems on 137.92MHz facilitated tracking by STADAN, in particular by Carnarvon, Tananarive, Rosman, Santiago, Fairbanks, Canberra, Quito and Johannesburg.

Explorer 47 was turned off in October 1978 after six years, with eight of the 13 experiments still fully working. The mission cost was $14.5m, split between the cost of the satellite and the launcher. Its scientific outcomes are reviewed with the last in the series (Explorer 50), but Explorer 47's particular contribution was that it made it possible to estimate the speed of cosmic rays, which travel near to the speed of light. The success of the mission led to consideration of a trio of satellites operating in Earth-Moon space and this subsequently emerged as the International Sun Earth Explorer (ISEE, see Explorer 56) with the European Space Research Agency (ESRO).

Explorer 48 (SAS-B): Dome with a Telescope

Explorer 48 was a 185kg SAS astronomy satellite, launched by Scout D-1 on 16th November 1972. It was the second small astronomy Explorer after *Uhuru* (Explorer 42), the fifth launched from San Marco and the first-ever night launch from the platform, into a starry African sky. It was aimed at 555km. SAS-B was also the first gamma ray Explorer for eleven years (the first was Explorer 11 in 1961). A feature of the design, commonplace nowadays, was a standard base on which different instruments could be mounted on following missions.[45] Explorer 48 followed a decision by the National Research Council Space Science Board in *Priorities for space research 1971-80* to award priority to gamma ray astronomy, which was considered critical to understanding energy transfers in the universe. Explorer 48 had just one experiment, a 32-level, digitized spark chamber gamma ray telescope – built by Goddard and housed in a dome on top – to measure the intensity, energy and direction of gamma rays with a sensitivity ten times greater than anything previously. Although they had been flown on balloons, this was the first such telescope in orbit, with twelve times the sensitivity of the OSO-3 observatory. The chamber design permitted a three-dimensional electronic picture to be built of incoming gamma rays, with an accuracy of 1° and a spectrum of 25 to 200MeV.

The mission's objectives were to measure the direction of galactic and extra-galactic x-ray radiation, to measure the spectrum in the 25 to 200MeV range, to determine sources within and outside our galaxy, to look for short gamma ray bursts from supernovae and to search for pulsed gamma radiation from pulsars. Secondary objectives were to measure gamma ray intensity and how that differed across the sky, to better understand supernovae and to test the steady-state theory of the universe.

The origins of gamma rays were unclear. They have enormous power – 200,000 times the power of visible light photons – but most is absorbed in the atmosphere. Even above the atmosphere they are transformed into electron-positron pairs which must be detected and observed. Explorer 48 would start with an all-sky survey – expected to be ten times more sensitive than anything done previously – and then move on to discrete sources. It would focus on the galactic plane in general and the Crab nebula in particular, with the hope of shedding light on the dynamics of the Milky Way.

Explorer 48

The spacecraft was 1.29m tall and 55cm diameter, with four 3.96m wide solar panels supplying 27w power to recharge nickel-cadmium batteries. It was stabilized by electromagnets. Data were transmitted in real-time at 1 kbps and downloaded for five minutes every 96-minute orbit by VHF transmitter. The cost of the mission was $9m, with $1.45m for the Scout and $600,000 paid to Italy for launch costs. Tracking was done by Quito, Ecuador; U.S. stations in the Seychelles and Ascension; and the French stations in Guyana and Brazzaville, Congo; with data going to Goddard. The project manager was Marjorie Townsend, an engineer from George Washington University. She was the first woman to hold the post of project manager at NASA and represented a significant breakthrough for women in space science. Her achievements were rewarded with the NASA Exceptional Service Medal, the Knight of the Italian Republic for her earlier role in *Uhuru* and later the Federal Women's Award. The program scientist was Albert Opp and the principal investigator was Carl Fichtel.

On 20th November 1972, the telescope experiment was turned on, becoming operational a week later. In its first months, Explorer 48 observed the galactic plane, the galactic centre and x-ray sources such as the Crab nebula. Preliminary results were presented to the American Physical Society the following April. Rich gamma ray sources had been detected in the galactic plane; high energy gamma radiation from the Crab nebula (only low energy hitherto); and gamma rays not only far away from the galactic plane but from outside our galaxy, which could be of cosmological origin. Explorer 48 then picked up a pulsar in Vela, believed to be 10,000 years old, with two non-simultaneous bursts of gamma rays for every radio wave burst, a phenomenon never before observed.

Fresh outcomes were presented at the summer's second international symposium on gamma ray astronomy at Goddard. Explorer 48 was strongly indicative that most cosmic rays came from our own galaxy rather than further afield. They were most likely the outcome of supernova explosions, or pulsar remnants from such explosions. The theory that cosmic rays came from supernova explosions was not new – it dated to the 1930s – but this was the best supporting evidence to date. SAS-B found that high energy cosmic ray gas was unevenly distributed throughout the galaxy, with concentrations in different places. Gamma rays seemed to come from concentrations of cosmic rays in the spiral arms of our galaxy, principally those richest in dense molecular clouds. The detection of high energy, short wavelength radiation from inside the galaxy, its centre and the Crab nebula had far-reaching implications for cosmologists, for it suggested the remnants of the fireball of the big bang. Explorer 48's power supply failed on 8th June 1973 and the satellite had to be closed down, ending the scientific return.

Explorer 49 (RAE-B): Last American Moon Probe for 20 Years

Explorer 49 was the second Explorer to orbit the Moon, taking a Delta 1913 (nine strap-on boosters) from Cape Canaveral on 10th June 1973 on a direct ascent path with one course correction. It was the second Radio Astronomy Explorer, similar to its predecessor Explorer 38 seven years earlier, but this time placed in lunar orbit to avoid Earth's radio noise and designed to function on slightly different wavelengths. The idea of lunar-orbit-based radio astronomy came from John Clark, director of Goddard, in 1968, but could probably be traced back to a near-namesake, *2001* author Arthur C. Clarke. Explorer 49 was to probe low frequency radio noise, searching five categories of radio emission: galactic and extra-galactic; the Sun; Jupiter; Earth's magnetosphere; and cosmic sources (Cygnus A and Centaurus A being specifically identified).

Like its predecessor, Explorer 49 was a small cylinder 92cm across and 79cm tall, its 38.5w power coming from four solar panels with 3,850 solar cells for a nickel-cadmium battery. Its total weight was 328kg (200kg probe with 128kg retro rocket). Explorer 49 carried two tape recorders with a capacity of 225 minutes on two tracks and playback time of 45 minutes. Data return was via a low power UHF transmitter in real-time and via a high power UHF transmitter in stored form from the two tape recorders. Explorer 49 had three transmitters: two at 400.950MHz (high or low power) with tracking data at 136MHz, while uplink was sent on a 148MHz VHF system. There were three engines: a hydrazine course correction package; a cold gas attitude control system; and a solid-fuel lunar insertion motor.

Explorer 49

Explorer 49 over the Moon

The experiments comprised two Ryle-Vonberg radiometers (nine channels each) and three swept frequency burst receivers (32 channels each), with an impedance probe for calibration. The antennae formed an X configuration: a 229m V pointed away from the Moon and a similar one pointed towards it, each made of 5mm beryllium copper alloy strips. There was also a 37m dipole to be deployed parallel to the lunar surface and a 129m boron damper boom. The antennae would scan from 25kHz to 13.1MHz in 32 steps over 144 seconds. Attitude was determined by solar, horizon and panoramic sensors accurate to 1° and the spacecraft was orientated to the gravity gradient, toward the Moon. It was built at Goddard and the mission cost $11.1m (the Delta was $8.6m). The program scientist was Nancy Roman and the project manager was John Shea, with Robert Stone as the PI at the Goddard Laboratory for Extraterrestrial Physics.

A low energy flight path to the Moon was constructed over five days, with small trajectory corrections available at the ten-hour and 95-hour point (one was made). Explorer 49 reached the Moon on 15th June 1973 after 113 hours, its retro-rocket firing for 20 seconds to achieve an original orbit of 1,053–1,063km, 221 minutes, 58.7°. The motor was then dropped. Deployment of the antennae took place in two phases, starting with the 37m dipoles and then the others in stages, finally reaching their full 228m length in November 1974. This was a record at the time and possibly remains so today. Two small cameras were placed on the tips to follow the deployment and the X-shape of the spacecraft measured almost half a kilometre between the tips (458m) when completed. On 20th June, the first data were received in the form of the 'crackle, squeeze and buzz of celestial low frequency radio noise,' as reported in *Goddard News*. The mission proved the viability of using lunar orbit as a base for radio observations, although Earth's radio noise was readily apparent whenever Explorer was on the Earth side of the Moon.[46] Explorer 49 was the last U.S. Moon probe until Clementine (1994) and the Moon was not used again for astronomical observatories until the Chinese Chang e 3 in 2013.[47]

Explorer 50 (IMP J): Last IMP Transmits From 1973 to 2006

Explorer 50 launched on a Delta 1604 (six strap-on boosters) from Cape Canaveral on 26th October 1973. The IMP J, the tenth and last of the series, was placed in a more elliptical orbit than planned (190,000 by 244,000km) and the second furthest out. IMP J carried 12 experiments to study solar wind, solar rays, magnetic fields and their effect on the weather. By this stage, IMPs had provided ten years of data going back to 1963 and had followed Earth's radiation environment over a whole solar cycle, starting from a quiet Sun period. The intention for this mission was to enter an orbit similar to Explorer 47, IMP H, of 192,620 by 234,000km, about halfway to the Moon but 180° apart, with J at R38 and H at R32. This would enable data on, for example, solar flares to be correlated from the opposite ends of Earth. Their data would also be matched against Pioneer 10 and 11 en route to Jupiter and Mariner 10 en route to Venus and Mercury.

Explorer 50 was a 398kg 16-sided drum, 1.35m diameter, 1.57m tall, with its instruments at the top and its motor below. There were two 3m long experiment booms, eight antennae and a cold gas Freon-14 monopropellant for attitude control. Its rotation was scheduled at 23rpm. IMP J carried a solar and Earth sensor and its telemetry was 1.6kbps on 137.98MHz using two transmitters. Tracking was from stations in Carnarvon; Tananarive; Rosman, North Carolina; Santiago; Fairbanks, Alaska; Canberra; Quito; and Johannesburg, sufficient to give 100% coverage. The program scientist for Explorer 50 was L.D. Kavanagh, with Norman Ness at Goddard. The spacecraft was constructed at Goddard and cost $8m, with launch cost of $7m. It carried 12 experiments for energetic particles (cosmic rays (four telescopes), ions and electrons, electrons and isotopes, charged particles); plasma (low energy particles); fields (magnetic fields, electrical fields); and engineering (data multiplex unit and solar panel test with improved cells).

In addition to Goddard, the institutions involved were NOAA; Los Alamos Scientific Laboratory; and the universities of Chicago, Maryland, Johns Hopkins, MIT and Iowa. IMP had grown from an original specification of 62kg for nine experiments weighing 19kg, to one of 13 experiments on a spacecraft of 395kg, with 81kg now available for experiments. The power supply had risen from 35w to 120w and the data rates were up from 25 bits/sec to 1,600.

The rocket motor was fired at 2.5 days, but the orbit achieved was R25 to R45, rather than circular at R32. Its orbital inclination was perturbed and ranged from 0° to 55° with a periodicity of several years. Explorer 50 was in the solar wind for 7–8 days of every 12.5-day orbit, giving a high rate of data return although the telemetry level varied, from 90% in the 1970s, falling to 65% in the 1980s, but back to 90% in the 1990s. IMP J was still operating in 1992 when it provided radiation measurements in advance of the Galileo probe's journey through the geomagnetotail on its way to Jupiter. The mission celebrated its 25th anniversary in October 1998. The magnetometer failed in June 2000, but the Charged Particle Measurements Experiment and the Energetic Particle Experiment were still going strong. The mission was formally ended in October 2001, but the Canberra station continued to follow it until 7th October 2006 – a remarkable achievement.

Overall, the IMPs provided an initial dataset of ten years. They brought numerous firsts: the first accurate measurements of the interplanetary magnetic field; the first mapping of the shock front boundary and turbulent transition region; the first detailed information on the tail region; the first evidence of the neutral sheet; the first evidence of energetic electrons in the neutral sheet which may feed aurorae and replenish the radiation belts; the first operational solar flare warnings; and data on the large-scale but negligible magnetic fields of the Moon and the solar wind cavity void behind the Moon. Their achievements included the:

- First extensive observations of the interaction of the solar wind and magnetosphere;
- First determination of the geometry and structure of the bow shock;
- First mapping of the geomagnetic tail beyond lunar orbit;
- First measurements of the interplanetary magnetic field; and the
- First observations of the interaction of the magnetized solar wind with the Moon.

The extended geomagnetic tail plasma sheet may have been the most important discovery of the series, but there were others where quite specific details were obtained. IMP plasma measurements were able to distinguish helium, hydrogen and oxygen ions. A solar x-ray detector made the first independent confirmation of cosmic gamma ray bursts, peaking in the 150keV region between the x-ray and gamma ray domains. CalTech scientists determined that less than 20% of low energy electrons had positive charge and less than

Explorer 51 (AE-C): Dipping into the Atmosphere

Table 2.2 IMP operations

Explorer	Apogee	Radii	Inclination	Transmissions
A. Explorer 18 S-74	202,000km	31.7	33.3	27 Nov 1963 - 6 May 1964
B. Explorer 21 S-74a	101,300km	15.9	33.5	4 Oct 1964 - 9 Apr 1965
C. Explorer 28 S-74b	264,000km	41.5	34	29 May 1965 - 12 May 1967
D. Explorer 33	456,000km	71.6	24.1	1 Aug 1966 - 15 Oct 1971
E. Explorer 35	Lunar orbit			19 Jul 1967 - 16 Jan 1971
F. Explorer 34	225,000km	35.4	67.2	24 May 1967 - 3 May 1969 re-entered
G. Explorer 41	182,800km	35.4	86	21 Jun 1969 - 23 Dec 1972 re-entered
I. Explorer 43	205,800km	32.3	39.9	13 Mar 1971 - 2 Oct 1974 re-entered
H. Explorer 47	247,000km	37	28.6	22 Sep 1972 - 31 Oct 1978
J. Explorer 50	288,000km	45	68.1	26 Oct 1973 - 7 Oct 2006

1% of cosmic ray electrons from the Sun were positrons. The intensity of low energy electrons was found to vary from day to day. Strong low frequency electromagnetic noise was found in the distant plasma sheet. In the solar wind, two different ion streams were found travelling at different speeds in double streams. The solar wind was hot, but could become saturated, at which point waves were generated to scatter the flow. Data on fluxes of protons made it possible to compute hourly averages. Finally, there were engineering advances in the form of digital data processors, metal oxide silicon transistors and magnetically clean spacecraft. Table 2.2 summarizes the span of the series, from 1963 to 2006.

Explorer 51 (AE-C): Dipping into the Atmosphere

Explorer 51 was an Atmospheric Explorer (AE), launched on a Delta 1900 (nine strap-on boosters) from Vandenberg on 16th December 1973, to continue experiments begun by Explorers 17 and 32 and pave the way for two more in 1975 (Explorer 54, 55). The most remarkable feature of the mission was that it could lower its perigee to only 115km to dip into the atmosphere, followed by a re-boost to a safe height. No previous attempts had been made to dive a satellite into the atmosphere like this and only two others would do so (Explorers 54, 55), although many years later, MAVEN made what were called 'deep dips' over Mars. The hydrazine-fuelled motor used on Explorer 51 was called the Orbit Adjust Propulsion Subsystem (OAPS). The mission was designed to use real-time data as much as possible, utilizing a computer called the Sigma 9 to enable rapid response, so that the orbit could be changed quickly to measure the impact of a solar flare, for example.

The aim of the mission was to explore the atmosphere at its lowest accessible altitudes – 115km to 300km – where important energy transfer, atomic and molecular processes and chemical reactions take place critical to the heat balance in the atmosphere. Up till then, this area had been probed only by sounding rockets for a few minutes at a time. The aim of the mission was to examine particles, fluxes, airglow intensities, temperatures, magnetic fields, heat balance, incoming solar ultraviolet radiation and energy conversion. It was to gather data on photochemical processes, matching them against incoming solar energy and energy transfer. The mission would study the composition and processes of the thermosphere, in particular ion and neutral composition, energetics of the ionized atmosphere, low energy electrons and air glow excitation, but also the global structure and dynamics of the atmosphere. The 14 instruments were designed to measure solar ultraviolet rays; the composition of positive ions and neutral particles; the density and temperature of neutral particles, positive ions and electrons; and the measurement of airglow emissions, photoelectron energy spectra and proton and electron fluxes up to 2keV.

The mission plan was to enter an initial orbit of 156km by 4,300km, lowered over several days to 120km. Explorer 51 would dip-and-reboost several times, back up to 390km and decaying to 150km before a dip (the terms used were 'excursions', 'perigee restorations' and 'apogee restorations'). When propulsion was almost exhausted, which was expected after six months, it would climb back to an orbit between 300 and 600km. The aim was to sample the upper atmosphere at different points: from 10°N to 68°N and then back down to 60°S. The spacecraft carried 168kg of fuel and it was anticipated that 1.5kg would be used for each firing. Skimming the atmosphere would create friction, so louvres and insulation were fitted to keep its internal temperatures in the range 10° to 35°C.

Explorer 51 was a multi-sided polyhedron, 1.35m diameter, 1.15m high and weighing 658kg, including 95kg of instrumentation. For orientation, it had horizon scanners and solar sensors. It was spun between 1rpm and 4rpm. There was a solar array for the nickel-cadmium batteries which powered the instruments, real-time telemetry and a tape recorder. Transmissions were on 137MHz, at a rate of 16kbps real-time and 131kbps on the tape recorder. The mission was commanded from the Atmosphere Explorer Operations Control Center (AEOCC), with commands issued through Rosman, North Carolina and Madrid, Spain. The program scientist at NASA was E.R. Schmerling, with Nelson Spencer at Goddard. The total cost of the mission was $20m.

Explorer 51 found that the weather of the upper atmosphere was constantly changing, with powerful winds ten times as severe as near the surface blowing west to east at 250km/hr, generally day side to night side. The upper atmosphere was dynamic and unpredictable, which was a big surprise.

Explorer 52 (*Hawkeye*): Last Van Allen

Explorer 52 was *Hawkeye*, a small 27kg solar wind satellite launched on 3rd June 1974 and a contributor to the International Magnetospheric Study (IMS). Its E-1 Scout headed directly south out of Vandenberg, with the fifth stage burn of its B-3A rocket motor over the south pole monitored by the National Science Foundation South Pole Station. *Hawkeye* was sent into a highly eccentric orbit (500–101,746km, 90°, 38 hours) to study the solar wind's interaction with Earth's magnetic field. The most striking feature about the mission was its unusual name and, as with Injun, the culprit was the University of Iowa, where *Hawkeye* was the name of the Iowa mascot and football team. It was the sixth and last of the Injun series as well as the last Van Allen mission in a series he had done so much to define.

The primary mission objective was to conduct particles and fields investigations of the polar magnetosphere of the Earth out to R21. It was tasked to investigate the interaction of the solar wind with the Earth's magnetic field, especially the north polar cap. The secondary objectives were to measure magnetic field and plasma distribution in the solar wind and to study emissions caused by solar electron streams in the interplanetary medium, aiming to locate magnetically neutral points or lines during both quiet and disturbed magnetic conditions.

Explorer 52 (*Hawkeye*): Last Van Allen 127

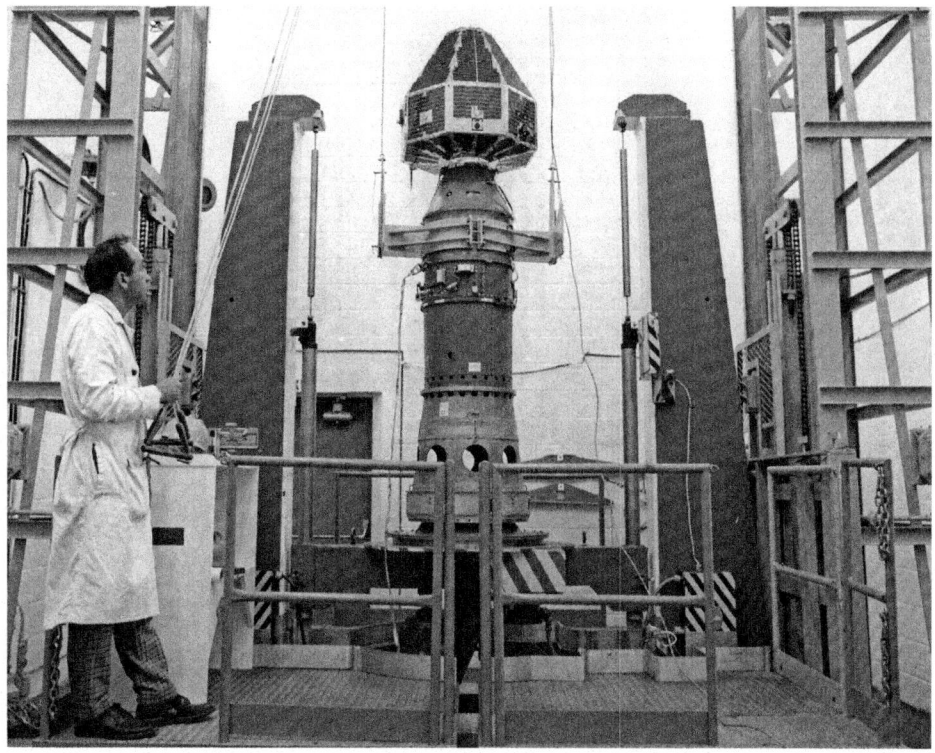

Explorer 52 *Hawkeye*

The top-shaped spacecraft was a 75cm tall cone set on an eight-sided base, with solar cells on all sides providing power of 22 to 36w. The spacecraft spun every 11 seconds and was controlled by the North Liberty Radio Observatory in the University of Iowa's Spacecraft Operations Control Center, with support from Goddard. Data were obtained in real-time only, at frequencies of 136MHz and 400MHz at 100 bps. *Hawkeye* was built in Iowa and cost $3.6m, but Langley Research Center had management responsibility. Its instruments comprised:

- Fluxgate magnetometer, 1.6m long, to measure magnetic topology and temporal variations in four sensitivity ranges (James Van Allen);
- Low Energy Proton Electron Differential Energy Analyzer (LEPEDEA) to measure the angular distribution of electron and proton intensities;
- Very low frequency, extremely low frequency electric and magnetic field receiver with two antennae extending to 23m.

James Van Allen remained active in the space program until his retirement in 1985, receiving many awards marking a lifetime of scientific achievements. Scientists the world over joined him to celebrate his 90th birthday in 2004. He lived another two years until 9th August 2006, with colleagues gathering one last time to commemorate his inspirational role in space science.

Early Explorers

James Van Allen

Explorer 53 (SAS-C)

Explorer 53, launched on a Scout F from San Marco on 7th May 1975, was a Solar Astronomy Satellite (the third) to study galactic x-ray sources. It was the last San Marco launch of an Explorer, aimed at a circular equatorial orbit at 486km, 96 minutes, for a one-year design lifetime. The planned night-time launch was to help the equipment stay cool before take-off.

The general purpose of the mission was to study major energy transfer mechanisms in the universe in the 'fast developing science of high energy physics'. Its specific objectives were to survey the celestial sphere for sources radiating in the x-ray, gamma ray, ultraviolet and other spectral regions. Explorer 53 was to measure the x-ray emissions of discrete extragalactic sources; monitor the intensity and spectra of galactic x-ray sources from 0.1 to 60keV; and monitor the x-ray intensity of Scorpio x-1, the first non-solar x-ray source, which had been discovered by sounding rocket in 1962. It was to determine bright x-ray source locations to an accuracy of 15 arcseconds and continuously search the sky for x-ray novae, flares and other transient phenomena. One of its functions was to return to objects first identified by *Uhuru*.

Explorer 53

The spacecraft weighed 195kg and was 1.452m high and 4.7m wide. Four solar paddles provided power for a 12-cell nickel-cadmium battery. The spacecraft could be spun slowly (0.1°/sec) or pointed at a particular object for 30 minutes. SAS-C incorporated nickel-cadmium batteries, a charge regulator and secure logic circuitry to allow commands to be stored and performed beyond the range of ground stations, all compressed by 100:1 in the spacecraft memory. The compression technology was later adapted for electronic implants for patients suffering neurological disorders. The mission was tracked from Quito, Ascension and the San Marco mobile Italian station and was managed by Goddard. The spacecraft was built at Johns Hopkins and the x-ray experiment, which included two star cameras, by MIT. The principal investigator was George Clark of MIT. The four experiments were:

- Galactic absorption experiment to estimate the density and distribution of interstellar matter;
- Scorpio monitor experiment, to view time variations in Scorpio x-1 from 0.2 to 50 keV;
- Galactic monitor experiment, to locate x-ray sources;
- Extra-galactic experiment, to collect data on weak x-ray sources outside the Milky Way in a 100^{o2} region, designed to take in the Virgo cluster of galaxies, the Andromeda galaxy, the Large Magellanic Cloud and the galactic equator.

The mission got off to a difficult start when the spacecraft wobbled due to a stuck nutation damper, but Explorer 53 went on to complete a four-year mission until April 1979. It mapped precise locations for 60 x-ray sources to within an arc minute and made a survey of the soft x-ray background (0.1-0.28 keV). Its first discovery was quasar 0241, the nearest of 600 quasars thus far identified at 800 million light years distant. Earlier, 0241 had been identified by SAS-A as an object, not necessarily a quasar, but obscured by thick dust. Explorer 53 also discovered the first highly magnetic white dwarf binary in Hercules and an isolated white dwarf in Algol. It discovered a dozen x-ray burst sources including a rapid burster (MXB1730-335), a neutron star pulsing every 3.6 seconds and located x-ray sources near the centre of globular clusters. It was the first to identify a quasar from x-ray emissions. The mission came at around the time of Britain's Ariel 5 x-ray satellite (15th October 1974) and both measured x-ray bursters which had short, periodic but intensive emissions, the shorter indicating the presence of the black holes, the longer a rotating neutron star. Explorer 53 was the last American x-ray mission for some time.

Explorers 54, 55 (AE-D and -E): Dipping into the Poles, Equator

The official Explorer numbering ends at 55 and the last two were Atmosphere Explorers (AE). Explorer 54 was sent into polar orbit from Vandenberg on 6th October 1975, while Explorer 55 was put into near-equatorial orbit from Cape Canaveral on 20th November 1975. Both were launched by Delta 2910. Their instrumentation was similar and the missions were timed closely together, with Explorer 54 carrying out a program in the polar regions while 55 operated around the equator and carried instruments to measure ozone and connections to its depletion by humans.

Explorer 54 was aimed at a polar 90° orbit of 250km to 3,800km, to be lowered as much as possible – consistent with its survival – as far down as 157km, repeatedly dipping down and reboosting but then circularized as its fuel neared depletion. Explorer 54 carried 12 instruments similar to those of Explorer 51: an ultraviolet nitric oxide photometer; a cylindrical electrostatic probe; an atmospheric density accelerometer; a photo-electron spectrometer; a retarding potential analyser; a visual airglow photometer; an extreme solar ultraviolet monitor; a magnetic ion mass spectrometer; a low energy electron spectrometer; an open source neutral mass spectrometer; a neutral atmosphere composition spectrometer; and a neutral atmosphere temperature spectrometer.

Like 51, the 676kg Explorer 54 dived into the upper atmosphere, but much further north to sample the polar atmosphere. The satellite carried a computer system designed to create a database that would make information quickly available to the principal investigators, one which, for the first time, was accessible from small terminals. There was a shared experimental analysis approach, with data for one investigator automatically available to others to encourage a team approach to selecting targets.

With the launch of the 721kg Explorer 55, NASA highlighted its role in studying Earth's protective ozone layer. Ozone is essential to protect the Earth from ultraviolet radiation and related warming and its reduction had been noticed from the early 1970s. Ozone appeared to be breaking down under the influence of industrial pollution, such as oxides of nitrogen and chlorine-based aerosols. At the time, consideration was first being given to banning chlorofluoromethanes (CFMs), because they were thought to destroy ozone. Nitric oxide had grown in importance because it was one of the main constituents of the ozone layer. In contrast to 54's polar orbit, 55 was aimed at the equator from 20°N to 20°S. It carried a backscatter ultraviolet spectrometer to measure ozone, an instrument originally flown on another Goddard project, Nimbus 4, from 1970. The PI for this instrument was Donald Heath, who spoke of the alarm amongst scientists at the depletion of ozone at 22–25km. Satellites were the only way to get global, seasonal and solar cycle data. As it dipped into the atmosphere, Explorer 55 tried to establish an ozone profile from 22km to 50km. Its 12 instruments were cylindrical electrostatic probes; an atmospheric density accelerometer; a photoelectron spectrometer; a positive ion mass spectrometer; an airglow photometer; a solar extreme ultraviolet spectrometer; an open source neutral mass spectrometer; a neutral atmosphere composition, to measure concentration and distributions of neutral gas; a neutral temperature instrument, to measure atmosphere kinetic temperature; a backscatter ultraviolet spectrometer (a new instrument); a planar ion trap, to measure ion temperature, concentrations and velocity of plasma flow (also new); and a solar extreme ultraviolet photometer.

Explorer 54 was able to collect samples at the polar latitudes in its first three months, but regrettably the solar panels failed in January 1976, bringing the mission to an end. It burned up a month later. Happily, it was possible to revive the old Explorer 51 the following month to partly replace what 54 would have done and that lasted until its re-entry on 12th December 1978. Explorer 55 worked between 130km and 400km and made six full latitude cycles before reboosting to a circular orbit of 390km on 20th November 1976. Whenever it decayed naturally to 250km, it was raised back to this height. It re-entered on 19th June 1981 after over five years and 31,268 orbits, crashing into the Caribbean east of Nicaragua. The last signals were received over Hawaii, but none came through over the next station (Ascension Island).

Finally, there was a failed Scout launch at the end of the year on 6th December 1975, when the 71kg Dual Air Density Explorer A and B failed to reach orbit.

Conclusions

Explorer 55 was the last of this stage of the program, effectively bringing the program developed in the 1950s (the 'S' series) and the 1960s (designated subsets) to a conclusion. Reviewing the series, Homer Newell described them in *Beyond the atmosphere - early days of space science* as 'cousins, stemming from a common, rather straightforward technology,' with similar structures, temperature controls, tracking and telemetering devices and antenna systems. Over the period, the technology advanced, better components and materials became available and improved housekeeping equipment was devised, but they retained many common features. By the end, the series had become much more capable and he cited the example of Explorer 35, which could outperform all the early ones and could do so as far away as the Moon. The early missions are summarized in Tables 2.3 and 2.4.

Table 2.3 Early Explorer missions ('S' series).

Designator	Type	Explorer number
S-1	Magnetosphere	Failure, 16th July 1959
S-2	Multi-purpose	6
S-1A	Magnetosphere	7
46	Geodetic	Failure, 23th Mar 1960
S-30	Ionosphere	8
56 (A)	Atmospheric balloon	Failure, 4th Dec 1960, 9
P-14	Plasma and magnetic fields	10
S-15	Gamma rays	11
S-45 (A)	Solar astronomy	Failure, 24th Feb 1961, 25th May 1961
S-3	Energetic particles	12, 14, 15, 26
S-55 (A, B, C)	Meteorites	13, 16, 23
S-6	Atmosphere Explorers	17, 32, 51, 54, 55
S^3	Small Scientific Satellite	45

Table 2.4 Early Explorer missions (sub-series).

Type	Initial	Explorer number
Energetic Particles Explorer	EPE	12, 14, 15, 26
Atmosphere Explorer	AE	17, 32, 51, 54, 55
Interplanetary Monitoring Platform	IMP	18, 21, 28, 33, 34, 41, 47 50
Atmospheric Density Explorer	ADE	19, 24, 39
Beacon Explorer	BE	22, 27
Ionosphere Explorer	IE	20, 25, 40, 52
Geodetic Earth Orbiting Satellite	GEOS	29, 36
Solar Radiation Explorer (Solrad)	Solrad	30, 37, 44
Direct Measurement Explorer	DME	31
Radio Astronomy Explorer	RAE	38, 49
Small Astronomy Satellite	SAS	42, 48, 53
Meteoroid Technology Satellite	MTS	46

Overall, the early missions up to the point reached here (1976) had carried out the full range of the program laid down by the IGY. The scientific haul had been substantial, most evident in space astronomy (for example, the SAS series), the magnetosphere and the physical environment between Earth and the Moon. The program ventured into areas of cutting-edge science, like gamma rays (Explorer 11, 48) and solar emissions (Explorer 30) and the level of knowledge of the ionosphere, magnetosphere and the radiation belts by the end of this period bore no resemblance to what had been known fifteen years earlier – it was a whole new world.

Over time, Explorer satellites provided accurate measurements of air density and micrometeorites. There were important discoveries, such as the great length of the Earth's magnetic tail, further out than the Moon and of how noisy Earth's radio environment had become (Explorer 38), while the SAS series opened the door to the extraordinary world of the x-ray universe. Important advances were made in knowledge of such distinct areas as the Earth's geode (Explorer 29) and whistlers (Explorer 40), not to mention exotic areas like red arcs and gravity waves (31, 32). The first sky maps were compiled.

The spacecraft improved in technical capabilities and the Thor Delta rocket made it possible for larger satellites to be launched, while the introduction of new launch sites made possible the use of polar orbits (Vandenberg) and equatorial orbits (San Marco). The program became more sophisticated, with double missions (Explorer 24/25) and satellites able to dive into the upper atmosphere (51, 54, 55). Imaginative use was made of high-altitude orbits to traverse the Earth-Moon system and even to put two satellites into lunar orbit (Explorers 35, 49).

Nevertheless, many of the principal scientific achievements came from the small satellites launched by the Scout. Equipment grew more reliable, missions lasted longer and the proportion of prematurely ending missions fell. The frustrating level of failed launches, always a feature of the introduction of new rockets, tailed off. The Juno was phased out in favour of the Scout, which over time became very reliable. Some of the most important improvements were in the ability to store and transmit ever larger volumes of data, again emphasizing the vital role of the tape recorder.

Chapter 1 had ended with concerns that space science might suffer from the huge diversion of effort and resources into the dominant space program of the 1960s, the manned space program and the race to the Moon. New administrator James Webb had given a commitment that space science should continue to retain priority, partly because he did not see this as an either/or discussion, but also because space science was a necessary complement to the manned program. The degree to which this was the case may be ascertained by examination of the budgets devoted to space science, which came under the heading of *Science and applications* in the NASA budget. The funding authorized under this heading rose from $62.2m in 1960 to a peak of $773m in 1966, falling back to $638m in 1968. The physics and astronomy heading therein rose from $27.6m in 1959 to peak at $160.1m in 1965, before falling back to $145m in 1968. This is very much in line with the space budget as a whole, which peaked in 1965-6. After that point, investment fell back, due to the completion of the primary infrastructure, as well as research and development giving way to operations, which are generally less expensive. At a much bigger political level, federal priorities also began to shift, with pressing demands from the Vietnam war and domestic social issues.

The early Explorers had their own budget line under the heading *Physics and astronomy* (code 85), as outlined in Table 2.5. This shows peak years in 1963 and 1965 ($64.9m and $50.5m respectively), consistent with the wider pattern, but over 1963-8 spending was consistently over $40m a year (as per the bottom line). However, this table does not capture the growth in space science *outside* the Explorer program, for example in physics and astronomy (top line). Homer Newell had campaigned for orbiting astrophysical observatories, an idea also championed by NRL and NASA physicist John Lindsay. It was Lindsay who called a NASA 'discussion group' on an Orbiting Solar Observatory (OSO) weighing over 140kg, far beyond the Explorer capacity of the time, attracting approval from NASA later that year. Eight were eventually flown, to be joined by the Orbiting Astronomical Observatory (OAO) series. These were costly, but productive missions, but it meant the end of 'Explorer' as *the* self-contained space science program.

NASA's budget rose more than tenfold from $305m in 1959 to peak at $5,249m in 1965, before falling back to $3,550m by 1976. Within that, space science (without applications) was allocated $617m in 1964, peaking at $664m in 1966 and falling back to $356m in 1969. According to Homer Newell, this was a far cry from the one or two millions per

134 Early Explorers

Table 2.5 Explorer budgets, 1959-68 [in US$ millions]

	1959	1960	1961	1962	1963	1964	1965	1966	1967	1968	Total
Physics and astronomy	27.6	14.3	44.6	88.2	148.6	146	160.1	141.1	134.3	145.3	1049.7
Explorer program											
Explorer (a)	0.25	1.86	3.92	5.74	10	20	16.6	19.1	10.1	11.3	99.764
Explorer (b)	6.2	12.9	19.9	4.4	32.8	15.5	21.5	18.6	18.2	17.5	
Energetic particles		0.6	1.2	0.9	6.8	0.9	0.9	0.4	0.5	-	7.3
Atmosphere		0.4	1.8	1.9	5	0.8	1	0.4	0.3	0.3	6.5
Ionosphere		0.2	0.8	0.9	0.8	1.1	0.5	0.2	-	-	2.9
Micrometeorite S-55											0.3
IMP						8.1	4.1	9.3	7.9	6.5	6.2	38.9
Air density						0.2	0.9	0.6	1.3	1.5	1.2	5.6
Electron density											0.6
Beacon		1.4	1.1	1.1	1.2	1.5	1.8	0.5	0.4	-	5.1
S-15						0.3		0.2			29.8
S-30											0.9
S-45											0.5
Radio astronomy							0.9	4.4	5.2	2.7	13.3
Small astronomy									0.7	2.7	3.4
Total Expenditure	6.45	17.36	28.7	14.9	64.9	45.2	50.5	43.6	43.4	41.9	

a= Explorer heading in budget; b = also includes projects funded under other headings that subsequently became part of the Explorer program and other Explorer-class projects. There was also a heading for small scientific satellites ($2.1m) and University Explorers Rice University ($3.7m). From: Ezell, Linda Neuman: *NASA historical data book, Programs and projects, vol II 1958-68; vol III, 1969-78*. Washington DC, NASA, SP-4012.

year available to the earlier Rocket and Satellite Panel and greatly exceeded budgets for ground-based science. He made the point that, spurred on by the growing number of exciting discoveries that the field had to offer, scientists were too quick to complain about not getting their fair share of the space budget. James Webb took the view that the Apollo budget helped to keep the other budgets up, including space science and within that the Explorer program, which may be a fair judgement.

Tensions between NASA and the National Academy of Science and its SSB, the site of the epic struggle of autumn 1958, never entirely went away. Some scientists in the academy argued against the observatory series – they were controversial at the time – preferring instead smaller spacecraft. They contested the amount of resources going into manned spaceflight and argued for scientist-astronauts to be included. They disputed the resources taken up by the upcoming Viking program to put a small laboratory on Mars, for which many small missions could have been flown instead (though Viking's success weakened their case). Although no one probably realized it at the time, these strains previewed a crisis in space science in general – and *small vs large* spacecraft in particular – that was to erupt in the 1980s.

What of the future? In early 1964, when newly appointed President Lyndon Johnson asked NASA to indicate its post-Apollo priorities, NASA took the opportunity to ask the SSB for its views, which it supplied on 11th August 1964 as *Future goals of the space program* (vol. 5).[48] This recommended a focus on astrophysics, principally arising from exciting developments in the mid-1960s such as the discovery of quasars; continued investigations of the upper atmosphere, magnetic fields and particle fluxes; and larger but standardized spacecraft designs, with double the payloads flown to date. Its views were elaborated in 1965, on the basis of its two-week summer study *Space research - directions for the future*, held that year in Woods Hole, Massachusetts. Meanwhile, within NASA, a *Future programs task group* had reported as early as January 1965. This gave a strong endorsement to space science in general and the Explorer program in particular, citing its many achievements and concluding reassuringly: "This program of launching relatively inexpensive Explorer class satellites to make measurements of specific phenomena will be continued." Explorer's future looked bright.

References

1. Wilson, A: *A look at Juno II.* Spaceflight, vol. 19, no. 5, May 1977. For the record, Juno III, IV and V never flew, but the Juno V became the basis of the Saturn V (LePage, Andrew: *Juno V - the early history of a superbooster.* Space Views, September 1998).
2. LePage, Andrew: *Vintage Micro - the second generation Explorer series; NASA's forgotten lunar program*, www.drewexmachina.com.
3. *The Explorer satellites.* NASA Facts.
4. Marcus, Gideon: *Earthbound Pioneer (Explorer 6).* Quest, vol. 19, §1, 2012.
5. 17th August (Pioneer 0); 11th October (Pioneer 1); 6th November (Pioneer 2); 6th December (Pioneer 3). Pioneer 4 passed the Moon in March 1959 at 60,000km.
6. Donald Le Galley (ed): *Space physics.* New York, Wiley, 1964.
7. Summaries of mission outcomes come from numerous sources such as Rosenthal, Alfred: *Venture into space - early years of the Goddard Space Flight Center.* NASA, U.S. Government Printing Office, 1968.
8. Trinklein, Frederic, & Huffer, Charles: *Modern space science.* New York, Holt. Reinhart & Winston, 1961.
9. NASA: *Juno II summary report, vol. I: The Explorer VII satellite.* NASA, Washington, 1961.
10. Information supplied by the Space Science & Engineering Center, University of Wisconsin Madison.
11. Jastrow, Robert: *Results of experiments in space.* NASA, Goddard, 1961.
12. Wilson, Andrew: *Scout - NASA's small satellite launcher.* Spaceflight, vol. 21, no. 11, November 1979.
13. Wilson, Andrew: *Delta digest.* Spaceflight, vol. 21, no. 10, October 1979. The first Echo launch, on 13th May 1960, failed on the second stage.
14. Furniss, Tim: T*he thirteenth Delta.* Flight International, 7th January 1989.
15. Corliss, William: *America in space - the first decade.* NASA, 1968; Donald Le Galley (ed): *Space science.* University of California, 1963; Jastrow, Robert: *Results of experiments in space.* NASA, Goddard, 1961; Wilmot Hess (ed): *Space science.* London & Glasgow, Blackie, 1965.
16. Kraushaar, William *et al*: *Explorer 11 experiment on cosmic gamma rays* Astrophysical Journal, vol. 141, §3, 1st April 1965; see LePage, Andrew: *The original gamma ray observatory.* www.drewexmachina, 15th September 2015.

17. Some of its findings are in *NASA in space - a pictorial review*. NASA, undated; and Donald Le Galley (ed): *Space physics*. New York, Wiley, 1964.
18. Donald Le Galley (ed): *Space physics*. New York, Wiley, 1964.
19. NASA: *Explorer 26 press kit*. Washington DC, NASA, 1964.
20. Hess, Wilmot: *The effects of high-altitude explosions*. NASA, Washington DC, 1962, Technical Note TN D-2402.
21. Gombosi, TI et al: *Anthropogenic space weather*. Space Science Review, November 2016.
22. Donald Le Galley (ed): *Space physics*. New York, Wiley, 1964.
23. Donald Le Galley (ed): *Space physics*. New York, Wiley, 1964.
24. Corliss, William: *America in space - the first decade*. Space Physics and Astronomy. NASA, 1968.
25. Wilmot Hess (ed): *Space science*. London & Glasgow, Blackie, 1965.
26. Day, Dwayne: *Lost and forgotten - the non-ruins of Slick 5*. Spaceflight, vol. 48, no. 6, June 2006.
27. Corliss, William: *America in space - the first decade*. Space Physics and Astronomy. NASA, 1968.
28. Thornton Page & Lou Williams Page (eds): *Space science and astronomy*. New York, Mac Millian & Collier, London, 1976.
29. *Explorers: searching the universe 40 years later*. NASA facts series. Greenbelt, Maryland, Goddard Space Flight Center. 1998.
30. The full record was Injun 1, 29th June 1961, fail; Injun 2, 24th January 1962, fail; Injun 3, 12th December 1962, fail; Injun 4, 24th November 1964, Explorer 25; Injun 5, 8th August 1968, Explorer 40; while Injun 6 was also a name for Explorer 52, 3rd June 1974.
31. Corliss, William: *America in space - the first decade*. Space Physics and Astronomy. NASA, 1968.
32. Norris, Pat: *Remembering Explorer 29*. Spaceflight, vol. 57, §11, November 2015.
33. Solrad designators are problematical, some being called Sun Ray, while Vanguard 3 and Explorer 7 were also considered undesignated participants in the program. The missions were Solrad 1 (22nd June 1960), 2 (30th November 1960, fail); 3 (29th June 1961, fail); 4A (24th January 1962); 4B (26th April 1962); 5 (not launched); 6 (15th June 1963); 7A (11th January 1964); 7B (9th March 1965); 8 (Explorer 30, 19th November 1965); 9 (Explorer 37, 5th March 1968); 10 (Explorer 44, 8th July 1971) and 11A and 11B (14th March 1976).
34. *UCL and Explorer 31*. Flight International, 23rd December 1965.
35. Dyson, PL; Newton, GP; Brace, LH: *In situ measurements of neutral and electron density wave structure from the Explorer 32 satellite*. Journal of Geophysical Research, 1st June 1970.
36. Madden, Jeremiah: *Interim flight report Anchored Interplanetary Monitoring Platform AIMP 1 - Explorer 33*. NASA, Goddard Space Flight Center, December 1966; *Second interim flight report, Explorer 33*. Do, May 1967.
37. Behannon, Kenneth: *Mapping of the Earth's bow shock and magnetic tail by Explorer 33*. NASA, Goddard Space Flight Center, July 1967.

38. Frolov, Ivan *et al: The Arctic basin - results from the Russian drifting stations.* Chichester, Praxis/Springer, 2005.
39. Kreplin, RW & Horan, DM: *The NRL Solrad 9 Satellite Solar Explorer B.* NRL Report 6800, 1969.
40. Gurnett, Donald: *Electric field and plasma observations in the magnetosphere* in ER Dyer (ed), Critical problems of magnetospheric physics, American Geophysical Union, 1972.
41. Corneille, Philip: *San Marco started Italy's space affair.* Spaceflight, vol. 53, no. 12, December 2011.
42. Louise Harra & Keith Mason (eds): *Space science.* London, Imperial College Press, 2004.
43. Hoffman, RA et al: *Explorer 45 (S^3-A) observations of the magnetosphere and magnetopause during the 4-6th August 1972 magnetic storm period.* Journal of Geophysical Research, vol. 80, §31, 1st November 1975.
44. *Magnetosphere Explorer launched.* Flight International, 28th September 1972.
45. *Gamma ray satellite launch; Explorer 48's gamma ray survey.* Flight International, 2nd November 1972; 16th August 1973.
46. Alexander, JK *et al: The scientific instruments of the Radio Astronomy Explorer 2 satellite.* Astronomy & Astrophysics, vol. 40, §4, May 1975.
47. For a context, see LePage, Andrew: *The original lunar observatories,* from www.drewexmachina.com, accessed 21st August 2014.
48. John M. Logsdon (ed): *Exploring the unknown - selected documents in the history of the US civil space program, NASA History series, 1995: vol. 5: Exploring the cosmos.*

3

Explorer and the Crisis in Space Science

The arrival in the White House of President Richard Nixon in January 1969 signalled a radical change in direction in the space program. Once the Moon landing was achieved, some change was inevitable and, as noted in chapter 2, NASA had taken the precaution of considering the post-Apollo future some years earlier. To those who chose to read the signals, like Homer Newell, the new Republican administration was committed to an attack on inflation and controlling or reducing federal spending. This would include the space budget – indeed, the big question in the minds of space planners was how far Nixon would let the budget drop.[1]

Nixon's decisions with regard to the space budget went through three stages.[2] First, he appointed a task force on space which reported to him on 8th January 1969, although its report was not published (*Report of the task force on space*). This took the view that space science was important, that 'its potential should be continuously developed through sound and stable programs' and that the U.S. should remain competitive with the USSR in this field. The focus of space science was seen very much around astronomy and astrophysics. This was encouraging.

Second, he appointed a Space Task Group, chaired by Vice-President Spiro Agnew, which issued *The post-Apollo space program – directions for the future* (September 1969). This presented an ambitious post-Apollo program of more Moon landings and onward to Mars in the 1980s. In the area of space science, it considered that the program to date had been 'successful and highly productive' and drew attention to its achievements in optical and radio astronomy, notably in the high energy x-ray and gamma regions. In *Astronomy, Physics, the Earth and Life Sciences*, it proposed that in each of these disciplines, the extension of existing or planned unmanned programs promised continued high science return. Work in astronomy, physics and the life sciences, as well as in the Earth sciences and remote sensing, would form an essential part of the foundation for the future.

However, the Spiro Agnew report, enthusiastically supported by new NASA Administrator Thomas Paine, won little favour with the President or the Congress. The third stage was the two critical judgements Nixon now made. The first of these, which probably showed an accurate reading of the public mood, was that the space program should be reduced from one of overriding national purpose to one that must take its place among many domestic priorities. NASA's budget fell from $5,249m in 1965 to $3,550m in 1976. The second was that the United States must maintain a *manned* space program – not to do so was unthinkable (though off-stage, Van Allen and some of his colleagues would have settled for a much smaller budget, $2bn, provided it *all* went to space science). The logic of both led President Nixon, through a tortuous process, to the approval of the Shuttle in January 1972 to replace the Apollo program, which would conclude with the Apollo-Soyuz rendezvous mission in 1975. In approving the Space Shuttle, Nixon selected the defining element of the U.S. space program until the 2010s. Although invited to do so, he did not support funding a space station, so the Shuttle had nowhere to shuttle (the space station was eventually approved by President Ronald Reagan in 1984). The Shuttle would become the sole rocket and the 'low-cost' means of getting all people and all cargo into space.

Richard Nixon, Thomas Paine and Spiro Agnew

In theory, none of this should have affected the well-entrenched role of space science, but in practice Nixon's decision was to have important consequences. Now, *all* missions had to be redefined around the Shuttle, both programmatically and physically – to fit into its huge cargo bay – creating a momentum for ever-larger spacecraft, rather than smaller ones (like Explorer). The Scout rocket managed to win a reprieve from the Shuttle-only decision, but the overall impetus to larger spacecraft was irresistible. For example, the production line of the Delta rocket, which had launched many Explorers, began to wind down toward closure in 1986. For the Explorer program, all missions now had to fit the Scout or the Shuttle.

The more serious consequence, though, was one that echoed the 1959 decision to set up the Mercury and manned space programs. Would investment in the Shuttle have a negative impact on resources for space science, like Explorer? It quickly transpired that the development costs for the Shuttle had been greatly underestimated, so that they began to consume the largest share – as much as two-thirds – of the NASA budget. The seductive appeal of lower operational costs also proved to be an illusion, since they came in at $500m a launch. A significant number amongst the science community began to fear that the Shuttle would siphon off funds that should have gone to space science. One historian's judgement was that although the President gave a strong commitment to astronomy and physics, in practice this had to be accomplished on a bare-bones budget.[3] NASA responded by trying to reassure the science community of the benefits of the Shuttle for space science and the Woods Hole SSB summer school in July 1973 was devoted to the topic. The scientists gathered were not appeased, because the Shuttle just did not seem to be appropriate for the kind of payloads represented by sounding rockets or small satellites, especially those that might have unusual orbits. There were mutterings that NASA might expect trouble.

Crisis in the Early 1980s

These fears were probably well justified. The rate of Explorer launches fell away, with only 12 between 1975 and 1992. In several years, there were none at all. The first year in which there were no launches was 1976, but this was followed by 1980, 1982 and 1983, a thin period for the program. There were no launches from 1985 to 1988, nor in 1990–1. Only two Explorer selections were made in this period, Announcement of Opportunity AO6 and 7 in 1974 and 1975, which approved Dynamics Explorer (DE), the Active Magnetospheric Particle Trace Explorer (AMPTE), the Extreme Ultraviolet Explorer (EUVE) and the X-Ray Timing Experiment (XTE) which did not fly till the 1990s (Rossi). The budget squeeze begun in 1972 with the Shuttle decision eventually worked its way through the system, with deep space missions most visibly affected: NASA sent no missions beyond Earth orbit from 1978 to 1989.

The high development costs of the Shuttle did indeed absorb the largest single part of the space budget, squeezing out everything else. By Financial Year 1977 (FY 1977), the Shuttle accounted for 59% of the NASA budget (with other launch operations a further 8%, science 15% and applications 8%).[4] This was in sharp contrast to the previous decade. Apollo spending did not come at a cost to space science, but the Shuttle did. Not only that, but there was a downward pressure on the NASA budget as a whole. For example, FY 1978 estimates for the NASA budget expected it to fall from $3.5bn in 1976 to $3.1bn by 1982 and there was further pressure for budget cuts during the early period of the Reagan administration. Space science became a smaller part of a smaller budget.

Crisis in the Early 1980s 141

President Richard Nixon, Space Shuttle and NASA Administrator James Fletcher

The Explorer program was still fortunate in that it had, in effect, its own annual budget line and that unlike lunar and interplanetary missions, individual mission start-ups did not require Congressional approval and its associated micro-management.[5] At the same time, Explorer came out of the broader space science budget that included high-profile planetary missions, so that if they were squeezed, Explorer was likely to suffer too. For the moment, however, Explorer was able to hold the line. Although FY 1978 projected a fall in the space science budget from $434m to $388m, the planetary missions were the hardest hit ($254m falling to $84m) with physics and astronomy actually rising ($159m to $235m) and life sciences also increased ($20m, rising to $68m). In relative terms, the allocations to space science did not fall greatly – they were 17.6% over 1959-68, down slightly to 17% over 1969-78 – but they were part of a smaller budget, so in absolute terms they did fall. The Explorer program found itself in a perfect storm in which space budgets overall were falling and almost two thirds of that declining budget was going to the Shuttle. This was squeezing space science out of the program.

The Carter administration (1977-81) was not blind to the situation it had inherited and its transition document warned that the Shuttle had become the end, rather than the means of the space program, with space science and applications getting what was left over. At one stage, President Jimmy Carter even contemplated the extreme measure of cancelling the Shuttle in order to save other programs. He issued a directive – NSC-42 *Civil and further national space policy*, actually written by Zbigniew Brzezinski – on 10th October

1978 which emphasized a commitment to space science, but in effect stated that it would only be possible to give it greater attention once the requirements of the Shuttle were phased down. The space science section spoke of a 'new era' in astronomy, a better understanding of the Sun and its interaction with Earth. In other words, less space science now, but more later. A year later, Presidential science advisor Frank Press wrote a memorandum to the President on 27th November 1979, expressing his concern about the lack of fresh starts to space science programs and lamenting that the expensive development costs of the Shuttle were driving out relatively inexpensive space exploration.[6] Meanwhile, the wish list of what *could* be done grew: The National Academy of Sciences, in its 1980s decadal study *Astronomy and Astrophysics in the 1980s* by Harvard astronomer George Field, made proposals for 'great observatories' like Hubble and the Gamma Ray Observatory to be launched by the Shuttle, alongside an augmented Explorer program and three new x-ray Explorers.

At this point, Goddard itself was also at a low ebb. The centre had got off to a flying start under Harry Goett, who was greatly admired and became known as a man who really got things done. He was criticized though, for neglecting the paperwork, which could create serious problems over contracting and many other things, but his downfall was his independent-minded way of running Goddard. James Webb – never a man to cross – would not tolerate Goddard, or any other NASA field centre, being run as an independent empire. Given Webb's management responsibility to deliver a Moon landing using all of the centres, Goett was eventually asked to go in 1965.[7]

The tradition of building in-house had also virtually gone from Goddard by the late 1970s, with almost all satellites being contracted out (until Thomas Young became director, when one of his first decisions was to resume satellite building in-house). The attractions of Goddard to engineers fell and morale at the centre plummeted. Goddard also suffered from two rounds of what were euphemistically called Reductions In Force (RIFs). The financial situation became so grim that, in the early 1980s, the closure of the Scout's main base at Wallops Station was proposed. NASA's clever response was to transfer Wallops from Langley to Goddard (to which it was geographically much closer) in 1982 and re-title it the 'suborbital projections and operations directorate', whose sonorous and less visible title might attract less attention from cost-cutters. The ruse appears to have worked.

These issues reached a demarche in the early 1980s. The battleground was not the Explorer program, but Halley's comet and the conclusion of this long, drawn-out battle took place in the early days of the Reagan administration. Comet Halley was due to return to the solar system in 1986 after an absence of 76 years and the main space-faring nations had prepared an armada of spacecraft in greeting – VEGA (USSR), *Giotto* (Europe) and *Suisei* and *Sakigake* (Japan). American scientists, though, had no success in trying to get approval for an American spacecraft, to their increasing embarrassment at international gatherings.

The Carter administration had hoped that the Shuttle might have begun to fly during his tenure and that the financial pressure might then ease, but this was not the case. The incoming Reagan administration inherited the same dilemmas and difficult choices, but these had now reached a critical stage. George Low, who had left NASA in 1976, wrote the transition advice for the new administration in the form of the *Report of the transition team, NASA* on 19th December 1980. This included a budget chart showing the decline of the proportion spent on science over time, while manned spaceflight and the planned Space Shuttle consumed the lion's share of resources. The problem was restated.

George Low (left), with Thomas Paine

Although the Shuttle finally flew in 1981, the financial pressures remained and the Office for the Management of the Budget (OMB) proposed to cancel the only two interplanetary projects in the pipeline, the Venus Orbiting Imaging Radar (VOIR) and the Galileo probe to Jupiter, that autumn. NASA's budget had diminished to the point that it even considered closing JPL and turning off the transmitters of the Voyager probes as they headed toward the outer planets.

In the end, the Congress mobilized to reinstate VOIR, Galileo and JPL. The Voyagers were also reprieved, but no new funding was added. An important element in this process was the Planetary Society – formed in 1980 by Carl Sagan and JPL's Bruce Murray – who organized a grassroots campaign. Whereas advocacy for space science had come from the lofty towers of the SSB in the 1950s, such advocacy now came from an improbable insurgency of enthusiasts, itself a tribute to the engagement of the space program far beyond its professional borders that, contrary to some predictions, did not fade after Apollo. The reprieve for VOIR, Galileo, JPL and the Voyagers was a turning point in the decline of space science, but brought no early relief for the smaller programs. There was still no dedicated Halley mission, but at least a line had been drawn beyond which there would be no more retreat. However, the situation affecting space science continued to be difficult and was officially and formally elevated to that of a 'crisis' with the publication by NASA's advisory council of a report entitled *The crisis in Earth and space science - time for a new commitment*. It concluded that there had been a transition, more due to circumstances than conscious policy choice, with fewer and fewer new starts, counter-productive competition for them and a decline of the smaller scale, less glamorous, less visible activities that had been the foundation of the program. Delays, overruns, cancellations, deferrals, erratic funding and a dependence on the Shuttle as the single launch vehicle had led to diminished flight opportunities. *The crisis in Earth and space science* report recommended reinstating the availability of expendable launch vehicles to enable broader choices in mission capabilities, especially for space science.

The roots of the crisis of the early 1980s, forced to a climax during the period of the Reagan administration and the subject of much criticism of the Carter presidency, can ultimately be traced to the downscaling of the program under Nixon and the concentration of more limited resources on one project (the Shuttle).[8] How that crisis was resolved is the subject of *Chapter 4: Faster, better, cheaper?* For the moment, the Explorer program during these difficult years is reviewed. This covers Explorers 56 to 67, which were known as ISEE 1 and EUVE respectively under the new NASA terminology.

ISEE 1 (Explorer 56)

The term 'Explorer' ceased to be used officially at this point, but in order to keep track of the numerical sequence, the 'Explorer number' is given in the subtitles. The first mission was the International Sun Earth Explorer (ISEE), an international collaborative mission involving three satellites.

ISEE arose from the 1976–9 International Magnetospheric Study, a collaborative study between the space-faring nations to examine the effect of the Sun on Earth's weather and the magnetosphere. ISEE was originally announced as a collaborative American-European program, the International Magnetosphere Explorer (IME) project in 1973. A Memorandum of Understanding (MOU) was agreed 17th March 1977, with the proposed satellites originally being given the names ISEE A mother (NASA), ISEE B daughter (ESA) and ISEE C (NASA, heliocentric), with mother and daughter to be launched together and the heliocentric one following. They were later renamed 1, 2 and 3. It was consciously intended as a sequel to the IMP series, with a mother-daughter spacecraft to make repeated passes through the Earth's magnetic field to measure fine scale and time variations. Because ISEE 1 was built by Goddard it counts as an Explorer but ISEE 2, built in Europe, does not form part of the Explorer number sequence. European participation came in the form of Imperial College London; Space Research Laboratory, Utrecht (protons); the Max Planck Institute (solar particles); and Paris observatory, Meudon (magnetic field lines) under the aegis of the newly-formed European Space Agency (1975).

ISEE 1 and 2 were launched together by Delta from Cape Canaveral's pad 17 on 22nd October 1977. The two satellites separated an hour after launch and both were placed in highly elliptical orbits. ISEE 1 (340kg) entered an orbit of 337km by 137,904km, R23, while ISEE 2 (166kg) flew from an almost identical path of 341km by 137,847km. It was anticipated that they would drift as far as 5,000km apart as a result of Sun-Earth-Moon perturbations. The idea was that ISEE 3 would later be sent to L1 halo orbit some 1.5 million km beyond Earth, so that it could see solar activity an hour before it reached ISEE 1 and 2 (the 'L' points were named after Joseph-Louis Lagrange – hence 'L' – who in 1772 determined the gravity-neutral points in Earth-Moon-solar orbit). The three satellites would carry out continuous measurements of the transient boundaries and discontinuities of the magnetosphere and the solar wind. A specific aim was to observe the 11-year solar cycle which had begun in June 1976 and monitor its effects on the magnetosphere, ionosphere and upper atmosphere, leading to improved knowledge of weather, climate, energy and ozone depletion. They would study the plasmapause, the magnetopause and the bow shock. In detail, they were tasked to observe particles and waves reflected from the bow shock; the neutral magnetic and current sheets in the solar wind sweeping past; wave-particle interactions; the velocity, composition and density of solar wind; the speed of the bow shock and its other features; the turbulence of magnetic field fluctuations; the nature of the magnetopause boundary; the flow of plasma and energetic particles up and down the tail; the structural features of the inner plasma sheet; the movement of the ring current; and the triggering of magnetic substorms.

ISEE 2 [from the author's collection]

ISEE 1 was a 16-side cylinder 1.73m across, 1.61m tall, painted green for thermal and electrical protection, while ISEE 2 was a circular cylinder, 1.27m diameter, 1.14m tall, with solar cells on three curved panels. They were both spun at almost 20rpm. ISEE 1 was built at Goddard and ISEE 2 at the European Space Technology Centre (ESTEC), under what was called the STAR consortium of Belgium, Denmark, France, Spain, Germany, Italy, Netherlands, Sweden, Switzerland and Britain, under the guidance of Dornier in Friedrichshafen in southern Germany, the old Zeppelin base nearby. Different European companies took responsibilities for different parts – including high-level companies such as Fokker in the Netherlands and BAC in Britain – with the solar arrays coming from Germany, for example. The missions would use a Satellite Situation Center in Goddard and data exchange offices in Meudon, France and Boulder, Colorado. ISEE 1 transmitted data at between 4kbps and 16kbps. The principal instruments covered the following fields:

- Fast plasma (ISEE 1 and 2), with instruments to determine electron and ion velocities, from Los Alamos Laboratory in New Mexico and Max Planck Institute in Germany;
- Low energy protons and electrons (1 and 2), with a Low Energy Proton and Electron Differential Energy Analyzer (LEPEDEA), built by L.A. Frank at the University of Iowa;
- Fluxgate magnetometer (1 and 2), developed by C.T. Russell of the University of California Los Angeles;
- Plasma waves instrument (1 and 2) 1Hz to 200Hz for electric fields and 1Hz to 200Hz for magnetic fields (University of Iowa);
- Plasma density (1 and 2), a University of Paris instrument;
- Energetic electrons and protons (1 and 2), with instruments by NOAA and the Max Planck Institute, using a magnet to separate them, in the 20keV to 2MeV (protons) and 20keV to 1MeV range (electrons);
- Electrons and protons (1 and 2), with two solid state detector telescopes from University of California Berkeley (electrons 8–200keV, 30–200keV; protons the same ranges plus 200–380keV);
- Fast electrons (1 only) (Goddard);
- Low energy cosmic ray and gamma ray burst detector (1 only) (Max Planck Institute);
- Quasi-static electric fields (1 only), to measure electric fields with a 50m wire boom (University of California);
- Electric fields (1 only) (Goddard);
- Ion composition energetic ion mass spectrometer (1 only), 0–17keV, (Lockheed);
- VLF wave propagation experiment (1 only) for the injection of waves into the magnetosphere, modelled on a transmitter operated by Stanford University in the Antarctic;
- Solar wind ions (2 only), an Italian instrument from Frascati, Italy to measure flow direction and energy spectra in the positive ions of the solar wind.

This was a big collaborative effort, with 117 principal investigators from 35 universities in ten countries. The project cost $45m. The ISEE 1 program scientist was Erwin Schmerling, while Alistair Durney was project scientist on ISEE 2.

ISEE 1 and 2 re-entered the Earth's atmosphere at the same time, during their 1,518th revolution on 26th September 1987, with 17 of the 21 on-board experiments still operational. To the wider world, though, they were eclipsed by the adventures of ISEE 3 (Explorer 59).

IUE (Explorer 57): 19 Years of Dazzling Science

The second mission of this phase of the program was also multinational, the International Ultraviolet Explorer (IUE), intended for 1976 but eventually launched on 26th January 1978. Originally there was to have been a fourth small Astronomy Satellite, SAS-D, but it was subsumed into this larger international project. The European Space Research Organization (ESRO) had proposed a large orbiting astronomical observatory as far back as 1964, to fly an ultraviolet spectrometer which would obtain spectra of quasar 3C273 in particular. When ESRO funding fell short, the British astronomer Robert Wilson approached Goddard's Albert Boggess for an international project that would share funding and outcomes. This approach worked, for they now obtained funding from the new European Space Agency (ESA) – which had replaced ESRO – as well as from Britain and the United States. The project was re-titled the International Ultraviolet Explorer (IUE). The satellite was put under the joint use of NASA, ESA and Britain's Science Research Council (SRC), with observing time shared between them according to their contribution: NASA, 66%; ESA, 17%; and SRC, 17%. ESA made available its tracking station in Spain, the European Ground Observatory. Such a management system was quite normal for ground observatories, but had not been applied before for space-based instruments. It was seen at the time as the most important astronomical project until the Space Telescope, then scheduled for 1983.[9] IUE built on the experience of the two Orbiting Astronomical Observatories and ESRO's TD-1, but with much higher sensitivity, 24-hour operations and greater altitude from a quasi-synchronous orbit. Before it was launched, 200 astronomers from 17 countries were lined up to use it, the idea being that they would come to Goddard or Villafranca, Spain to do so (nowadays, thanks to the internet, they would work from their home or office computer).

IUE was designed to investigate stars, quasars and galaxies in the ultraviolet band between 1,150 and 3,200Å, short wavelengths blocked out by the atmosphere and the origins of emissions of the fundamental elements of the universe such as hydrogen, helium, carbon, nitrogen and oxygen. There was a special interest in quasars (because of their enormous energy), Seyfert galaxies, hot stars and the hot outer atmospheres of cool stars, like our Sun. The study list comprised stars akin to our Sun, the material where stars were formed and – further afield – sedate galaxies, strange objects (e.g. emitting x-rays) and violent quasars. Specific objects of interest were pulsars, neutron stars, black holes, binary stars, the intergalactic medium and other matter.

The eccentric 25,000km by 46,000km, 29° elliptical, quasi-synchronous orbit was designed to make possible long-term observations. IUE would drift back and forth from 71° W, with a hydrazine engine used for station-keeping, so that it remained in view of Goddard all the time but the easterly swing would take it over Madrid for ten hours a day. IUE had an open access policy, making time available to guest astronomers putting in proposals of merit, with one or two sessions for between six and eight observers per day for a week. IUE was simple for guest observers to operate directly. Ground controllers from both sides of the Atlantic could simply command IUE to point toward the guests' objects of interest. In the end, about half the world's astronomers used the satellite, including those from the USSR and China. Although the telescope was small, it was highly efficient because of its stable location, 24-hour-per-day observation periods and the capacity to make exposures up to 14 hours long. It could also be pointed quite quickly at new or sudden areas of interest, like novae or comets.

The octagonal, 4.3m tall and 1.3m diameter spacecraft housed a telescope on top and two 4.3m solar arrays at the side. It weighed 671kg including its motor. The instruments were a reflecting telescope with a spectrograph to form the light into spectral displays and two cameras to transform them into readable images. The optical system was a 45cm diameter Cassegrain telescope of 675cm focal length. The NASA program scientist was

Nancy Roman and the project scientist at Goddard was Albert Boggess. One of the scientists involved was Alan Stern, who subsequently went on to the Extreme Ultraviolet Explorer and achieved global fame for the New Horizons pass of Pluto in 2015.

Getting IUE off the ground was challenging, as a short-circuit was found only two weeks before launch. The satellite was disassembled on New Year's Eve to find the errant circuit and then brought to the launch pad within two weeks. The launcher was a Delta from Cape Canaveral, which injected IUE into a transfer orbit, raised to over 45,000km by the firing of the apogee boost motor.

The first results were reported by Herbert Gursky and Andrea Dupree of the Harvard Smithsonian Center for Astrophysics, Cambridge, Massachusetts.[10] The ultraviolet instrumentation was able to find a globular cluster 150,000 light years away, with ten to 20 bright blue stars orbiting a hot, bright object at the core, likely to be a black hole. In its first 15 months, IUE took 9,000 images and obtained spectra of stellar sources to 12th magnitude and extragalactic to 17th magnitude, the latter requiring 14 hours of continuous observations. IUE had observed hot dwarfs, cool stellar winds, supergiants and novae, while its discoveries included gold in A-type stars, hot circumstellar shells in stellar x-ray sources, ultraviolet sources in the centre of globular clusters, acetylene in the atmosphere of Saturn and limb brightening on Jupiter. It also made the first ultraviolet spectra of novae and their remnants.

IUE became well known for its work on comets, observing Comet IRAS-Araki-Alcock when it approached four million km from Earth in May 1983 and finding, for the first time in a comet, diatomic sulphur.

Despite the absence of an American spacecraft going to Halley, IUE was able to make up some of the deficit, as 150 hours of IUE observing time were allocated to Comet Halley from April 1985, with 60 hours of that time allocated for closest approach. IUE first picked up Comet Halley on 11th September 1985, taking its first pictures. Spectra were taken on 15th December, showing the comet clearly heating up as it approached perihelion passage, five times more active than it had been in September. The dust coma was estimated at 20,000km. IUE instruments calculated the outflow of Halley's comet – 10 tonnes of water per second – suggesting that these bodies were highly dynamic and not merely inert dirty ice balls. By this stage, IUE had generated 350 scientific papers.

An early discovery of IUE was of a halo of hot (100,000°C) material made of oxygen, sulphur, iron, silicon and carbon, surrounding our galaxy to a distance of 50,000 light years. These first results were discussed at a scientific conference held at Goddard in May 1980, during which 92 papers were presented. A second discovery was that some twin quasars were not really twins after all, but a single object whose light waves were so distorted by a gravitational lens that they appeared to be double to initial observations.

In June 1985, IUE found a variable cloud of gas around β Pictoris, a star in the southern skies constellation of Pictor, such gas clouds being considered at the time as indicators of planetary formation. On 24th February 1987, IUE was quickly – within 14 hours – on to the blue supergiant supernova 1987a in the Larger Magellanic Cloud, 163,000 light years away. This was the brightest supernova since one spotted by Kepler in 1604. IUE's scheduled program was quickly interrupted to obtain the first images and there were always contingency plans for such targets-of-opportunity events. It turned out that Supernova 1987a was an intense source of ultraviolet radiation and the first 15-second exposure image was badly over-exposed. Going back over previous imaging of the region, it was found that the supernova originated in the area of supergiant Sanduleak 69 202 and two adjacent fainter blue stars, one of which must be the candidate.

IUE also pointed its telescope at all the planets, except Mercury for fear of the telescope being blinded. Turned toward the planets, its observations suggested water on Mars and aurorae on Jupiter. It made the first intensive study of Jupiter's auroral belts. Observing Venus, it indicated a decline in sulphur monoxide and sulphur dioxide, suggesting their diminution in the aftermath of a volcanic eruption, an impression confirmed by Venus Express in the new century.

IUE celebrated ten years in January 1988, although it had an original design life of only three. Its achievements by then included the discovery of galactic halos and a moon of Jupiter, as well as the first photographs of Comet Halley from space and tracking volcanic activity on Io. The list of scientific papers had now reached the 1,400 mark, making IUE the most productive telescope ever. The typical number of guest astronomers at Goddard reached 800, with another 750 at Villafranca, Spain. IUE was by no means the largest or most powerful telescope – it was only 45cm – but it observed in ultraviolet and above the atmosphere. To mark the anniversary, Goddard convened a week-long conference, titled *A decade of ultraviolet astronomy with the International Ultraviolet Explorer*, attended by 300 astronomers from 23 countries. Topics discussed included supernovae, planetary nebulae and stellar activity. That same year, IUE won the Presidential award for design excellence.

The remarkable discoveries of its first decade would be matched, if not surpassed, by those of its second. In May 1993, IUE joined a 55-day program with ROSAT and Hubble to observe Seyfert galaxies, in order to model their physical conditions. It followed the variable star W Serpentis for eight days to see how the gas from one star was drawn into the accretion disk of its companion. IUE spent June 1993 studying Sun-like stars for 16 days to compare their magnetic fields, active regions and spectral lines with our own Sun. In October 1993, new software was introduced that enabled more detailed examination of IUE data, using techniques not available 15 years earlier. Goddard's head of observations Andrew Michelitsianos used it to re-analyse the gravitationally-lensed binary quasar 0957+561 (whose light was so bent as to present two images of one object), finding extremely hot x-ray gases which were undiscovered in the original observations. Later that year, IUE was redirected toward the Cassiopeia 1993 nova.

In March 1994, IUE observed Comet McNaught-Russell, returning to the inner solar system for the first time since it was observed by Chinese scientists in the year 574, suggesting a period of 1,440 years. In April, IUE observed SN1994I in M51, faint in ultraviolet but still important for modelling supernovae, then observed an eclipse in binary HR2554 that comprised an A-type and adjoining larger G-type star. In May, IUE observed the nucleus of active galaxy PKS 2155-304, its brightness likely due to the energy falling into a massive black hole at its centre. It also observed Jupiter, in anticipation of the arrival there of Comet Shoemaker-Levy in July. In the event, IUE was the only telescope able to follow all 21 impact events, which took place over 24 hours, following which IUE observed Jupiter's aurorae and the moon Ganymede.

In June, IUE focussed on FK comae, a cooler and more rapidly rotating star than our Sun. In September, IUE began a month's monitoring of binary UX Aries, 160 light years away, whose larger star orbits the smaller every 6.4 days, with each star rotating every 6.4 days due to tidal forces. It then began a program of observing young stars with substantial disks of gas and dust around them. In October, IUE observed hot stars in the Large Magellanic Cloud 150,000 light years from Earth in an area full of x-ray-emitting hot bubbles. In November, it observed three massive stars in the stellar system HD 5980, in the Small Magellanic Cloud, whose combined brightness was 115,000 times greater than our Sun and had increased 50% since IUE was launched, principally because one had tripled in brightness and seemed to be in an advanced state of evolution. In June 1995, IUE observed an eclipse of the exotic binary system V342 Aquilae where a high rate of mass transfer was taking place, with the outer envelope of one star streaming onto the other as they orbited one another every 3.4 days.

In May 1995, IUE took advantage of an event that takes place only once every 14.7 years, when Saturn presents its rings edge-on – the rest of the time they are tilted, up to 30° – making it possible to view both hemispheres entirely, as well as their polar aurorae. In September, IUE observed spiral galaxy NGC 1512 some 13 megaparsecs or 42 million light years distant, along with its companion galaxy NGC 1510. Observers were on the look out for interactions or even collisions between the two which could trigger starbursts or the formation of new stars.

Long after its demise, IUE's observations were engaging scientists. One focus for IUE was η Carinae, one of the most puzzling, erratic stars in the Milky Way, 8,000 light years from Earth and visible in our southern skies.[11] It was identified as a variable star in the 19th century, but over 1841–3 brightened like an 'almost-nova' to become the brightest in the sky. By 1868 it had faded and could no longer be seen, except through telescopes. It gave indications that it might blow any time, like a volcano, so watching it could yield clues to the processes that led to stellar explosions. Observations in the 1940s indicated that it had left behind a debris cloud with a complex structure from a hundred years earlier. Observations in the 1970s suggested that it was an unstable binary star system, one of which was 90 times the size of our Sun (making it a supergiant) and the other 30 times (subgiant), orbiting one another every five and a half years. Ejections were most turbulent and rich in nitrogen, oxygen, magnesium, iron, sulphur, silicon and calcium, all indications of nuclear burning. All these reasons made it an important target for IUE observations.

IUE was designed to last a minimum of three years but had consumables for five. In the event, its gyros determined its lifetime. IUE had six gyros, of which two were spare, but needed three at any given time. One was lost in the first year (it was turned off and would not turn on again), a second failed in March 1982 and a third that July, so it had run out of spares. The fourth failure on 17th August 1985 was critical, but in anticipation Goddard engineers had devised a plan to use the Sun sensor as a substitute gyro and reprogrammed the orientation system accordingly. This was an original, creative, lateral-thinking fix that had never been considered before this mission. Eventually four gyros failed, but IUE worked on its two remaining gyros for eleven years and on only one, combined with a star tracker, by 1995. Due to budget reductions, NASA almost decided to close the mission down, but ESA stepped in and increased its control share to 16 hours per day. As its resources reduced, NASA relinquished IUE operations in October 1995 and transferred all to ESA, although Goddard continued to provide 'limited operational support' which included an education project to give access to high school students to make observations. IUE's final observations were of Comet Hayutake in March 1996, which it tracked for five days taking five-hour exposures. ESA's funding for the mission ran out a year later, so IUE was finally shut down on 30th September 1996 after 19 years. This was 14 years more than designed and it could have continued to operate had the funds been available. The signal to empty its fuel effectively sent it spinning and made it useless, saving operating costs of $1m to $2m a year. Out of control, IUE has drifted off course, but is so high up it is likely to remain in orbit for some time.

In the end, IUE achieved a huge scientific haul. The statistics alone are dazzling. An astonishing 4,000 guest scientists had participated in observations and 3,600 papers were published, with 250 PhDs awarded on the strength of IUE data. It was the focus of attention of nine symposia of the International Astronomical Union. Before the arrival of the new 'great observatories', it was considered the most successful astronomical mission ever. In 1997, Goddard set about creating an IUE archive to store 23 GB worth of 104,470 observations and 44 spectra data, to be made available worldwide, for which it received a design award for excellence.

IUE's main discoveries were the detection of aurorae in Jupiter, Saturn and Uranus; strong magnetic fields in stars; hot dwarf companions to Cepheid variables; and high velocity winds in stars other than the Sun. IUE imaged starspots, took temperature and density gradients in stars, found gas streams around binary stars, identified oxygen neon magnesium novae, found a ring around supernova 1987a and discovered galactic halos and extragalactic symbiotic stars. IUE imaged not only bright quasars, but many fainter ones. It took spectra of the Sun, hot stars, variable stars and Active Galactic Nuclei (AGNs), only one of which had been observed in ultraviolet before. It made exposures for numerous hours at a time and obtained the first ultraviolet spectrum of a star in another galaxy, the first ultraviolet recording of a supernova and globular clusters over 15,000 light years away. IUE measured a hot gas halo of oxygen, sulphur, iron, silicon and carbon up to 100,000°C around the Milky Way and found that the star Capella had a chromosphere like the Sun. IUE observations found stellar wind driven away from hot stars, shock waves from supernovae interacting with interstellar material, a coronal halo around own galaxy and x-ray binaries in which a normal star orbits with a white dwarf, neutron star or black hole. IUE was particularly important for studying objects in the hard-to-reach ultraviolet range, notably gas clouds and the halo of hot gas outside our own galaxy. IUE obtained high-resolution data on stars and planets in the UV spectrum, from which it was possible to learn about their temperature, behaviour and density. Its achievements included the discovery of sulphur emission from a comet, the measurement of inter-stellar wind, supernova observations and the mapping of low-density gas bubbles around the Sun and other stars.

HCMM (Explorer 58)

The Heat Capacity Mapping Mission (HCMM) spacecraft was intended as the first of a series of Applications Explorer Missions (AEM). The mission objective of the HCMM was to provide comprehensive, accurate, high-spatial-resolution thermal surveys of the Earth's surface. HCMM was launched by Scout F from Vandenberg on 26th April 1978. The HCMM comprised an instrument module with its single instrument, a heat capacity mapping radiometer and its supporting gear, as well as a service module containing the data handling, power, communications, command and attitude control subsystems to support the instrument module. The spacecraft spun at 14rpm and transmitted real-time data only to seven ground stations.

HCMM was sent into a circular sun-synchronous orbit, crossing the equator at 14:00 every day so that it would sense surface temperatures near the maximum and minimum of the diurnal cycle. It crossed the same pathway between 85°N and 85°S every 16 days, drifting by about an hour each year. Because it was drifting to unfavourable Sun angles for the solar panels, its orbit was adjusted downward from 620km to 540km in February 1980. This poorly-known mission finished on 30th September 1980.

ISEE 3 (Explorer 59): Halo, Comet and Return

ISEE 3 was the third companion of ISEE 1 (Explorer 56) and ISEE 2, but launched separately later. It was to have a long and famous history, being repurposed to fly to a comet and then make a return to the vicinity of the Earth in the new century.

ISEE 3 was put into what is called a halo orbit at L1, R335, 1.5 million km on the sunward, inner side of the Earth from the Sun on 12th August 1978, the first spacecraft to be stationed there. The L1 position is where solar, Earthly and lunar gravity are in balance, placing the satellite in a perfect position to report on solar radiation heading toward Earth. ISEE 3 was the first spacecraft to study the constant flow of solar wind from the Sun from this stable position. The purpose of the mission was to investigate solar-terrestrial relationships at the outermost boundaries of the Earth's magnetosphere; examine the structure of the solar wind near the Earth and the shock wave that forms the interface between the solar wind and Earth's magnetosphere; investigate motions and mechanisms operating in the plasma sheets; and continue the investigation of cosmic rays and solar flare emissions in the interplanetary region. It interrogated the solar wind every 40 minutes – slow by contemporary standards – and it was one of the first spacecraft to use s-band communications, which until then had only been used for military satellites.

ISEE 3 was a 469kg spacecraft, but carried as much as 94kg of propellant, in case the Delta's injection was inaccurate. ISEE 3 had 13 scientific instruments, of which three came from Europe (France, Germany and the Netherlands). One was to investigate protons in outer space (Space Research Laboratory, Utrecht and Imperial College London); the second to study the composition of solar particles (Max Plank Institute); and the third an experiment to examine magnetic field lines (Paris Observatory, Meudon). The remaining instruments were designed to study solar wind plasma; plasma composition; magnetic fields (magnetometer); plasma waves; energetic protons; radio waves; x-ray and low energy electrons; low energy cosmic rays; medium energy cosmic rays; high energy cosmic rays; cosmic ray electrons; and gamma ray bursts.

By August 1981, the three ISEE spacecraft had completed their primary mission. By then, time had finally run out on any prospects for NASA funding for a mission to Comet Halley, an aspect of the financial shortfall discussed at the start of the chapter. At Goddard, Fred Scarf (PI for ISEE 3) and Dr Robert Farquhar (orbital engineer) came up with the idea of using ISEE's considerable spare fuel to redirect it toward Halley or another a comet. By January 1983, 64kg of its original 94kg propellant remained, enough to achieve ΔV of 300m/s.

Robert Farquhar

Although Halley was not out of reach, another comet offered a closer intercept sooner. The idea had the additional merit, in the competitive spirit of the space race, of getting the United States to a comet first. Comet Giacobini-Zinner had been discovered by Michel Giacobini in Nice in 1900 and was rediscovered by Ernst Zinner in 1913, hence the double name. It was a short-period comet orbiting the Sun every 6.6 years, flying between 1AU and 6AU, 31.9° to the ecliptic and associated with the Draconid meteor shower. It had last passed Earth in 1972 and was known for a distinctive one-million-km-long ion tail. It was observed to have a nucleus of 1,000 to 2,500m, spinning every 1.66 hours. The comet had provided meteor displays in 1933 and 1946, but little since. This next would be its eleventh apparition. Moreover, a rendezvous with Giacobini-Zinner also opened up the possibility of later making distant, upwind observations of Comet Halley in October 1985 and March 1986.

A feasibility study was conducted in 1981 and presented to the ISEE science team in February 1982. Reluctant to lose data from the L1 vantage point, the team rejected it, but Scarf and Farquhar did not give up and presented the proposal further up the line to NASA's Comet and Asteroid Working Group of the Solar System Exploration Committee (SSEC). The SSB then weighed in with its support. At this stage, the Director of Space Sciences at Goddard, George Pieper, changed sides from 'against' to 'for' and announced that he would pursue the comet interception unless instructed not to. The matter ended up at NASA Headquarters where Burton Edelson, Associate Administrator for Space Science and Applications, gave his support at the end of the month. Of ISEE's 13 instruments, the first six would be useful for the comet, especially the vector helium magnetometer which would measure the way in which the interplanetary magnetic field interacted with the comet. The only regret was that ISEE had not also been equipped with a camera.

Making a virtue of the situation, they renamed ISEE the International Cometary Explorer (ICE). The cost of the redirection was about $5m, considerably less than the $200m cost of a dedicated mission. It was a challenging profile, however, as the redirection involved five lunar flybys and 37 course corrections: ISEE had enough fuel to leave L1 but not enough to reach the comet unless it used the Moon for gravitational assist. The biggest risk was ISEE's battery, which had died, meaning that it could function only in direct sunshine. One lunar flypast involved 28 minutes of darkness, when it might freeze. The mission also presented a signals challenge, as ISEE was designed to transmit from no further than 1.6 million km, but was now expected to do so from 120 million km, so JPL upgraded its Deep Space Network (DSN) tracking system accordingly. Transmission rates, normally 2kbps, would be down to 1kbps at the 70 million km range of Giacobini-Zinner interception. An unexpected spin-off of the mission was the establishment, for the fast interception, of an electronic link between Goddard and the European Space Operations Centre (ESOC) in Darmstadt, Germany, the first internet-type connection between the two space agencies.

The adventure began. JPL commanded ISEE 3 out of L1 on 20th June 1982 to begin a series of five swings through the Earth-Moon system, one which also provided opportunities to study Earth's magnetic tail at numerous points. No less than 15 firings of its engine were required. ISEE's first loop around the Moon, 21,500km out, was on 21st September 1983, the second on 21st October and the third at a grazing 112km over Smyth's Sea on 23rd December, each generating more and more momentum to hurtle it out of the Earth-Moon system toward the comet. By the first week of January 1983, it was so far from Earth that tracking had to be transferred to the 64m Goldstone dish of the Deep Space Network. Meanwhile, Comet Giacobini-Zinner was picked up by Earth's telescopes on 3rd April 1984.

Four course corrections were necessary to ensure a good intercept. Conducted on 5th June 1984, the first involved two thousand pulses of the hydrazine engines over 4.5 hours to turn what would have been a 200,000km miss into a 10,000km intercept. This achieved a closing velocity of 21km/sec and also put the spacecraft within range of the Arecibo dish in Peru, which now joined Goldstone, Madrid, Canberra and the 64m Japanese Usuda dish. The next month, IUE picked up Giacobini-Zinner some 139 million km out, obtaining multicolour spectograms that revealed molecules of radical hydroxyl and carbon monosulphide.

The International Cometary Explorer (ICE), as it was now known, finally crossed the tail of Giacobini-Zinner on 11th September 1985, the first spacecraft to intercept a comet. ICE first detected the comet one million km out, much sooner than expected. The big fear was that the dusty debris of the comet would destroy it, disable it, or knock the antenna off course, which, in the absence of a tape recorder, would be the end of live data. A 2.3m/sec ΔV trim manoeuvre on 8th September had put ICE on course for close interception, a delicate decision, for coming too close could risk the spacecraft, while too much distance would risk the loss of useful data. Energetic ions were detected on the 10th and a bow shock the next day. Before interception, ICE detected plasma electrical waves from 2.3

million km out and then high speed, heavy ion beams of electrified atomic or molecular particles at 1.8 million km out. The latter had never been seen before and had possibly escaped from the comet. ICE entered the 22,530km-long tail some 7,800km behind the nucleus at 74,030km/hr, passing through the narrow neutral sheet and emerging from the tail 20 minutes later. The interception began at 09:10 UT and ended at 12:22. ICE passed through a region of turbulence between cometary ions and the solar wind before entering the comet's cold plasma tail. The shock boundary was 140,000km on the inward side, 100,000km on the outward side. There were no dust impacts, but micron-size particles in the plasma tail were recorded. ICE was within the comet's electric influence for about an hour beforehand and after, with 20 minutes in the plasma tail. The flyby distance was 7,862km, sufficiently near to the nucleus to obtain useful information and sufficiently far into the tail to learn about its properties.

Goddard held three press conferences, on the 9th, 11th and 13th September, to mark the first-ever spacecraft interception of a comet, upstaging the Russian, European and Japanese interceptions due the following year. However, the public impact was limited and was overshadowed by the later spectacular images of Halley by VEGA and Giotto. Nevertheless, Goddard was quickly able to announce significant scientific results. Its instruments were able to pick up the bow shock of the comet, the comet's electronic ions, the region of turbulence between the solar wind and the comet's own magnetic forces and the levels of cometary dust (less than feared, giving hope to later comet interceptors). Even without a camera, it was able to pick up useful data about the comet's tail and was able to notice its bow shock wave and turbulence.[12] The comet did not have a sharply defined bow shock, but a broad, turbulent u-shaped region where the solar wind was heated, compressed and slowed. The tail comprised hairpin-shaped magnetic field lines of dense, cold electric gases. Blue imaging showed up dense, narrow, cool plasma in the tail. In the comet itself, the ion composition experiment found mainly water vapour ions (H_2O+). The tail comprised two parallel lobes threaded by a magnetic field of opposite polarity. Amazingly, there was no damage to the spacecraft from the encounter. Goddard considered ICE to be a triumph and NASA opened a *Three lives of ICE* exhibition in the National Air and Space Museum in Washington DC that summer, with a one-tenth model, audio tapes of its signals and a proclamation of its 'firsts' (e.g. first to a Lagrange point, first use of multiple lunar swing-bys, first visit to a comet). On the first anniversary of the interception, Goddard hosted a conference on the mission in the museum on 11th September 1986.

ICE then went on to pass a distant 28 million km from Halley in March 1986, further than Earth had been from the comet in 1910. Some members of the ISEE science team remained sore about losing their spacecraft at the L1 point, because it was another nine years before another spacecraft reached the point. The ever-inventive team now devised a third mission: on 7th April 1986, Goddard commanded the thruster for a 40m/sec ΔV reduction to bring ICE back to Earth for August 2014. Hopes were expressed, however fanciful, that the ever-versatile Shuttle might be able to collect it.

Meanwhile, ICE sailed in a distant solar orbit, with communications becoming slow (down to 64bps) and infrequent because it was so far away. Its orbit was 0.927AU by 1.034 AU, 0.06°, slightly ahead of the Earth. Although operations officially ended, its carrier signal was not turned off. As a gesture to history, NASA formally transferred ownership to the Smithsonian Institution during a ceremony in Washington DC on 11th

September 1986. Acquiring hardware that was still in outer space was not a new idea, as the cash-stricken Russians sold their Moon rovers to entrepreneurs or collectors – but on a 'collect it if you can' basis.

ISEE (it now reverted to its original name, though both were still used) was due back in the vicinity of Earth on 9th August 2014 at a distance of 397,000km, before heading back into outer space, so some kind of capture manoeuvre would be required. Signals were duly picked up again early that year, and its beacon indicated that it was in the orbit expected.

In advance of its return, a group of ISEE enthusiasts called Rocket Hub planned to get in contact with the spacecraft again and re-capture it – if they could find the old command codes and get past legal obstacles. What became known as *Project Reboot*, also called the ISEE *Citizen Science Mission*, operating out of a disused McDonalds in Los Gatos, California, raised $159,000 in crowd funding in a project with NASA to regain control over the probe.[13] The signal indicated it still had fuel for a ΔV of 150m/sec, enough for recapture. The Shuttle had been retired by this point, so alternate options were to re-station it at L1 or design a mission to fly to another comet, but they had to reacquire signals and commands first.

Goddard found that its communications equipment for ISEE had been decommissioned in 1999, but the project was able to find the documentation and replicate the old electronics and they planned a new mission control in Morehead State University, Kentucky. On 29th May 2014, the amateurs successfully resumed command of ISEE. On 2nd July they fired the thrusters, which had not been used since 1987 and spun up the spacecraft for the capture manoeuvre.

ISEE 3 duly entered the Earth's field of influence again on a 42.5° trajectory that brought it 178,400km past Earth and 15,938km past the Moon on 10th August. At this stage, the luck of *Project Reboot* ran out as the pressurizing gas failed on the first of seven planned burns on 8th July, making future manoeuvres no longer possible. ISEE passed L1 on 17th August to enter a new 1.012 by 1.19AU, 1.17°, solar orbit, period 422 days, further out and 70 days longer than before.[14] The amateurs shut down all but those systems necessary to enable the continued return of scientific data, for five instruments were still working. Contact was lost on 16th September as ISEE headed away from Earth. Although due back to Earth in 2031, it is unlikely that contact will be re-established again, both because its trajectory may be insufficiently well known and because it may have frozen.

SAGE (Explorer 60): Aerosols and Ozone

The Stratospheric Aerosol and Gas Experiment (SAGE) spacecraft was the second of the Applications Explorer Missions (AEM), following HCMM (Explorer 58). Weighing 147kg, it was launched from Wallops Island on a Scout on 18th February 1979. As with HCMM, it comprised an instrument module and a standard service module. The purpose of the mission was to obtain global stratospheric aerosol and ozone data to better understand Earth's environmental quality and its radiation budget over a year. Power problems developed after 15th May 1979, but operations continued until 19th November 1981. With battery failure, the last signal was received on 7th January 1982. The mission's outcomes were published in the *Bulletin of the American Meteorological Society*.[15]

MAGSAT (Explorer 61): The Geodynamo

MAGSAT was launched on 30th October 1979 on a Scout from Vandenberg, for an intentionally short mission to provide the most accurate measurements ever of the global magnetic field.[16] It was a small satellite, weighing 158kg, which entered a low, near-polar, sun-synchronous dawn-dusk orbit, of 94° inclination, 355km perigee and 562km apogee. It was a joint venture between NASA and the United States Geological Survey (USGS) to measure near-Earth magnetic fields. Its specific objectives were to obtain a description of the magnetic field to inform updates and refinements to world and regional magnetic charts, the compilation of a global crustal magnetic anomaly map and interpreting that map to model the geology and geophysics of Earth's crust.

The spacecraft was built from leftover hardware from the Small Astronomy Satellite (SAS-C, Explorer 53) and comprised an instrument module with a precision vector magnetometer and a caesium-vapour scalar magnetometer at the end of a six-metre-long graphite epoxy scissors boom, as well as a service module for data-handling, power, communications, command and attitude-control. The magnetometer was able to collect 16 vector measurements and eight scalar measurements every second with a resolution of 0.5γ. The magnetometers were designed to be accurate to 2γ, while elaborate systems were in place to use star cameras and Doppler tracking to ensure accuracy within 70m. MAGSAT was expected to make a fresh picture of Earth's magnetic field every 12 hours.

The results of the mission were compiled with those of the Danish Oersted satellite.[17] Between them, they were able to map the dynamo of the Earth's liquid core that shapes its magnetic field to a scale hitherto impossible. The dynamo had changed over the previous 20 years, a little under the Pacific, but substantially under the poles and below southern Africa.

After a mission that lasted 75 days beyond its expected 1,250, MAGSAT re-entered on 11th June 1980 between Norway and Iceland. The chief project scientist Robert Langel described it as having provided the most accurate model of the Earth's magnetic field to date, a tenfold improvement over anything before and with an accuracy of within 0.03%. Its investigations covered geophysics, geology, field modelling, marine studies, magnetospheric, ionospheric and core and mantle studies. Data collected over the Pacific, for example, were able to map crustal anomalies along a 3,000km stretch covering the Marianas Trench from an altitude of 187–191km. Its data were expected to be of considerable importance in the search for minerals, as MAGSAT flew as low as 200km compared to previous such surveys at over 400km. MAGSAT confirmed that the Earth's magnetic field was decreasing in intensity by 0.09% annually, likely to lead to a reversal of north and south magnetic poles in about 1,200 years (the last reversal was 700,000 years ago).

Dynamics Explorer (DE) 1, 2 (Explorer 62, 63): High and Low

Dynamics Explorers 1 and 2 were launched together on a Delta 3913 from Vandenberg on 3rd August 1981, into different orbits, to study electric currents and fields around the Earth. This was the first Delta to use the powerful Castor IV strap-on booster. The mission was designed to extend knowledge of the hot, tenuous, convecting plasmas of the magnetosphere and the cooler, denser plasmas of the ionosphere, plasmasphere and upper atmosphere. The aim was to put DE 1 into a highly elliptical orbit 25,000km out for auroral imaging and the measurement of auroral field lines, while DE 2 would fly in a lower, co-planar polar orbit (205km–1,300km) to make composition, temperature and wind measurements. The mission cost $54m, with a further $23m for the launch.

The idea was that their polar co-planar orbits would permit simultaneous measurements at high and low altitudes in the same field-line region. Their purpose was to study the environment of the Earth, from the upper atmosphere to far out in the magnetic field, with the intention of shedding light on the magnetosphere, the northern lights, radio transmissions and weather. They complemented the three Atmosphere Explorers which studied the upper atmosphere down to 130km and the three ISEEs which studied the interaction of the solar wind with the magnetosphere. Specifically, DE1 and 2 were expected to obtain information on the coupling of energy, electrical currents, electric fields and plasmas in the magnetosphere, ionosphere and atmosphere. Later, there would be a four-satellite mission called Origin of Plasmas in Earth's Neighbourhood (OPEN). The two spacecraft were similar, drum-shaped, 16-side polygons, 1.36m in diameter and 1.14m tall. Their weight was also similar, with DE 1 weighing 403kg and DE 2 weighing 415kg. Their solar cells covered 4.64m^2 and both had a 1.5m S-band antenna. The project scientist was Robert Hoffman of Goddard and the signals would be transmitted to the computer facility at that centre, which would then compile archives on microfilm or microfiche. The facility was accessible to the user community.

Dynamics Explorer 1 was to take photographs of the northern lights, to show visually how energy was transferred from the magnetosphere to the upper atmosphere while other instruments detected ionized particles; undertake global auroral imaging of wave measurements in the heart of the magnetosphere and the crossing of auroral field lines at several Earth radii; and make measurements along a magnetic field flux tube. This required long antennae of 200m. The spacecraft was set to spin at 10rpm. Transmissions were both realtime or by tape recorder at 16kbps. On DE 1, the plasma wave instrument comprised two 4m antennae, two 100m wires and a 5.9m boom, while the magnetometer was on a 5.9m boom opposite. DE1 carried six 10m antennae and a 6m boom.

Lower down, at 350km by 1,000km, DE 2 would skim the atmosphere from pole to pole, moving rapidly through the polar regions where these disturbances were most intense, with similarly long antennae to measure the voltage or strength of these current lines and electrical fields as well as their speed, up to 1,600 km/hr. DE 2's perigee was set sufficiently low to permit measurements of neutral composition, temperature and wind, while the apogee was set high enough to permit measurements of the interaction regions of supra-thermal ions. It attempted to capture electron beams shooting from space into the atmosphere, radio noise bursts thought to emanate from a region between 4,800km and 13,000km. On DE 2, the vector electrical field instrument had antennae 23m long end to end and a 5.9m long magnetometer, while the Langmuir instrument had two probes 70cm long. DE 2 carried an optical interferometer from the University of Michigan and University College London to measure the wind and temperature structure of the upper atmosphere. This had triaxial antennae 23m long. The spin rate for DE 2 was one per orbit and the solar array charged two 6amp nickel-cadmium batteries. The instruments for the two spacecraft were:

Dynamics Explorer (DE) 1, 2 (Explorer 62, 63): High and Low

Dynamics Explorer 1, 2

Dynamics Explorer 1

- Magnetometer
- Plasma Wave Instrument
- Spin Scan Auroral Imager
- Retarding Ion Mass Spectrometer
- High Altitude Plasma Instrument
- Energetic Ion Composition Spectrometer

Dynamics Explorer 2

- Magnetometer
- Vector Electric Field Instrument
- Neutral Atmosphere Composition Spectrometer
- Wind and Temperature Spectrometer
- Fabry Perot Interferometer
- Ion Drift Meter
- Retarding Potential Analyzer
- Low Altitude Plasma Instrument
- Langmuir Probe Instrument.

The satellites found themselves in lower orbits than intended due to the fuel tanks being 200kg short of fully-fuelled. The filler thought the tank had overflowed, so stopped filling without having checked the gauges, but there was also a new monitoring system in operation about which he had not been told. DE 1 reached 22,400km and DE 2 reached 1,006km, rather than the higher altitudes intended, which reduced the mission duration. DE 1 stopped taking commands in October 1990 and was retired on 28th February 1991. DE 2 re-entered the atmosphere on 19th February 1983.

The missions had a big visual impact, providing the first global images of the aurorae. The University of Iowa spin-scan auroral imager took extraordinary pictures of the auroral oval over the polar regions, Greenland, Alaska and northern Canada, the first such global image.[18] It took global pictures of the auroral region every 12 minutes, showing substorms and the electrical currents increasing in intensity as they glowed between the magnetosphere and ionosphere. Examining the pictures taken by DE 1's ultraviolet imager, Dr Louis Frank of the University of Iowa found dark spots or atmospheric holes 20–30km across on the sunlit atmosphere. They had a high water content and he controversially speculated that they were 'caused by small comets breaking up in the upper atmosphere.'[19]

Solar Mesosphere Explorer (Explorer 64): Global Ozone

The Solar Mesosphere Explorer (SME) studied the ozone layer. The mission was first proposed by Charles Barth of the University of Colorado in 1974, who became its PI. It was not the first ozone satellite – these were Explorers 55 and 60 (SAGE) – but, by going into polar orbit, SME would cover the whole globe, including the especially important Arctic and Antarctic regions. SME was launched on a Delta II from the Western Test Range (WTR) on 6th October 1981 into a 540km sun-synchronous polar orbit.[20] It was one of the first satellites intended to use the Tracking Data Relay Station (TDRS) network of 24-hour satellites to relay its transmissions.

The general objective of the mission was to investigate the processes that created and destroyed ozone in the Earth's mesosphere and upper stratosphere. Its specific objectives were to determine the nature and magnitude of changes in mesospheric ozone densities arising from changes in the solar ultraviolet flux; the interrelationship between solar flux, ozone and the temperature of the upper stratosphere and mesosphere; the interrelationship between ozone and water vapour; and the interrelationship between nitrogen dioxide and ozone. It was expected to provide a comprehensive picture of the processes that affected atmospheric ozone; follow changes in ozone distribution, especially in response to incoming solar radiation; measure temperature, pressure, water vapour, nitrogen dioxide and near-infrared airglow; determine solar terrestrial correlations with ozone density; and examine other changes in ozone abundance and their connections, if any, to the Sun. A particular point of interest was how ozone was transported from 30km altitude to 90km.

The spacecraft was 1.7m tall, 1.25m in diameter, with a 2.2m wide solar array of 2,156 cells. It comprised an instrument module and a service module and was built by Ball Aerospace. SME was managed by JPL, with the Laboratory for Atmospheric and

Solar Mesosphere Explorer (Explorer 64): Global Ozone

Space Physics in the University of Colorado responsible for its science instruments, operations and data. Its communications were via Goddard. The five experiments carried were:

- Ultraviolet ozone spectrometer to measure ozone at between 40km and 70km altitude;
- 1.27 Micron spectrometer to measure ozone and hydroxyl from 50km to 90km;
- Nitrogen dioxide spectrometer to make measurements between 20km and 40km;
- Four-channel infrared radiometer to measure temperature and pressure between 20km and 70km; and ozone and water vapour from 30km to 65km;
- Solar proton alarm detector to measure the integrated solar flux in the range 30-500 MeV.

Explorer 64

SME was an unusual shape, with its instruments mounted on an incomplete flat disc whose reverse acted as solar cells. It looked like a box, with a flat ring underneath and the infrared radiometer in the shape of a small telescope on top. Power was stored in two nickel-cadmium cells and there were two tape recorders, each able to hold 16MB for up to 96 hours and transmit at 8.192kbps from a 5w transmitter. SME was spun at 5rpm, with its axis always at 90° to its path so that it could scan the Earth to the horizon.

SME suggested that the variation in low-altitude nitric oxide was due to variation in the solar output of soft x-rays. Nitric oxide density at 110km varied with the 27-day solar rotation period and the 11-year solar cycle. The mission finished in December 1988.

AMPTE Charge Composition Explorer (Explorer 65)

AMPTE stood for Active Magnetospheric Particle Trace Explorer and aimed to study how the solar wind entered and behaved in the Earth's magnetosphere and its tail, with trace ions of lithium and barium being released into the magnetospheric stream to identify the patterns. It would be the first Explorer to create an artificial comet. AMPTE was a three-part project between the United States and Germany, with Britain joining later. There were actually three satellites: the American one (Charge Composition Explorer (CCE) or AMPTE 2, closest to Earth), the German one (Ion Release Module (IRM), or AMPTE 1, releasing the tracers) and its British sub-satellite called UKS or AMPTE 3. Although the American satellite was technically AMPTE 2 or CCE, the single term 'AMPTE' was often used. It was built at Johns Hopkins, managed by Goddard and controlled by JPL, while IRM was operated from the European Space Operations Centre in Oberpfaffenhofen, Germany and UKS from the Rutherford and Appleton laboratories at Chilton, Oxfordshire. The principal investigator was S. M. Krimigis and the project manager was Gilbert Ousley.

AMPTE was launched by Delta 3924 on 16th August 1984, a three-year gap since the previous Explorer. Soon after launch, during the first orbit, CCE made a sharp plane change from 28.7° to 4.8°and climbed to an apogee of R8, while IRM and UKS went on to R18. IRM and UKS were in a similar high elliptical orbit outside the magnetosphere and in the solar wind, 100km apart, while AMPTE's orbit was lower, equatorial and in the magnetosphere. AMPTE was a real-time interactive experiment to study how solar wind ions entered the magnetosphere; the convective-diffusive transport and energization of magnetospheric particles and the interaction of plasmas.

The American AMPTE was an octagonal prism, 1m high, 2m diameter, weighing 242kg and orbiting at 1,113km–49,667km, 939 minutes, 4.83°. It had a 140w array and was spin-stabilized at 10rpm. In comparison, IRM was a 2m cylinder, 2m diameter, weighing 605kg and orbiting at 383km–114,570km, 2,657 minutes, 29°. The UKS sub-satellite was a multifaced prism of 1.5m diameter, 50cm long and 77kg, operating in the same orbit as IRM with two booms to study natural magnetic particles, fields and waves. The role of the IRM was to provide multiple ion releases in the solar wind, the magnetosheath and the magnetotail, with diagnostics of each, while the UKS used thrusters to keep station near the IRM to provide two-point local measurements. The function of the American AMPTE was to detect lithium and barium tracer ions from the IRM releases and see how they were transported into the magnetosphere.

The lithium and barium clouds were released at four intervals. They were extremely tenuous, with 1kg generating a cloud 30,000km in diameter. The German spacecraft made the first releases, of lithium, on 11th and 20th September 1984. With UKS in the neighbourhood, it was measured by the American satellite and they established that less than 1% of solar wind was entering the magnetosphere on those dates. Solar wind sped toward the Earth at up to at 1.6 million km per hour and then hit the magnetosphere typically 64,000km sunward, extending four million km night side.

The second release, of barium ions, was intended to be a 160km diameter comet-shape, released 113,000km over Lima, Peru on Christmas Day, not for religious reasons but because it was a new Moon, so it would be visible over the western U.S. The Sun was 18° below the horizon so the artificial comet would be visible for ten minutes. The idea was that the German satellite would release four 10kg canisters of barium vapour, to be analysed by the British and American satellites travelling behind. In the event, the release took place on the 27th December, high over the Pacific, making an artificial comet which solar winds then drove away from the Sun's direction. The artificial comet was quickly eroded by the solar wind and ground observations were bedevilled by cloud, but it was seen by aircraft-based observers. There were two surprising results. The barium cloud expanded perpendicularly to the solar wind before it blew in its direction and developed complex motions and structures. In contrast, the lithium disappeared and no traces of it were found.[21]

The third phase of the mission, from 21st March to 13th May, involved two barium and two lithium releases by the German satellite into the plasma sheet on the opposite flank of the magnetosphere, with the precondition of good viewing conditions for at least one of the ground stations (Kitt Peak, Arizona; White Sands, New Mexico; and Mauna Kea, Hawaii). No tracer ions were found by the American satellite. In the fourth stage, another release was made on 17th July, at 113,000km from Earth into the solar wind on the flank of the Earth's magnetosphere. Two aircraft were involved in the airborne observations off the coast of Mexico. The first, an American Convair 990, blew a tyre on take-off, crashed and burned out, but not before the 19 crew and observers escaped. Luckily the second, an Argentinian Boeing 707, got good observations, as did ground observers in Phoenix, Arizona.

The British satellite failed on 16th January 1985 due to a power system malfunction. The American spacecraft encountered power supply problems from early 1989, failing on 12th July that year.

COBE (Explorer 66): The Big Bang

The COsmic Background Explorer (COBE) was one of the most important of the Explorer series. It found the relict radiation from the big bang and its scientists were later awarded the Nobel prize. The 2,260kg satellite was launched on 18th November 1989 on a Delta II 5920 from Vandenberg into an 888–897km orbit, 99°, 102.9 minutes, circling Earth 14 times a day. It had been a record five years since the previous Explorer, AMPTE, an indicator of how space science had thinned out in the post-Nixon, Shuttle era. COBE was an example of how the Explorer program had become bigger (the satellite was over two tonnes) and took many years to develop (15 years from proposal to launch), although the outcome more than met expectations.

The cosmic background was first theorized by George Gamow, Ralph Alpher and Robert Hernan in the 1940s, building on the earlier work of Albert Einstein and Edwin Hubble. Cosmic background radiation had first been detected by Arno Penzias and Robert Wilson of Bell Telephone Laboratories in 1964, using a sensitive microwave receiver. They discovered a faint background hum of radiation at 7.35cm, the microwave range of the spectrum, coming in uniformly from everywhere, in effect the signature of the first light. They guessed that it was the remnant radiation left over from the 'big bang', which had big cosmological implications. If it could be proven, it would oust the traditional 'steady state' model of the universe. Cosmic background radiation was confirmed by subsequent sounding rockets and balloons and was again found to be smooth and uniform. But if the universe was as violent as we now know it to be, then it would not be perfectly smooth, so COBE was designed to carry instruments sufficiently sensitive to determine any variations. The product of COBE was intended to be multi-colour sky maps in 100 microwave and infrared wavelengths, showing the traces of the big bang; irregularities, called 'anistropies'; as well as the locations of interstellar dust clouds. They would help to determine the rotation and expansion of the universe.

COBE was put forward in a 49-page proposal by John Mather in October 1974 as *Cosmological Background Radiation Satellite*, suitable for an Explorer-class satellite on a Delta. John Mather was a young Goddard scientist who had written his PhD on cosmic background radiation. He was a self-taught child prodigy scientist who came from a scientific family – his father had reorganized the dairy industry – and had entered science fair projects from the age of nine. Six people were involved in the original proposal. They probably had no idea that the COBE project would eventually involve 1,500, cost up to $400m and take 15 years. John Mather became the PI on the satellite and later wrote a first-hand account of the mission, making it one of the best documented of the series.[22]

John Mather

The proposal was an understated text, defining the task of making definitive measurements of the radiative relics of the earliest stages of the universe at 2.7K. Nobody had done this kind of mission before. COBE was an in-house project at Goddard and was not contracted out. Mather's view was that no private contractor had the capacity and that, if done in-house, it would be a useful training exercise.[23] NASA approved COBE for funding, with a Delta launch, on 19th July 1982.

COBE required 6,000 rivets, took nine months to assemble and was rolled out in August 1985. The principal challenge was to keep its telescope cool, the task of a 600l vacuum flask called a dewar (named after James Dewar, its inventor) which had to keep the instruments at a temperature of 1.4K (-271.6°C) for a year. Using a conventional analogy, this was equivalent to keeping coffee hot at the same temperature for 30 years! The dewar was the heaviest single item on the spacecraft at 642kg and was duly completed in November 1985.

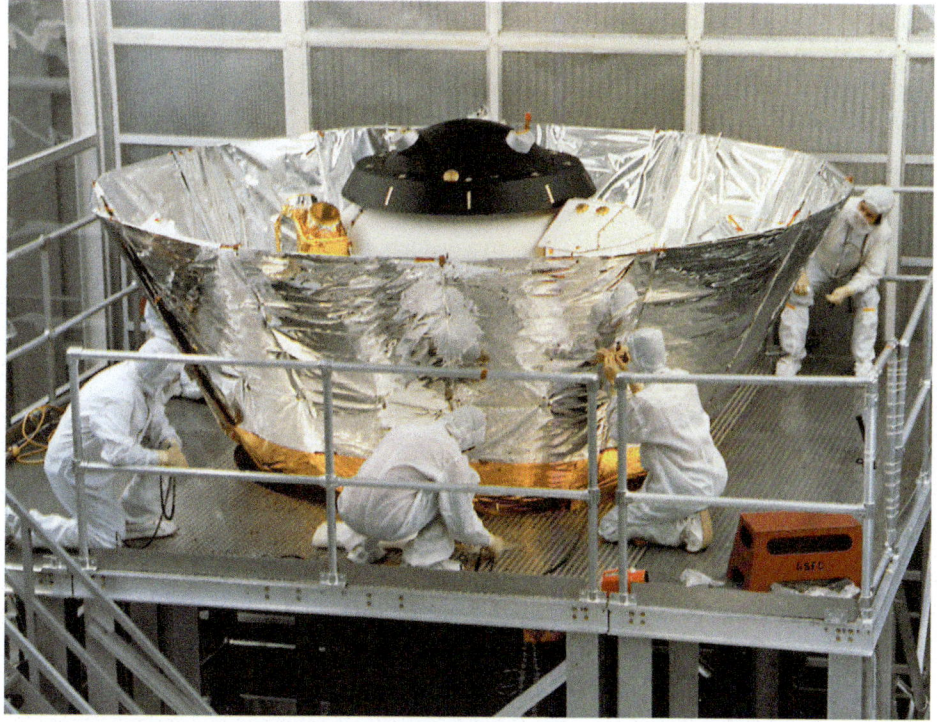

COBE preparations

The new instruments were the slowest and most difficult part of the project and building them was one of Goddard's biggest challenges. The mission was guided by a Science Working Group of 20 members. COBE used technology originally developed by the American-Dutch-British Infrared Astronomical Satellite (IRAS) in 1983 and had a 20cm telescope. It had three instruments, developed by Charles Bennett:

- FIRAS (Far Infrared Absolute Spectrophotometer);
- DMR (Differential Microwave Radiometer); and
- DIRBE (Diffuse Infrared Background Experiment).

The role of DMR was to determine whether the original explosion was equally intense in all directions and was set to make measurements in the 3.3, 5.7 and 9.6mm wave bands. FIRAS would survey the sky for a year twice, in the 0.1 to 10mm waveband, to determine the spectrum of the background radiation with a hundred times greater precision than ever before. The aim of DIRBE was to search for diffuse glow in the 1–300 micrometre range.

The mission as originally proposed was to be launched on a Delta rocket, but with the Shuttle now designated the universal launcher, it was re-assigned to a Shuttle launch from Vandenberg Air Force Base in 1987, later slipping to 1989. Although this would require 860kg of propulsion fuel to raise it from the Shuttle's altitude of 300km to its intended 900km sun-synchronous orbit, it also enabled the spacecraft to carry more sophisticated instrumentation. By 1986, the satellite was ready, except for its instruments and was on target for a 1989 launch.

The loss of the Space Shuttle *Challenger* and its crew of seven on 28th January 1986 was not only a national disaster but brought despair to the COBE team. The only two viable alternatives were the Soviet Proton, which in the cold war atmosphere of the time was ruled out; or the European Ariane, which NASA headquarters also ruled out because the tricky politics of negotiating with the French were too problematical. This left the team members with the only alternative: downsizing the spacecraft back to its original launcher, the Delta, if one could be found. They did track down the very last Delta, although it was corroded and patched. NASA gave the COBE redesign a priority, because the agency badly needed some good news, so Goddard brought in Delta engineers, working together in a cubicle-free, open plan office. This arrangement in effect forced everybody to work together in the same room and intensified personal contact to the point that formal staff meetings were scrapped. The redesign involved 300 people working 24/7 in three shifts, which added $1m a month to the bill.

This was a serious re-design that involved cutting the weight in half from 5,000kg to less than 2,500kg and the diameter from 5m to less to 2.5m, as well as addressing vibration tolerances. It meant a redesign of the primary structure and reconfiguration of the solar arrays, thermal shield and microwave radiometer. The easy part was to drop the propulsion unit, which was not needed because the more versatile Delta could get COBE straight into its intended orbit, but changing the dewar was not negotiable.

The proudly-made-in-Goddard new COBE first went on display at a COBE open day at the centre on 30th October 1988, along with displays and video of its mission, so it was on course for its May 1989 deadline. Then the spectrophotometer failed and its thermos cooler had to be opened to fix the instrument. The instrument team insisted on a two-week electrical clean room test which delayed things further, but nobody later contested their wisdom in running such a check. The satellite was finally shipped to Vandenberg on 4th October 1989.

COBE was eventually launched at the end of Pacific darkness on 18th November. Separation took place 61 minutes into the mission for a 14-day checkout and 16-day calibration. Data were dumped once a day for over nine minutes at 655 kbps, with three ten-minute passes a day over Goddard's Processing Flight Facility at Wallops, before being forwarded to the COBE Science Data Center at Goddard HQ. COBE had two tape recorders, with data sent out through the TDRS system. Its spin rate was 0.8 rpm and the instruments could scan the sky every six months.

COBE in orbit

The first results came through early in the new year.[24] They were presented at the American Astronomical Society in Crystal City, Virginia and confirmed the idea of the big bang. John Mather and George Smoot presented the first 20 days of data, including sky maps, over the course of which the instruments had been pointed at the north galactic pole. They found the sky to be extraordinarily smooth but expected that over two years they would detect lumpiness, or 'anistropies' to use the scientific term. Later, the team was able to identify primordial hot and cold spots in the big bang radiation that would grow into huge clouds of galaxies and giant empty spaces.

They confirmed that COBE had discovered large quantities of carbon and nitrogen between the stars, thought to be the primordial remnants of the big bang 15 billion years earlier. The FIRAS data were so unambiguous regarding the uniformity of the post-big bang radiation as to astonish the project scientists and quickly made all the effort worthwhile. FIRAS was able to obtain a cosmic background spectrum; DMR made a map at 31, 53 and 90GHz, indicating that the background was equally bright in all directions and that the universe was very smooth; while DIRBE obtained maps of half the sky at 1.2, 12 and 240 microns, revealing stars and dust. Their presentation to the American Astronomical Society in January 1990 received a standing ovation from normally staid scientists.

COBE completed its all-sky survey on 18th June 1990 and ran out of helium on 21st September 1990, as expected. This made the FIRAS no longer operable, but the other two instruments would continue to function. COBE lost its C-gyro in September 1993 and radar tracking showed 14 pieces of debris orbiting close by, which was mysterious. Science operations concluded on 23rd December 1993. The mission concluded in January 1994, the control centre closed down and its staff went off to write up end-of-mission reports. In March 1991, Goddard had already opened the Cosmology Data Analysis Center, where up to 80 scientists could work on COBE data. According to John Mather, it had done everything expected of it. We now knew that 99.97% of the energy of the universe was released in the first year of the big bang. Since then, its temperature has cooled over the following 15 billion years, from trillions of degrees to 2.72K. COBE found the hot and cold spots that dated to 300,000 years after the big bang.

More results were announced at the 1992 meeting of the American Physical Society. The DMR confirmed the temperature of the universe at 2.725K (±0.002K) with faint but observable differences of temperature, some as small as a three-millionth of a degree, suggesting not only the glow left over by the big bang but dark matter. Relict radiation was found because of the identification of plasma ripple variations of 1:100,000 parts. The DMR and DIRBE instruments refined the results to pick out different radiation from early stars and galaxies. In addition, COBE found ten new galaxies emitting infrared and a resonance ring of dust in Earth's orbit. The plane of the Milky Way was obvious, slightly warped and a barred spiral in shape. It was calculated that the Sun was 8.6 kiloparsecs from the galactic centre. An important finding was that a large amount of starlight could not be seen, suggesting that it was hidden by dust. DIRBE detected ten new far-infrared light-emitting galaxies missed by IRAS.

In January 1993, four new COBE images were published at the American Astronomical Society in Phoenix, Arizona, showing the universe in the 1–240 micrometre range across the s-shaped plane of the Milky Way. The success of the mission was celebrated at Goddard by naming one of its connecting streets as 'COBE road'.

COBE led to Nobel Physics prizes for Dr George Smoot and Dr John Mather (2006), perhaps the most prestigious of a shower of awards that came their way. This was no surprise, as COBE was one of the most important contributors to modern understandings of cosmology.[25] Stephen Hawking described the COBE findings, confirming the big bang as the origin of the universe, as 'the discovery of the century.'

There were two footnotes. First, scientists were nevertheless puzzled at how small the temperature differences were and devised a follow-up mission (WMAP, see next chapter). Second, Russian scientists were greatly disappointed, for their *Relikt* spacecraft, Prognoz 9, had found and measured the big bang radiation and published a galactic map accordingly in 1983, six years before COBE even left the ground. They got no recognition, never mind Nobel prizes and their findings were even disputed in the west. That was the cold war.

EUVE (Explorer 67)

The idea of an extreme ultraviolet mission came from the July 1975 Apollo-Soyuz Test Project, whose high-profile political role overshadowed the valuable science carried on board. Apollo 18's telescope had found four Extreme Ultra Violet (EUV) sources, contrary to earlier expectations that it would not be possible to see this end of the spectrum and that such rays would be absorbed by interstellar hydrogen. Apollo 18's SAG telescope, named after the Space Astrophysics Group at University of California Berkeley led by Professor Stuart Bowyer, had 20 hours of observation time and was aimed at 30 targets. On 22nd July 1975, Apollo 18 rolled toward Coma Berenices, where it picked up a hot dwarf, HZ43, some 300 light years away and with it a new field of astronomy. The Apollo 18 observations enabled the scientists to estimate its temperature at 110,000K – more than 20 times hotter than the Sun – and its radius at 7,800km (about the same as Earth), emitting powerful amounts of extreme ultraviolet radiation.

Years later, the number of extreme ultraviolet sources discovered in the universe was still small: only nine had been found by the end of the decade (Jupiter, Saturn, Titan and six stellar objects). The SAG group from Apollo 18 prepared a proposal for an extreme ultraviolet mission by the end of the following year, sending a 300-page dossier to NASA. This was eventually approved as the EUV Explorer (EUVE).

The project was assigned to JPL in 1982 and the instrumentation to Stuart Bowyer's Berkeley group, with the expectation of a 1988 launch. The mission was to make a six-month all-sky survey with three 40cm telescopes, the first in extreme ultra-violet. The program manager was John Paulson, who had directed SME, then Ron Ploszaj. Stuart Bowyer was the PI and Roger Malina the project scientist, with Robert Stachnik the program scientist at NASA. Like COBE, it was built in-house at Goddard, where it was managed by Frank Volpe.

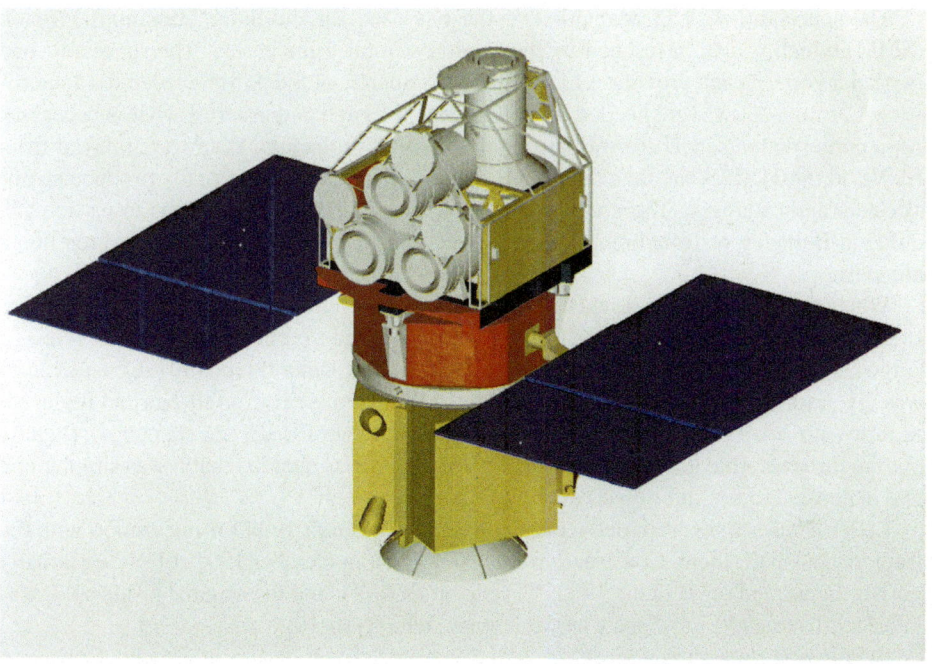

EUVE

Designing instrumentation that could detect extreme ultraviolet radiation was challenging, requiring steep and smooth mirrors (called grazing incidence mirrors) and detectors. Although this was not as difficult as the Hubble Space Telescope, it required new standards of manufacture and design. There were originally three grazing incidence all-sky scanning telescopes clustered together, as well as a pointing deep sky telescope that was designed always to operate either in shadow or pointed away from the Sun. These 188kg Wolter-Schwarzschild scopes were the most powerful of their kind ever built at that point. A fourth 336kg spectrometer was added to the three telescopes in 1983 to make an energy distribution chart to explain the chemistry and structure of these sources and enable follow-up studies with individual stars. The plan was for the three telescopes to make an all-sky survey in the first six months, while the fourth would survey individual selected stars below 100Å. Getting a spectrum of a standard source was expected to take 20 minutes, but up to a week for very faint objects. Two of them worked in 70 to 400Å, the other 400 to 760Å, below the 1100Å to 3200Å range of IUE (Explorer 57).

This spacecraft's general objective was to carry out a full-sky survey, followed by a deep-space survey and pointed observations. Its specific objectives were to discover and study ultraviolet sources and their relationship with the interstellar medium. It was hoped that this mission would add hundreds of new sources, such as white dwarfs. Although EUVE's time was focussed on the 'local bubble' of 250 light years around the solar system (an area of 12 major and well-known stars), it was intended to look further to white dwarfs, red giants, novae, young hot stars, cool stars and other objects of opportunity. This was NASA's first dedicated mission into extreme ultraviolet.

The spacecraft was an adaptation of the bus used for the Solar Maximum Mission (SMM) which had been rescued by the Shuttle several years earlier. The spacecraft had two solar panels, each providing 1,000w, about a quarter of the power needed for a microwave. Of this, 300w were needed for the science package, a quarter of what is necessary for a domestic toaster. There was substantial thermal protection. Data were relayed from EUVE to the TDRSS and then down to Berkeley, which would eventually produce an all-sky map and catalogue. There was a guest observer program for scientists to participate, either at Berkeley or from home. Viewing time was divided into 200 blocks of ten hours annually.

When construction began in 1983, EUVE was scheduled for Shuttle launch in 1988. Like COBE, it was delayed by the loss of *Challenger* and the end of the Shuttle's role as a cargo carrier. EUVE was rescheduled for one of the new Delta IIs for 1991. Consideration was given for the Shuttle to rendezvous with EUVE, remove it from its bus and replace it in turn with two future Explorer satellites (XTE and then FUSE, see chapter 4), thereby getting three satellite uses from one bus. In the end, it was decided that it was simpler and less expensive to use three buses and expendable rockets.[26]

EUVE's telescopes were delivered to Goddard in March 1990 for integration with the Explorer bus built there. Construction was completed in October 1991. EUVE eventually got off the ground on 7th June 1992, 17 years after ASTP and the original proposal. It was launched from pad 17 at Cape Canaveral on a Delta 6920-10.

EUVE began its survey work on 24th July 1992. Even during its checkout phase, it detected 20 known and two unknown EUV sources. By the time it had been in orbit

150 days, it had observed an energetic elliptical galaxy two billion light years away which had as much energy as a trillion suns and might have a black hole at its centre. The University of California, Berkeley, released its first list of bright EUVE sources, including hot white dwarfs, planetary nebulae, flare stars and cool star coronae. By January 1993, EUVE had detected seven white dwarfs and by the following month its bright source list had risen to 135. With the completion of the all-sky survey, the guest observer program was introduced. The first year of findings was presented at the 182nd American Astronomical Society national meeting in June 1993, highlighting the detection of ionized helium in local interstellar gas, rare extragalactic objects and the elements that blanketed light from white dwarfs. The last gaps in the all-sky survey were filled in August 1993, cataloguing 801 objects. EUVE eventually identified 900 stars and 11 galaxies with radiation in the ultraviolet range.

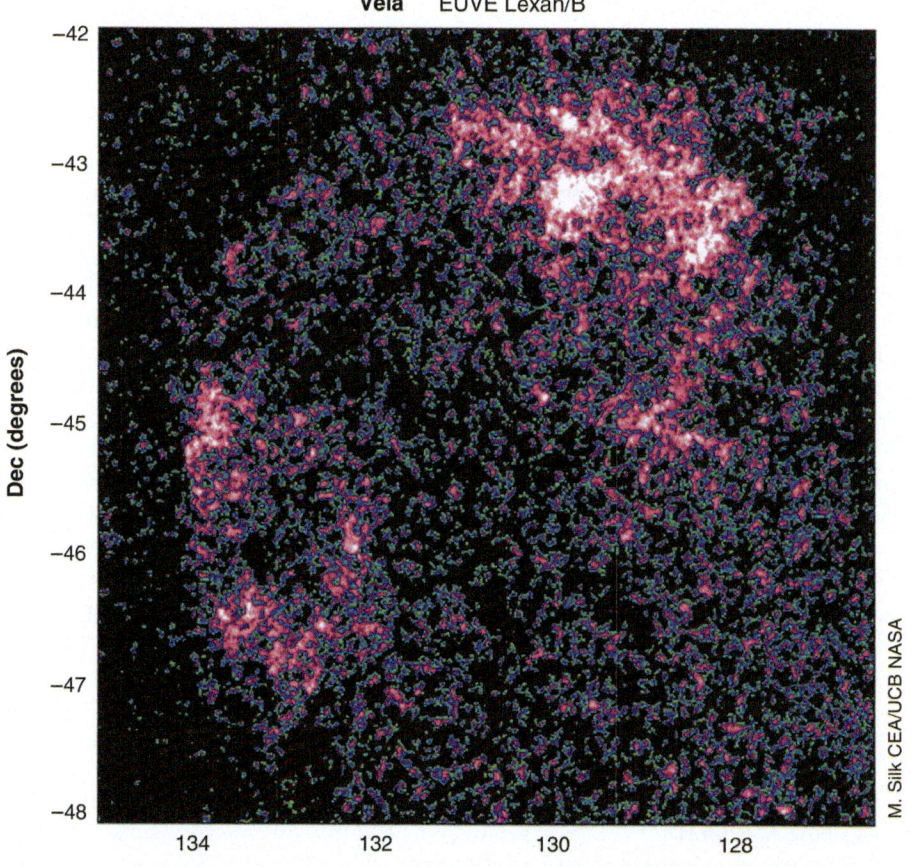

EUVE supernova remnant

EUVE next focussed on a target of opportunity, an outbursting cataclysmic variable, SS Cygnus. In April 1994, EUVE turned to the brilliant B-type star Hadar in the southern constellation of Centaurus. EUVE detected thermal extreme ultra violet emissions from neutron star surfaces, made the first measurement of stellar coronal abundances and imaged hot, hydrogen-rich white dwarf atmospheres. It discovered a new class of massive white dwarfs, high-density coronal plasmas, EUV emission from Comet Hyakutake, helium outgassing in the Martian atmosphere and EUV emissions from neutron stars. It detected photospheric emissions from stars (e.g. ε Canis Major) and a Quasi Periodic Oscillation (QPO) in Cygnus.

The EUVE mission was extended twice, but because of costs and diminishing scientific return the instruments were turned off in December 2000. The spacecraft was made safe and its transmitters turned off on 31st January 2001. EUVE burned up over central Egypt a year later on 30th January 2002.

Conclusions

This necessarily short chapter looked at the Explorer program in the aftermath of the revision of the space program by President Nixon. The pace of the Explorer program slowed from 55 in the seventeen years 1958-75 (over three a year) to only 11 in the seventeen years from 1977 to 1994 (much less than one a year). Overall, though, it is hard to avoid the conclusion that with the space program smaller and with resources concentrated on the Shuttle, space science was squeezed, confirming the fears of scientists and contrasting with the Apollo period. The use of the Shuttle as principal carrier also set in train a logic whereby science missions had to be redesigned around the Shuttle's cargo capacity, COBE being the prime example. Although the Scout was spared the 'all-aboard-the-Shuttle' rule, in practice it was used only three times (Explorer 58, 60 and 61).

Despite the low numbers, the missions flown in this period were highly successful. COBE was nothing less than a cosmological breakthrough, while EUVE was a pioneer of extreme ultraviolet astronomy. IUE worked for 19 years and brought in one of the biggest hauls of scientific observations before the Hubble Space Telescope. Other missions were at the frontier of atmospheric science, especially those that dipped into the high atmosphere (Dynamics Explorers 1 and 2), targeted ozone research (SME) and mapped the geodynamo (MAGSAT). Honouring NASA's tradition for improvisation, ICE was first into the tail of a comet.

References

1. Newell, Homer: *Beyond the atmosphere - early days of space science*. Washington DC, NASA, SP-421.
2. For an analysis of the key decisions of this period, see Logsdon, John: *After Apollo - Richard Nixon and the American space program*. New York, Palgrave Macmillan, 2015
3. Ezell, Linda Neuman: *NASA historical data book, Programs and projects, vol. II 1958-68; vol. III, 1969-78*. Washington DC, NASA, SP-4012.

4. MacNeil, Nick, Carter-Mondale Transition Planning Group: *NASA recommendations*, 31st January 1977.
5. Logsdon, John: *The survival crisis of the U.S. solar system exploration program.* 2012, Palmgrave McMillan. Originally an unpublished paper for the NASA History Office, 1989.
6. John M. Logsdon (ed): *Exploring the unknown - selected documents in the history of the U.S. civil space program, vol. V, Exploring the cosmos.* NASA, SP 4407, 2001.
7. Bizony, Piers: *The man who ran the Moon - James Webb, JFK and the secret history of project Apollo.* London, Icon Books, 2006.
8. Roland, Alex: *A spacefaring people - perspectives on early spaceflight.* Washington DC, NASA, 1985, SP-4405.
9. *International Astronomy Satellite; IUE: ultraviolet telescope in space.* Flight International, 8th June 1972; 11th February 1978.
10. *Black holes and globular clusters*, Spaceflight, vol. 21, no. 1, January 1979.
11. DeVorkin, David: *Mystery star.* Air & Space, January 2016.
12. Carter, L.J: *Comet fly-by first results.* Spaceflight, vol. 28, no. 3, March 1986; Le Page, Andrew: *ICE - the first comet flyby.* www.drewexmachina.com, 12th September 2015.
13. Jonathan's Space Report, 696, 19th April 2014; Dunham, David: *The return of the ISEE-3/ICE spacecraft.* Paper presented to the International Astronautical Congress, Toronto, 2014; *The Tortuous path to a comet's tail* in John M. Logsdon: Exploring the unknown vol. VI, NASA 2004
14. Other figures are also given, such as 0.927 by 1.035AU, 0.1°.
15. McCormick, M.P et al: *Satellite studies of the stratospheric aerosol.* Bulletin of the American Meteorological Society, vol. 60.
16. Both upper case (MAGSAT) and lower-case (Magsat) forms were in use, MAGSAT being the most common.
17. Hulot, Gauthier et al: *Small-scale structure of the geodynamo inferred from Oersted and Magsat satellite data.* Nature, vol. 416, 11th April 2002
18. *Auroral oval from space.* Spaceflight, vol. 28, no. 5, May 1986.
19. *Earth sprayed by small comets.* Spaceflight, vol. 39, no. 8, August 1997.
20. Wilson, Andrew: *The Solar Mesosphere Explorer.* Spaceflight, vol. 24, no. 1, January 1982.
21. Gombosi, TI et al: *Anthropogenic space weather.* Space Science Review, November 2016.
22. Mather, John C. & Boslough, John: *The very first light.* London, Penguin, 1998. His colleague George Smoot wrote *Wrinkles in time* with Keay Davidson (New York, Morrow, 1993).
23. NASA: *Redesigning the Cosmic Background Explorer.* Monograph, NASA, undated.
24. *Early COBE results in accordance with big bang theory.* Spaceflight, vol. 32, no. 3, March 1990; *NASA's COBE spacecraft detects structure of early universe.* Spaceflight, vol. 34, no. 6, June 1992.
25. Webb, Stephen: *Planck's new view of the universe.* Patrick Moore & John Mason (eds): *Patrick Moore's yearbook of astronomy, 2014.* London, MacMillian, 2013; Milstein, Michael: *News from the dawn of time.* Air & Space, vol. 28, §7, Feb/Mar 2014.
26. *Satellite telescopes.* Spaceflight, vol. 31, no. 4, April 1989.

4

Faster, Better, Cheaper?

The small number and declining pace of Explorer missions increasingly concerned space scientists in the 1980s. What we might call the 'Explorer constituency' of Goddard and university-based scientists was a small part of the space science community, which was focussed on more visible and high-profile programs like large astrophysics projects and of course interplanetary missions. Ultimately, the crisis forced a rethink in NASA's approach, challenging it to use smaller budgets more smartly.

The loss of *Challenger* in 1986 prompted the biggest re-consideration in the space program since President Nixon authorized the Space Shuttle in 1972. The use of the Shuttle as a universal launcher was seen to be a mistake and the reintroduction of expendable rockets, notably the Delta II, reopened the door to a more flexible range of smaller missions. In 1987, the USAF invited proposals for a new medium launch vehicle, to which McDonnell Douglas responded by putting the Delta II back into production.

The *Challenger* accident provided some breathing space for a reconsideration of priorities, but indicators of the role of space science in general and Explorer-type programs in particular were ambiguous. The Ride report (1987), undertaken in the aftermath of the disaster, re-iterated the importance of scientific research, but its focus was on space transportation (Shuttle), the Moon and Mars.[1] The science areas attracting attention were not natural Explorer roles, but Hubble and new priorities such as 'mission to planet Earth' and life sciences. The early failure of Hubble led to further soul-searching and the *Report of the advisory committee on the future of the U.S. space program* (1990).[2] This stated, as a top recommendation, a science program 'which enjoys highest priority within the civil space program and is maintained at or above the current fraction of the NASA budget.'

An emerging subtext within this debate was the *size* of mission to be undertaken, with scientists coming to the view that the Explorer program should return to its original vocation of smaller, less costly spacecraft launched more frequently, a course preferable to the infrequent – albeit successful – missions of the 1980s. On 26th March 1987, Glenn Mason, Associate Professor of Physics at the University of Maryland, wrote to Dr. Lou Lanzerotti of Bell Telephone, commenting that 'what we had now' was no Explorers, stretch-outs and risky launch vehicles. He did not make much headway. On 1st May, Dr. Martin Weisskopf, chief of x-ray astronomy at NASA, who had been passed this letter, told him that the X-ray Timing Explorer (XTE) was under study and 'in the Explorer queue'. He attacked the 'vague feeling' among scientists that smaller was better, more productive and of higher scientific yield. Spreading resources to smaller, less risky missions was like building many merely adequate 16-inch telescopes rather than a single, more useful large one because of concerns that the large one might fail.

NASA eventually woke up. In 1988, Lennard Fisk, the new head of the Office of Space Science and Applications (OSSA), acknowledged the trend towards large and complex missions and voiced the need for a more strategic planning process. In 1986, NASA's own Space and Earth Science Advisory Committee had already noted a growing sense of disquiet and frustration at the slowing pace of space science, counterproductive levels of competition for infrequent missions and approvals of new starts whose chance of flying was questionable. In OSSA's *Strategic plan 1988*, Fisk looked back at the evolution of space science from the exhilarating pace of the 1960s to the fewer but more sophisticated missions of the 1970s and expressed the need now for 'increased opportunity with small missions'. They were vital to the program because they could be accomplished relatively inexpensively, allowed more innovative ideas, ran on a short time scale and offered quick turnaround. They trained the next generation of scientists and engineers, were of a scale suitable for the universities and could be developed over the same timescale as earning a degree. There should be a fresh mission start for a new small mission each year. The *Strategic plan 1991* repeated the commitment. Reassuringly, Fisk said, 'we are planning to augment the Explorer program to allow for more small missions which can be launched on Scout-class expendable launch vehicles. These missions are sufficiently small that they can be built and launched within three years, yet they are sufficiently capable to accomplish first-class scientific objectives in astronomy, space physics and upper atmosphere physics.'

Within Goddard from 1989, the Director of Engineering, Tom Huber, began to advocate a new line of smaller satellites, a return to Goddard's roots in innovative, small, cheaper (less than $30m) and quickly-produced spacecraft, while explaining how this could be done by incorporating advances in miniaturization, fibre optics and more powerful solid state computers that had more memory and used less power.

The SSB, never far away, then weighed in. Its committees on Space Astronomy & Astrophysics and on Solar & Space Physics recommended one new Explorer each year, with new management to ensure timely implementation. Their comments were thoughtful and incisive, for they appreciated the way the planning process had now evolved. Their analysis was that Explorer funding had lost both priority and capital, with the prospect that what was left would have to go to complete EUVE and XTE by 1993. Nobody wanted to drop these missions because they were considered of high scientific value, but their combined cost was $235m. Explorer was budgeted at $65m to $70m a year and these missions were using up more than two years' budget alone. There were also 43 proposals for new missions, but the evaluation process had been halted while the program was reviewed. They could begin new missions but that would further delay EUVE and XTE, which were already ten years behind.

In 1988, support also came from the Congress, which proposed 'a set of smaller, lower-cost missions designed to answer a variety of questions' with a constant level of activity in the 1990s.[3] Although late in the day, the Office of Technology Assessment then carried out a review of processes, priorities and goals in space science and applications.[4] This was very critical of the lack of flight opportunities for space science and pleaded for more smaller missions and a better balance between small, medium and large missions. It questioned why small missions were given as much scrutiny as large ones and subjected to Congressional micromanagement. Earlier, the Congressional Budget Office had made a special study in 1988 of the NASA program for the 1990s.[5] That study admitted that 'the resources required to build the space transportation system necessarily limited the resources available for space science activities and hence the flow of new experimental results.' It also expressed interest in the great observatory program (Hubble, Gamma Ray Observatory), but nevertheless recommended, under 'Explorer program', that a set of smaller, low-cost missions at a constant level of activity be included for the core program of the 1990s.

This new consensus paved the way for the arrival in the Explorer program of a new approach. The 'faster, better, cheaper' philosophy (there are variations on this combination of words) is actually primarily associated with the interplanetary program – strictly speaking, it applied only to the *Discovery* program (Near Earth Asteroid Rendezvous, Mars Pathfinder, Lunar Prospector) – but the approach took wider hold.

'Faster, better, cheaper' was the outcome of a number of factors. First, there was a diminishing profile for space science in the early 1990s, which challenged the scientific community to find solutions. Second, there was a desire for more frequent missions to plug gaps, educate the next generation and respond to targets of opportunity (Halley had shown the program's inability to do so). Third, there was criticism of NASA for over-investment in big projects, sometimes uncharitably called 'Battlestar Galacticas', where single point failures could lead to disaster. Such was the case with Mars Observer, which exploded on entry to Mars orbit and nearly the case with Hubble, where a mirror error would have ruined the whole project had astronauts not been able to fly to the telescope to fix it.[6]

At one level, there was nothing new about this. As far back as 1971, NASA Associate Administrator Homer Newell had written to his superior, Administrator James Fletcher (3rd December 1971) to complain that the science community was being shut out, that there was a huge emphasis on large-scale programs (Viking, Grand Tour) and that there was a need for smaller-scale projects. The science program had to strike a balance between disciplines and between large and small projects. There was a need for smaller projects, spacecraft and shorter lead-in times of the type that sounding rockets and the Explorers could offer, generating a steady flow of results.[7] In other words, small *vs* large was and would continue to be an enduring issue in designing space exploration in general and spacecraft in particular, but the issue was now in exceptionally sharp focus. Newell pointed out that small spacecraft were intrinsically cheaper. Citing the Comptroller's Office, Explorer was the cheapest of all American spacecraft to make.

Initially, there was scepticism about the possibility of running missions at much lower costs, with $250m to $300m cited as ballpark figures. Whatever the figures, NASA documentation began to speak more of higher returns for more modest costs and the standardization of design. Already, some more detailed thought was being applied, with future missions divided into 'expensive', 'moderate' and 'less expensive' and having to make more room for astronomy and astrophysics.

The use of the term 'faster, better, cheaper' is personally associated with Dan Goldin, who was appointed NASA Administrator by President George Bush on 1st April 1992, but was primarily associated with the Clinton presidency (lasting its full duration to 2001). Goldin publicly embraced 'faster, better, cheaper' at the World Space Congress in 1992. He established a Program Management Council to implement the 'faster, better, cheaper' approach (sometimes termed by its initials, FBC) and announced that the Cassini probe to Saturn, due 1997, would be the last big scientific mission for some time. NASA now had to stop 'putting all its eggs in one basket' but he also acknowledged that lower costs might mean occasional failures. Goldin said bluntly that NASA must focus on smaller missions and devolve responsibility, while also arguing that 'it's permissible to fail.' He had read the signs that tight funding for space exploration meant that existing resources had to be used more intelligently. Goldin also faced a problem similar to NASA administrators in the 1970s, namely that one particular project was consuming a significant percentage of resources. This time, it was not the Space Shuttle but the still-unbuilt space station *Freedom* and its subsequent replacement, the International Space Station. His reforms were timely, taking place during a period of aggressive cost-cutting by the Congress, especially in the late 1990s.

Dan Goldin

Goldin's philosophy was only half as controversial as its execution. Many found his personality abrasive and during his time (1992–2001), NASA's workforce fell from 25,300 to 19,000, although the number of scientific spacecraft increased fourfold on a third less budget. When he moved on, he claimed he had left the agency better than he found it and argued that you could only be loved if you never did the right thing. However, he was not well served by the Congress, which responded to his savings by insisting that he make even more. By the late 1990s, NASA was being praised for having saved substantial amounts for the government – the figure cited being $35bn over seven years, reducing its budget by a third in real terms. The Office of Federal Procurement Policy praised NASA for being 'the most aggressive' of all federal agencies, at what *Goddard News* gave the impression was a better procurement rally held on 13th November 1996.[8] The X-ray Timing Explorer (XTE) was especially praised for innovative procurement policies that had achieved eight months of schedule savings. In 1997, Goddard stated that Explorer missions run along the new lines had saved $80m, citing Rossi and the Advanced Composition Explorer which had both come in $30m under budget.[9]

Despite the praise, the Congress shocked both NASA and the scientific community in 1999 when the House of Representatives cut the agency's budget by 11%, or $1.4bn on $13.5bn for its 2000 request. Space science was the most affected, losing 29% or $640m. Within that, the Explorer program, one of the smallest, was cut from $151m to $47m. A betrayed Goldin was furious and vowed to fight the cuts, which threatened the closure of three of its ten field centres, such as JPL. He called the 11% cut a 'dagger in the heart of employee morale.' NASA had done more with smaller budgets, was a model of good government and had reduced its workforce by a third, but it had still increased productivity. It was doing the impossible and yet this was its reward. The Planetary Society warned that the Explorer program could not survive such Congressional budgetary butchery.

In 1995, the House had ordered the closure of Goddard, Marshall and Langley, but backed down when the Senate would not agree.[10] With the new cut, their future was again on the line. NASA Goddard was still a substantial facility, with 27 buildings, 154 laboratories and 200ha of land, costing $2.7bn out of NASA's $14.5bn budget. Staff levels peaked at 3,987 in 1993 but along with other NASA facilities, the centre was now required to do more with less. At Goddard, this was called 'refocussing' and involved program retraining and lateral transfers. Staffing fell to 3,900 the following year and 3,575 in 1995. The FY 1998 budget projected Goddard staff falling from 3,573 to 3,275, but somehow, the home of the Explorer program managed to hang on. After the 11th September 2001 terrorist attacks, space budgets became even further constrained and NASA's advisory council again made proposals to slash the field centres like Goddard, devolving tasks out to the universities and the private sector. Goldin called this 'throwing the children to the wolves.'

Goldin left only a few weeks later. During his time, NASA launched 171 missions, which was a considerable improvement in rate. Criticism of the 'faster, better, cheaper' philosophy came to a head when the two 1998 Mars missions failed, two of eleven failures during this time. A subsequent review panel considered that they had been 30% underfunded. Esteemed space commentator Jim Oberg criticized the policy as leading to the use of 'overworked, underpaid and inexperienced workers' in downsized teams, with NASA the poster-child of Al Gore's campaign to reinvent government. Older, experienced, but more expensive workers were driven out so that younger, enthusiastic ones – with good

Goddard Space Flight Center

qualifications, at least on paper – could make mistakes. The mantra underestimated the value of quality engineering, whose absence was no bargain at any price. Oberg called it inappropriate sloganeering and the approach remained controversial.[11]

The 'Faster, Better, Cheaper' New Explorers

The first record of the 'faster, better, cheaper' mantra at Goddard comes from a strategy day held there in autumn 1994. Some saw it as a threat, but Vanguard inventor Bob Baumann who was present made the comment that they had been doing 'smaller, faster, better, cheaper' since they were first set up.[12]

In reality, the re-design of the Explorer program had been under way for almost ten years. An influential person in the new philosophy was Charles Pellerin, head of the NASA Astrophysics Division which included Explorer.[13] He recognized the need for major changes in the program, especially to shorten the time from mission selection to flight. Pellerin effectively made the wake-up call internally on the slowing of Explorer missions, which were averaging eight years from approval to launch. Writing to his colleague Laurence Peterson, Assistant Director for Science, on 23rd October 1986, he pointed out tellingly that the last Explorer selections had been as far back as 1974-5 (see chapter 3). In Pellerin's view, costs were being driven upwards because of schedule delays caused by inadequate resources, which then attracted attention and recommendations to cancel to save money. The current situation was discouraging for the scientific community, but it deserved an honest and truthful assessment, without raising false expectations. Even new mission studies would drain limited resources and his main concern was that with the number of missions already in development, any new announcement would possibly close the Explorer opportunity to new missions to the end of the 1990s. Pellerin decided to hold a competition for what were called 'new small Explorer' missions and to concentrate on launching the existing backlog of approved missions. 'Large' Explorers would come to an end with the 1995 X-ray Timing Explorer (XTE).

Goddard convened a 'small payloads workshop' in February 1987, which led to the formulation of what was called the Small Explorer Program (SMEX) that June. The first AO was issued in May 1988 and 51 proposals had been received by the end of September. It was anticipated that they would be called 'Small Explorer' 1-6, with launch dates set over June 1991 to June 1993, covering the thematic areas of astrophysics, space and upper atmosphere science. Ron Adkins was appointed Small Explorer Program manager. The intention was to build and fly each mission within three years. They would involve young professionals in the scientific community, use a standard Goddard-built bus, be no larger than 227kg (the limit of the Scout) and cost less than $30m. The SMEX program involved technology transfer to industry and the SMEX bus design was handed over to an industry team called Swales in 1999, being first used in the Triana satellite.

The first successful missions and their funding were announced in October. The four selected were SAMPEX (1992), Submillimetre Wave Astronomy Satellite (SWAS, set for 1994), Fast Auroral Snapshot Explorer (FAST, 1993); and Total Ozone Mapping Spectrometer (TOMS).[14] The last took the form of both an instrument – flown on the Nimbus 7 and Russian Meteor 3 weather satellite – and a spacecraft, the TOMS Earth Probe, but was subsequently moved out of the Explorer program. Later, these small Explorers were joined by the Transition Region And Coronal Explorer (TRACE, 1998) and then the Advanced Composition Explorer (ACE, 1997) and the Imager for Magnetopause-to-Aurora Global Exploration (IMAGE, 2000).

In September 1994, the NASA Office of Space Science formally announced that the new 'faster, better, and cheaper' Explorer program would comprise three classes: Medium Explorers (MIDEX), with cost capped at about $70m; Small Explorers (SMEX), capped at about $35m; and University Explorers (UNEX), capped at about $25m (all excluding the launch costs). Just how restrictive these cost caps were became apparent when the cost for the upcoming Far Ultraviolet Spectroscopic Explorer (FUSE) looked like hitting $300m. It was ordered to be 're-baselined' to $125m, or cancelled.

The next stage, the MIDEX selection, was announced by the NASA Office of Space Science in April 1996 and then handed over to Goddard to manage. MIDEX was designed to be larger than SMEX but not sufficiently so to require the Delta rocket, instead expecting to fly on a new 'medlite' launcher than had not yet been developed. In the event, FUSE and IMAGE did both fly on the Delta. The idea of MIDEX was to focus on astrophysics and space physics, with a higher cost cap than the small Explorers ($70m) and the aim of one mission a year. The first two selected were the Microwave Anistropy Probe (MAP) and IMAGE. The next group of MIDEX selected for feasibility studies were Swift Gamma Ray Burst Explorer (SGRBE); Next Generation Sky Survey (NGSS); Full Sky Astrometric Mapping Explorer (FAME), Auroral Multi-scale Mission (AMM); and Advanced Solar Coronal Explorer (ASCE). The missions were given a budget envelope of $130m to $140m.[15]

That was only the beginning of the changes. One Principal Investigator (PI) was now made responsible for managing each project, from the bid to the end of the mission. This meant a significant shift in the role of PI, from literally being a principal investigator to being responsible for proposal, concept, cost and schedule – very much a management role. The PI took responsibility for proposing a mission and its experiments, working with a team of co-investigators – often drawn from a number of universities and who might not have worked together before – and aiming to achieve a cohesive, efficient team effort. NASA recognized that it should step back from its tendency to micro-manage and that its supervisory regime should be lighter, but with distinct, clear mileposts. PIs could come from government, laboratories in industry centres, or universities. In practice, the highest take-up was from the universities and 'faster, better, cheaper' decisively moved Explorers out of house. Educational and outreach programs would be encouraged, but the design had to be precisely costed, with fixed contingency (30%). If it exceeded the cap, then the mission would be de-scoped or even scrapped.

The 'faster, better, cheaper' approach duly took off – being most active from 1995 to 2003 – with six MIDEX and five SMEX selected (although two were subsequently cancelled). The cost was $1.7bn, or $200m a year. There was then a gap, with no missions selected from 2003 to 2009 and the effective budget falling to $100m a year, but the program recovered in 2010 with projections in the order of $150m annually for five years.

The UNEX class was short-lived. It began as the Student Explorer Demonstration Initiative (STEDI), managed by the Universities Space Research Association (USRA) and was intended to illustrate the possibilities of low-cost missions with research outcomes and high rates of graduate and undergraduate student involvement. These missions were designed to have low, fixed budgets ($10–15m) and rapid turnaround in the order of two years. STEDI was advertised on 12th May 1994, drawing in 66 proposals, of which six went for further study and two were approved: TERRIERS (which failed) and SNOE (which succeeded). The University of New Hampshire Cooperative Astrophysics and Technology Satellite (CATSAT) was an alternate and was funded from 1996, but was terminated when a suitable launch vehicle could not be found.

New, Small Launcher: Pegasus

The final part of the jigsaw was a new, small launcher to replace the aging Scout. It was the first entirely new American launch vehicle since the introduction of the Space Shuttle. In a radical, lateral-thinking, cost-saving measure, its launch base was not an expensive fixed facility, but an old bomber.

Pegasus had its roots in military programs, a response to a request for proposals by the Defense Advanced Projects Agency for a launcher for small advanced military satellites called 'lightsats'.[16] Orbital Sciences Corporation (OSC) and Hercules Aerospace responded with the only proposal, promising launch costs of less than $10m by using two old B-52s at Edwards, one of which had flown the X-15 rocket plane from there in the 1960s. The essence of the concept was that a conventional aircraft should do some of the heavy lifting, bringing a spacecraft up through the heaviest layers of the atmosphere to an altitude of 12,000m where it would launch. In charge of Pegasus test flights was Shuttle astronaut Gordon Fullerton. The development team comprised 80 engineers, with an emphasis on simplicity of design but using recent advances in the development of solid propellant engines and computer guidance. OSC, based in Fairfax, Virginia was formed in 1982, building upper stages for the Titan IV rocket and Mars Observer; while Hercules, of Wilmington, Delaware built solid rocket boosters for the Delta II in Salt Lake City, Utah. Development began in 1987 and cost a modest $50m. Pegasus first flew on 5th April 1990 when it was dropped from a B-52 flying out of Edwards Air Force Base to put a military satellite into polar orbit. NASA's use of Pegasus came later, after Goddard announced its selection as the launcher of choice for the new Small Explorer program in 1991. It was first used for FAST in August 1996.

B-52 and Pegasus

The Pegasus was a small, winged rocket able to lift 275kg into polar orbit of 500km. Its payloads were limited to 1.8m length and 1.2m diameter. The cost per launch was about $25m, making it ideal for 'faster, better, cheaper'. The original Pegasus was 16.9m long, 1.3m in diameter, had a triangle of wings with a span of 6.7m and weighed 19 tonnes. Like the Scout, it used solid propellant, in this case three stages of 726kN, 196kN and 36kN, burning for 70 seconds, 73 seconds and 68 seconds respectively. It resembled a cruise missile. Although the idea of air-launched rockets had been around for some time and was the subject of many studies in the USSR and Russia, the Pegasus was the first and it remains the only air-based launcher. OSC had its own mission control in Dulles, Virginia. Pegasus used several launch locations, for example Bucholz Army Auxiliary Airfield at Kwajalein Atoll Missile Range (1958), one of the world's largest coral atolls in the Ralik Chain of the Marshall Islands, half way between Hawaii and Australia. One launch even took place from the Canary Islands in Spain. Theoretically, they could take place from any airport on the planet.

After having made six launches, the B-52 was withdrawn and in 1993, OSC bought and converted a Lockheed L-1011 Tristar from Air Canada. The Tristar was called the *Stargazer* after Jean Luc Picard's flagship in *Star Trek Next Generation*. This had a slightly different launch method – under the fuselage rather than the wing – and enabled launches of the larger Pegasus XL. Pegasus XL was 17m long, 1.27m diameter and weighed 23 tonnes, with a payload capacity of 341kg to an orbit of 740km, still at the very small end of the launcher market.

L-1011 Tristar with Pegasus underneath

Pegasus was a different televisual experience from a conventional launching. It began with the L-1011 Tristar airliner taking off from an ordinary runway with the missile-like Pegasus underneath. If launched from Cape Canaveral, the L-1011 would fly off the coast to an altitude of 11,900m and release the Pegasus, in the same way that the B-52 used to drop the X-15 experimental rocket. The Pegasus would drop for five seconds before the 73,937kg thrust rocket ignited, beginning an initial ascent at 45°. Dropping the Pegasus would cause a severe jolt to the plane, which would shoot up when relieved of its load with a loud bang as its airframe flexed with the shock. As the timer counted to five seconds, the pilot had to make a sharp turn to get out of the way in case the Pegasus collided with it or exploded. Thankfully, this has never happened, but the crew's survival would by no means be certain if it did. The crew of the plane had no trouble hearing the roar of the now rapidly-accelerating Pegasus, as it hit Mach 2 straight ahead in a 35° climb on a 30m pillar of orange flame. The crew was normally able to see the ignition of the second stage at 88,087m.

Launching the Pegasus was not as easy as it seemed and the launch of a Brazilian satellite in summer 1993 was so chaotic as to require formal investigation.[17] This was worrying, as the Pegasus XL had been assigned a number of Explorer missions. The first launch of the XL from an L-1011 on 27th June 1994, carrying an experimental air force satellite, went off course and had to be destroyed. The second Pegasus XL was lost a year later on 22nd June 1995 when the interstage and second stage failed to separate properly, causing the rocket to loop wildly and losing another air force test satellite 90km off Monterey, California. This was later blamed on incorrect installation. However, FUSE was already booked on an upcoming Pegasus XL, with the Submillimetre Wave Astronomy Satellite (SWAS) to follow and both were now delayed. People began to question the start-up company Orbital, but it was defended as it was trying to lead out one of the few new launchers, always a learning curve.[18] Goddard even issue a formal Request for Information to industry to see could it any offer an alternative to the troubled Pegasus. It was recommended to fly FUSE on a Russian Cosmos rocket, but NASA Headquarters said no. The desperate idea of flying on the Chinese Long March was even explored, but in the event, NASA chose to stick with Pegasus. The next three launches were successful, but the following one failed in November 1996 when the payload failed to separate. Of the first 30 Pegasus missions, five had failed, of which three were complete failures, while the other two achieved orbit but not the right one. By 2003, though, the XL had achieved 21 straight successes.

Pegasus

SAMPEX (Explorer 68): Anomalous Cosmic Rays

SAMPEX (Solar Anomalous and Magnetospheric Particle Explorer) was the first SMEX mission. On 3rd July 1992, it flew out on the last Scout in the Explorer program, the launcher's 188th mission, a rocket so reliable that it had never failed since 1976. The target of SAMPEX was anomalous cosmic rays but its main achievement was to confirm the existence of a radiation belt of trapped heavy nuclei from the interstellar medium within the Van Allen belt.

Following approval in 1988, SAMPEX went through design, development and preliminary design review in 1989, critical design review in 1990, integration and testing in 1991 and then mission readiness review and launch in 1992. The 82-page proposal, by PI Glenn Mason of the University of Maryland, listed ten co-investigators. Daniel Baker was project scientist and the young team included a mission manager, instrument manager, parts engineer and fifty undergraduates. It was three years from approval to orbit and the mission cost was $77m. This was the way things were going to be now.

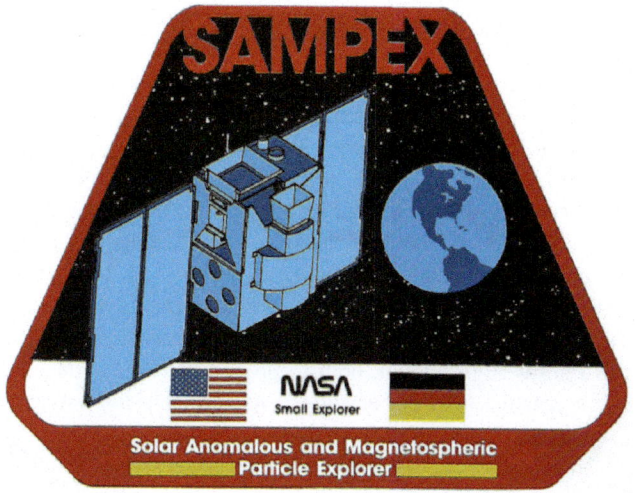

SAMPEX badge

SAMPEX was an international project with Germany, designed to study energetic particles reaching Earth from the solar atmosphere and interstellar space. There were four fields of work: low energy galactic cosmic rays; solar energetic particles; anomalous cosmic rays, to obtain their isotopic composition; and magnetospheric electrons, to determine the connection between events and the destruction of ozone. SAMPEX was launched in the decline to solar minimum. Its four instruments were:

- Low Energy Ion Composition Analyzer (LEICA), to measure the mass and kinetic energy of particles (University of Maryland);
- Heavy Ion Large Telescope (HILT), with its own supply of isobutane, to measure galactic cosmic rays and solar energetic particles near the magnetic poles, as well

as the energy and elemental composition of anomalous cosmic rays. It was provided by the Max Planck Institute for Extraterrestrial Physics in Garching, Germany;
- Mass Spectrometer Telescope (MAST), to determine the direction, energy and isotope of atoms travelling between 12% and 75% of the speed of light (CalTech, Pasadena);
- Proton Electron Telescope (PET), to measure the energy of electrons, protons and helium nuclei, especially electrons moving close to the speed of light and a threat to the ozone layer (CalTech).

Two of the instruments (HILT and LEICA), had already flown before as 'getaway specials' on the Shuttle, while MAST and PET had been cannibalized from the cancelled American part of the International Solar Polar Mission (Europe's part flew as *Ulysses*). External interest focussed mainly on Anomalous Cosmic Rays (ACRs), which had intermediate energies and were suppressed during solar maximum. Their existence had been suggested in the early 1970s by Fisk, Koslovsky and Ramaty in the *Journal of Astrophysics* and were believed to be comprised of helium, carbon, nitrogen, oxygen, neon and argon.[19] Anomalous cosmic rays were defined by NASA as atoms of local and interstellar gas that entered the solar system, were ionized and then accelerated to cosmic ray intensities at the shock wave of the solar wind and were likely to penetrate Earth's magnetic field at lower latitudes. Due to their relatively low velocity, anomalous cosmic rays did not reach Earth during periods of solar maximum.

The box-shaped SAMPEX was 1.4m long, 85cm diameter, with two solar panels and could only just be fitted within the Scout's nose fairing. Its orbit was 515–691km, 82°. Most of the data collection would be done over Earth's magnetic poles and would be downlinked twice a day to Wallops for onward transmission to Goddard and then the University of Maryland Mission Operations Center. There was student involvement through the Cooperative Satellite Learning Project involving a number of Maryland schools, such as Laurel High School, both to process SAMPEX data and to encourage careers in space science. SAMPEX developed its own mission badge and sticker with 'Anomalous Component Team' on it.

SAMPEX lived up to its high hopes and the promise of the new small Explorers. It found the precise location of trapped ACRs in the magnetosphere and determined their composition (C, N, O and Ne), spotted the early return of ACRs for the 1992 solar minimum, found a new radiation belt with helium, oxygen and nitrogen and discovered high energy deuterium trapped in the magnetosphere. Fluxes in electrons in geostationary orbit on 20th January 1994 were measured by SAMPEX and found to be strong enough to cause the loss of control on satellites (Intelsat and Anik were affected), a process called dielectric charging. Earth's magnetosphere was considered to be a cosmic electron accelerator of substantial strength and efficiency.

The formal discovery of the new radiation belt was announced at the American Geophysical Union meeting in Baltimore, Maryland in 1993, crediting data from SAMPEX and providing a headline in *Goddard News* of 'SAMPEX scientists locate radiation belt around Earth'. SAMPEX found the belt when it was orbiting 600km high, revealing that it was composed of anomalous particles and was at its most intense over the South Atlantic Anomaly. Anomalous cosmic rays became trapped in this field for weeks or months, where they bounced back between the north and south pole but got trapped between south America and Africa. Their intensity built up, more so during solar minima.[20]

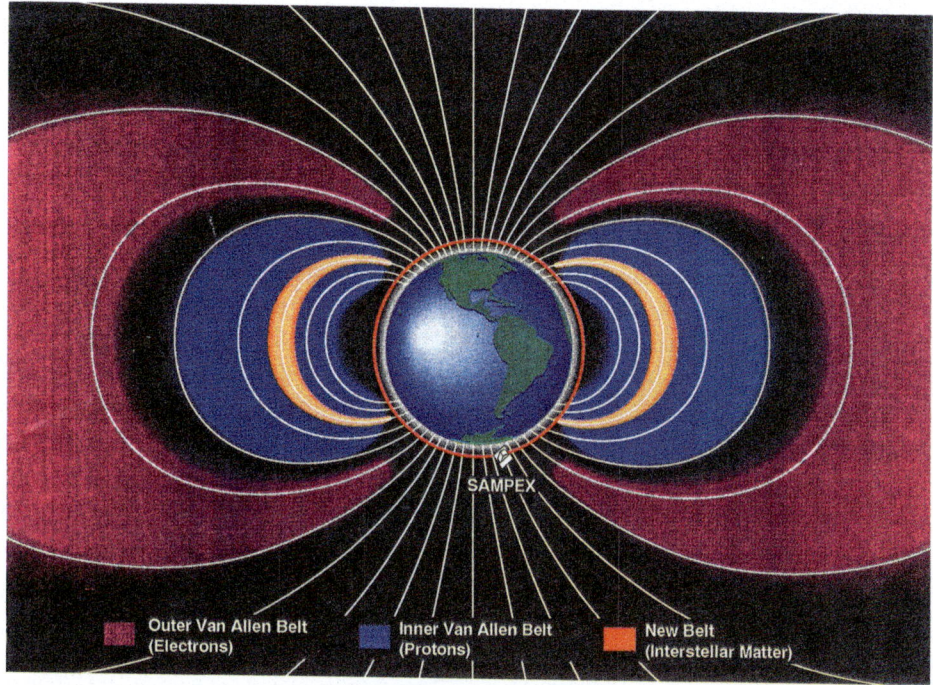

SAMPEX new radiation belt

This sparked off another '*Who found it?*' contest that echoed the discovery of the cosmic background radiation. As far back as 1977, leading Russian physicist Naum Grigorov had speculated in *Soviet Physics* that there was a third, small radiation belt around the Earth, his views being published as *Possibility of the existence of a radiation belt around the Earth consisting of electrons with energies of 100MeV and above*. He and his colleagues worked hard to locate the belt, using instruments on the Salyut 6 orbital station and Cosmos satellites. Grigorov announced his discovery of the belt in *High energy electrons in Earth orbit* in 1985. Early on, the SAMPEX team credited the Cosmos team with the first evidence for the belt, but believed that it was the first to pin down the location. This time there was an amicable sharing between the two sides, with none of the acrimony as there had been with COBE and no Nobel prize to complicate things. In 1998, SAMPEX was turned over to the University of Maryland as a training exercise for students. The mission concluded after twelve years on 30th June 2004 and burned up on 13th November 2012.

XTE/RXTE/Rossi (Explorer 69): Magnetars and Frame-dragging

The X-ray Timing Explorer (XTE) was the last of the pre- 'faster, better, cheaper' missions, but it did not get airborne until the age of the new Explorers. It was launched on 30th December 1995 on a Delta II, entering a 23° orbit at 580km.

XTE was an ambitious and, by the standards of 'faster, better, cheaper', large (3,035kg) and expensive mission ($195m, though in reality it came in at $40m less). It planned to use a large x-ray telescope to study white dwarf and neutron stars, binary star systems and x-ray novae; search for black holes and other mysterious distant objects for two years; and scan 80% of the sky every 100 minutes. It was hoped that it would find collapsed stars and massive black holes in quasars and galaxies. XTE carried the largest x-ray detector ever flown, the Proportional Counter Array. It was aimed at both our own galaxy and further afield, concentrating on compact x-ray objects, making timing and spectral studies. 'Timing' refers to measuring the duration of the emitting sources, which can be from sub-milliseconds to years, while the spectral instruments aimed to identify their location. XTE also benefitted from a newer, smaller, more precise and reliable guidance system called an Interferometric Fibre Optic Gyroscope. It had a 1GB solid state memory, an unprecedented capability at the time. The PI was Richard Rothschild of the University of California San Diego (UCSD), with Jean Swank of Goddard and Hale Bredt of MIT. XTE had three instruments:

- All Sky Monitor (MIT), to view 70% of the x-ray sky once per orbit, search for old or new sources and suggest targets for further investigation;
- Proportional Counter Array (Goddard), five nearly-identical proportional counter detectors to provide large area, low background and high time resolution of x-ray sources in 2–60 keV;
- Higher Energy X-ray Timing Experiment (University of California San Diego), for sensitive spectroscopy from 20–200 keV.

Planning for XTE began in 1991 and integration was completed in October 1994. It was a big spacecraft, 2.36m by 2.41m by 6.08m when stowed and arrived at Cape Canaveral on a huge C-5 Galaxy transport plane in June 1995 for two months of check-out and tests. It was brought to the pad two weeks before launch for final checks and simulations. This was one of the first Goddard satellites to transmit direct by TDRSS, which then transmitted first to TDRSS ground control and then on to the Science Operations Center in Goddard (SAMPEX, by contrast, transmitted straight to the ground during overhead passes). After reaching orbit, XTE was renamed Rossi, after Professor Bruno Rossi of MIT, an advocate for and pioneer of x-ray astronomy. The satellite was re-titled RXTE. Bruno Rossi was involved in sounding rockets which discovered the first x-ray source outside the solar system in 1962 (Scorpio x-1).

The focus of RXTE/Rossi was black holes at the heart of Active Galactic Nuclei (AGN) which emitted x-rays because of their highly compressed nature. Rossi made the longest-running study ever of AGNs. Its instruments picked up x-ray emissions lasting from microseconds to several years across the energy span of 2000 eV to 250,000 eV. Rossi found the smallest-ever black hole, only 25km in diameter, about the smallest postulated to be possible. The outcomes were announced at a press conference on 1st April 2008. The two scientists involved, Nikolai Shaposhnikov and Lev Titarchuk, focussed on binary XTE J1650–500 in the southern constellation of Ara, discovered by Rossi in 2001. They found that the mass of black holes was related to their x-ray intensity, known as Quasi-Periodic Oscillation (QPO). They knew the QPOs and masses of fifteen other black holes and thus they were able to calculate this one. They advised travelling spacefarers not to go too close, as small black holes could be more dangerous than larger ones, able to 'swallow you up and turn you into a strand of spaghetti,' as Shaposhnikov warned.

XTE/RXTE/Rossi (Explorer 69): Magnetars and Frame-dragging

XTE/Rossi

One of the first actions for Rossi in February 1996 was to study binary star 4U 1728-34, already known as a frequent source of powerful x-rays. It found three binaries with the fastest oscillations ever measured, one as much as 1,130 times a second. They could be waves from a dying neutron star or one spinning at an extraordinary rate. On 6th November 1997, scientists announced that they had observed a black hole 'frame-dragging', a concept postulated by Einstein in 1918 whereby a black hole dragged time and space around it as it rotated. This proved to be one of Rossi's most significant outcomes.

In 1997, Rossi was directed to Cassiopeia A, a supernova remnant 11,000 light years away, the star having exploded in the year 1680 but whose shell of hydrogen and helium was still blowing away. Scientists making an x-ray analysis believed that what they were looking at was 'synchrotron radiation' like that produced by accelerators on Earth. Cassiopeia A was accelerating cosmic ray electrons, protons and other nuclei, providing a possible explanation for the source of cosmic rays.

On 26th January 1998, Rossi scientists announced the discovery of a fast-spinning pulsar, rotating 60 times a second, formed 4,000 years ago and associated with supernova remnant N157B in the Larger Magellanic Cloud (LMC). On 21st May 1998, Rossi confirmed the existence of magnetars, a special class of ultra-magnetic neutron stars with a magnetic field a trillion times the strength of Earth, based on observations of SGR 1806-20 which had first been found in 1979. Rossi found a new magnetar, XTE J1810-197. At the time, only ten were known. The observer was Alaa Ibrahim, who believed that the magnetar was turned on in spring 2003, although a precursor could be found in old star maps to 1990. On 23rd July 1998, Rossi discovered the first known accretion-powered millisecond pulsar, SAX J1808.4-3658, emitting x-rays and spinning rapidly, 400 times a second. This was 'the holy grail of x-ray astronomy', a companion star from which gases are drawn off and accreted, causing pulsars to heat up to the point that the companion disappears. Pulsars ultimately vaporize their companions, which might explain why they were generally found on their own, evidence of their previous companion having disappeared, the otherwise perfect stellar crime.

XTE/Rossi impression of a black hole

Rossi found re-appearing bursting pulsars and QPOs in neutron stars, picked up a blast from a magnetar 20,000 light years away and measured the period of low-mass x-ray binary Cygnus x-2 (it was 77.7 days). Rossi was important for its work in gamma ray bursts (GRBs). These were extremely powerful, as bright as a billion trillion suns, first spotted by U.S. Air Force Vela satellites in 1967. These were not reported until 1973 because Vela's purpose was to detect nuclear testing in the USSR and the United States did not want the Soviet Union to know how it knew these things. Meanwhile, Soviet scientists measured GRBs from the 1970s using their own Konus detector. The satellite which did most to study GRBs was the Compton Gamma Ray Observatory, launched in 1991, which detected 2,700 and found that they came from all over the universe and had enormous energies. Rossi was able to extend the study of GRBs. On 27th August 1998, it detected a huge release of magnetic energy from soft gamma repeater SGR 1900 + 14 in Aquila, the star's entire crust being blown away. Rossi characterized a new kind of x-ray burst, a superburst, a thousand times more powerful than any known before.

Rossi was especially valuable in observing gas clouds.[21] The team detected 12 cloud events when gas clouds moved across the line of sight of x-ray objects, an average cloud being 6.5 billion km across, the width of the solar system out to Pluto. Only four had been discovered before. Cloud transits varied in duration from as short as five hours to as long as 16 years. There seemed to be several clouds moving along the central plane of our galaxy. There was one extraordinary event in the barred spiral galaxy NGC 3783 in Centaurus, 143 million light years away, when the AGN dimmed twice in 11 days, suggesting that the black hole was tearing the cloud apart.

Rossi operated until 2011, when it failed a review board and was decommissioned the following January. It led to 2,200 papers and 92 PhDs. At a late stage, students from Shenandoah Valley school were invited in for a hands-on experience of running the mission. Rossi's value lasted well beyond its closing date, for searches through its archives three years later found a rare middle-size black hole, M82 X-1, the brightest x-ray source in galaxy Messier 82. Examining its Quasi Periodic Oscillations, astronomers found it pulsing rhythmically 5.1 and 3.3 times a second, a 3:2 relationship found in many other QPOs. Further research into middle-size black holes was expected later in the NICER mission (see chapter 5). Rossi is due to fall out of the sky some time before 2023.

FAST (Explorer 70)

FAST stands for Fast Auroral Snapshot Explorer and was designed to investigate plasma in the auroral oval. It was the second SMEX after SAMPEX and was launched on 21st August 1996 on a Pegasus from the L-1011 out of Vandenberg. It had been delayed by two years because of the failure of the first Pegasus XL. FAST was launched into a high inclination, 83° orbit between 351km – 4,165km. The concept of the mission was to put a satellite over the poles under precise lighting conditions in order to understand how electrons were accelerated, collided with gas molecules in the upper atmosphere and caused aurorae.

The purpose of FAST was to follow the acceleration processes in the northern and southern lights by taking snapshots of electric and magnetic fields, as well as energetic ion and electron distributions at 1,931–2,574km above 60°, journeying into the heart of the aurorae where charged particles were energized. It was designed to measure acceleration physics and plasma processes there, especially the microphysics of space plasma and the accelerated particles that caused them. FAST aimed to examine electron and ion acceleration by parallel E-fields; the wave heating of ions; electrostatic double layers; field-aligned currents; kilometric radiation; and general wave and particle interactions. Its scientific instruments were energetic electron and ion electrostatic analysers.

FAST weighed 180kg and was costed at $60m, including the Pegasus launcher ($15m), bus ($27m) and instruments ($18m). It spun at 12rpm. The PI was Charles Carlson of the University of California, Berkeley. Rob Pfaff was the project scientist and contributors came from the University of California (Berkeley and Los Angeles); University of New Hampshire; Lockheed's Palo Alto Research Laboratory; and the Max Planck Institute in Germany. The mission included the first transportable tracking station at Poker Flat, Alaksa, as well as tracking by McMurdo base in Antarctica and the European Space Agency's station in Kiruna, Sweden. FAST had a 1GB solid state memory and provided 1TB of data in its first two years.

FAST

ACE (Explorer 71): Solar, Magnetic Minimum

The Advanced Composition Explorer (ACE) was launched through intermittently cloudy skies from Cape Canaveral on 25th August 1997, a day late because a fishing boat had strayed into the launch zone the previous day. It was parked 1.4 million km from Earth at L1 sunward, 1/100th the distance of the Earth to the Sun, so that it could observe the Sun and determine and compare elemental and isotopic composition of emissions from the solar corona, interplanetary, interstellar and intergalactic points. Some of the particles it would measure would be very fast, up to 5.6 million km/hr. and ultimately, ACE would help us understand the origin of matter. Compared to the IMPs, the collecting power of its instruments had improved up to a thousand times.

ACE was invented at a meeting of scientists at Johns Hopkins University on 19th June 1983. At that time, it was called Cosmic Composition Explorer.[22] A proposal was sent unsolicited to NASA and resubmitted in a concept study program in March 1986. It went to outline design in 1988–9 and was then selected for detailed design in 1989 with a view to a 1993 launch. This stage lasted much longer, until 1991–4 and ACE did not begin construction until 1994 for a 1997 launch. ACE came in 24% under budget, saving $34.3m (on $141.1m) and beat its final launch date deadline by four months.[23] In line with the 'faster, better, cheaper' philosophy, it was the responsibility of the PI and CalTech to deliver the instrumentation under budget and on time with minimal supervision.

The spacecraft design was based on that of AMPTE. It was a set of octagonal boxes, with tanks and engine on the bottom and four windmill solar panels and the instrument package on the top. With its solar wings, it had a span of 8.3m and was 2m in diameter, 1.9m in length and spun at 5rpm. It weighed 752kg (including 156kg of instrumentation) but it also carried almost 200kg of propellant for station keeping at L1. The instruments were positioned to keep their fields of view clear from each other and to be well distributed for weight. Available power was 443w and the downlink capability was 7kbps. One end of ACE was always pointed to the Sun, one always toward the Earth, with data transmitted by a parabolic dish on the bottom during telecommunication sessions of three to four hours per day to the Deep Space Network operated by JPL, for forwarding to the ACE Science Center at CalTech. ACE had two solid state recorders, each of 1GB, which were an in-house development at Goddard and now replaced the failure-prone traditional recorders. To reduce costs, the science centre, the mission operations centre and the integration and testing centre used a common computer system, a common sense approach but one that was new in spacecraft development.

The four principal problems tackled by ACE were the elemental and isotopic composition of matter; the origin of elements and subsequent evolutionary processing; the formation of the solar corona and acceleration of solar wind; and particle acceleration and transport. The project was managed by Goddard and developed by the Applied Physics Laboratory of Johns Hopkins University, with instrument development by CalTech under contract to NASA. The PI was Ed Stone of CalTech. No single instrument was critical to mission success, but in the event nine were included, with sensors a magnitude more sensitive than anything flown before. ACE had six high-resolution spectrometers to measure the elemental, isotopic and ionic charge state composition of nuclei, from hydrogen to nitrogen and from solar wind energies to galactic cosmic rays, to compare the elemental and isotopic composition of the solar corona, interstellar medium and galaxy, as well as particle acceleration processes. NASA explained that the diverse populations of energetic particles from the Sun, solar system, our galaxy and further afield required a broad suite of instruments to capture data from 100eV to 500MeV – from solar emissions to anomalous cosmic rays and galactic cosmic rays. The instruments carried were:

- Cosmic Ray Isotope Spectrometer, 100 to 600MeV;
- Solar Isotope Spectrometer, 10 to 100MeV;
- Ultra Low Energy Isotope Spectrometer;
- Solar Energetic Particle Ionic Charge Analyzer, 0.2 to 3MeV;
- Electron, Proton and Alpha Monitor, 0.03 to 5MeV;
- Solar Wind Ion Mass Spectrometer;
- Solar Wind Electron, Proton and Alpha Monitor, 1 to 900eV and ions 0.26 to 35keV;
- Triaxial Fluxgate Magnetometer.

The collecting power of the instruments was ten to a thousand times greater than any that had gone before. The real-time solar wind monitoring system had the ability to transmit data in real time at the rate of 434 bps and thereby give an hour's warning of impeding solar storms.

ACE was launched at solar minimum, so it was able to follow the gradual increase in solar activity over its two-year mission, but with enough reserves to last up to five. It took four months to reach L1 halo orbit, the first Explorer to arrive there since ISEE 3. ACE was declared operational the following February. The L1 halo orbit, at between 75,000km and 150,000km, was ideal for picking up solar wind, solar and interplanetary particles and cosmic rays well outside Earth's magnetic environment. ACE was joined in 2015 by DSCOVR, whose origins lay in Triana, the educational Earth observation satellite conceived by Vice-President Al Gore but buried by the George Bush presidency. Revived under President Barak Obama, DSCOVR was launched in February 2015 and arrived at L1 in June to study aerosols, ozone and radiation balance.

ACE launch. [From the author's collection]

ACE led to a huge scientific return, vindicating those who had first presented the project in the 1980s. Its first observations, as early as August 1998, were of small to modest solar flares called 'impulsive' solar flares, sometimes also called 'sneezes'. ACE made a new determination of the composition of the Sun and also found that cosmic rays accelerated at least 100,000 years after they were synthesized in supernovae. They may spend 15 million years in our galaxy before leaving. By 2015, the total number of refereed published papers from ACE had reached 3,384.

These were some of the highlight events. On 6th April 2000, ACE detected an interplanetary shock wave heading Earthward, which reached Earth an hour later and caused auroral displays seen as far south in Europe as the Netherlands. ACE was able to detect the way in which the velocity of the solar wind accelerated from 375km/sec to 600km/sec. By 2004, ACE had been through the cone of the interstellar wind flowing into the solar system (technically the Local Interstellar Cloud) seven times. ACE estimated that it was rich in helium, hot (6,000 °C) and very thin (0.264 atoms/cm^3), but uneven, with gusts, ebbs and flows.

ACE's mission coincided with 2008–9, the sleepiest years for solar activity since 1913, with more sunspot-free days than for a hundred years.[24] The Sun's own magnetic field fell from 6-8nT (nanoTesla) to 4nT. The upside of this was that an unprecedented number of cosmic rays reached Earth, as they always do in periods of low solar activity. The bubble of the heliosphere protecting the solar system from the interstellar medium was smaller, with lower pressure. Solar wind speeds were in decline (1% a decade), while galactic cosmic ray counts were increasing. Studies of polar ice from a thousand years ago up to the year 1700 showed much more beryllium 10, an outcome of cosmic rays, so the space age up to 2008 may have seen atypically high levels of solar activity. ACE detected solar events enriched in helium 3 and iron, called hybrid events. Its analysis of cosmic rays prompted scientists to explain them in terms of galactic super-bubbles in regions where supernovae explode.

Something which puzzled scientists was that magnetic activity and aurorae around Earth did not reach their minimum until eight months later in 2009. The explanation, ACE data suggested, was in the Sun's coronal holes which had lingered at the Sun's lower latitudes all 2008 and continued to cause disturbances that year. Combined with the solar and then magnetic minimum, cosmic ray intensities hit a 50-year high in 2009, some 19% more than anything seen previously.

Using ACE's Cosmic Ray Isotope Spectrometer, it was possible to determine that most galactic cosmic rays (GCRs) reaching Earth came from nearby clusters of massive stars. By then ACE had detected 300,000 GCRs, but 15 of them had a particular and very rare radio-isotope of iron, 60Fe, with a half-life of 2.6 million years, which indicated a supernova explosion during that period. The announcement was made in *Science* by Robert Binns of Washington University in St Louis. Their likely sources were in Scorpius and Centaurus. Intriguingly, 60Fe has been found in 2.2 million-year-old samples of both Earth's crust and nine of the Apollo lunar samples, opening up the new discrete field of 60Fe research. ACE was still returning data many years later.[25] According to NASA at the time of the launch, ACE should be able to stay at L1 until 2019 when its hydrazine expires.

SNOE (Explorer 72): Students and Nitric Oxide

SNOE (pronounced 'snowy') was the Student Nitric Oxide Explorer, the first NASA Student Explorer Demonstration Initiative (STEDI), operated by the University of Colorado to measure nitric oxide by altitude in the atmosphere. It was a companion to TERRIERS, both funded at $4.3m for two years design and one year of mission operation. SNOE involved students working with the University of Colorado Laboratory for Atmospheric and Space Physics (LASP) where the mission was controlled, the first one from a university. The PI was Charles Barth of the University of Colorado.[26] One of the main functions of SNOE was to demonstrate that low-cost satellite missions could be undertaken with student involvement. To keep costs down, the satellite was expected to use instrumentation from earlier missions and minimize the complexity of data and communication systems. The mission's orbit was set at 550km, sun-synchronous, 97.5°.

SNOE

Nitric oxide is found unevenly in the Earth's atmosphere, being up to ten times denser in the polar regions above 65°N and 65°S, especially between 100km and 110km. Its density changes in the lower thermosphere, where it can be transported downward to react with and destroy ozone. In the polar regions, it was possible that nitric oxide was produced by the impact of auroral electrons. Uncovering how it is produced and why it changes were the challenges of the mission. Sounding rockets had already measured nitric oxide in the poles in the course of their ascent and descent, but this was the first satellite intended to do so. The main scientific functions were to measure thermospheric nitric oxide and its variability, solar irradiance at soft x-ray wavelengths and auroral energy deposition at high latitudes. The three science instruments were an ultraviolet spectrometer to measure nitric oxide, a solar soft x-ray photometer and a far ultraviolet photometer. The ultraviolet spectrometer was based on the one carried on the Solar Mesosphere Explorer (Explorer 64) and collected data continuously. The auroral photometer was designed to determine the energy deposited in the upper atmosphere by energetic auroral electrons and was based on instruments flown by the orbiting geophysical observatories in the 1960s. The solar x-ray photometer was intended to measure solar irradiance between 2nm and 31nm.

SNOE launched on a Pegasus on 26th February 1998. There were two main results: solar soft x-ray irradiance (summer) and auroral energy (winter) controlled nitric oxide abundance at mid-latitude; while solar soft x-ray irradiance and auroral energy were overall controllers globally. SNOE re-entered on 13th December 2003, crashing into the ocean west of Peru after more than five years on orbit.

TRACE (Explorer 73)

TRACE was the Transition Region And Coronal Explorer, with instruments to study the interaction between fine scale magnetic fields on the solar surface and the Sun's photosphere, chromosphere and the transition regions in between. TRACE was an early 'faster, better, cheaper' mission, developed in less than four years and launched only a month behind schedule. It was sent up on 2nd April 1998 on a Pegasus XL from Vandenberg. TRACE was the first solar mission since the Solar Maximum Mission (Solar Max) in 1980 and was complementary to the Solar and Heliospheric Observatory (SOHO). It did not carry as wide a range of instrumentation as SOHO, but did possess the ability to take detailed images of the photosphere, transition region and corona. TRACE had 10 to 25 times better resolution than SOHO. The mission's PI was Alan Title of Lockheed Martin.

TRACE was 2m tall, more than 1m wide and weighed 250kg. The purpose of the mission was to use an extreme ultraviolet telescope to examine solar light in different bandwidths from a sun-synchronous orbit, making almost continuous observations possible. Its objective was to explore the three-dimensional magnetic structures which emerged through the photosphere (the visible surface of the Sun) and to define both the geometry and dynamics of the upper solar atmosphere – its transition region and corona. Prior to this mission, no images had been collected that showed the temperature range simultaneously with both high spatial and temporal resolution. TRACE was timed for the increase in solar activity at the beginning of the new cycle.

Signals from TRACE were first picked up two hours after launch at Poker Flat, Alaska. The gyros were activated on the first day and the scientific equipment on the second. First light for TRACE was 20th April 1998. TRACE's telescope was ten times more powerful than any that had gone before and studied the Sun from the relatively cool 6,000 °C lower atmosphere to the 1.7 million °C corona. TRACE was able to provide imaging of the solar atmosphere where temperatures ranged from 6,000K to ten million K. On 26th September 2000, TRACE scientists announced the location of a source that heated the Sun's corona to 300 times more than its surface. TRACE observed 41 loops reaching out from 4,000km to 290,000km. Heating occurred in the first 9,000km of the loop, its base, which cooled as it extended further out. On 31st May 1998, TRACE made a video of a bright explosion in the atmosphere of the Sun, a solar flare 89,000km long, 322km wide and travelling at 3.2 million km/hr.

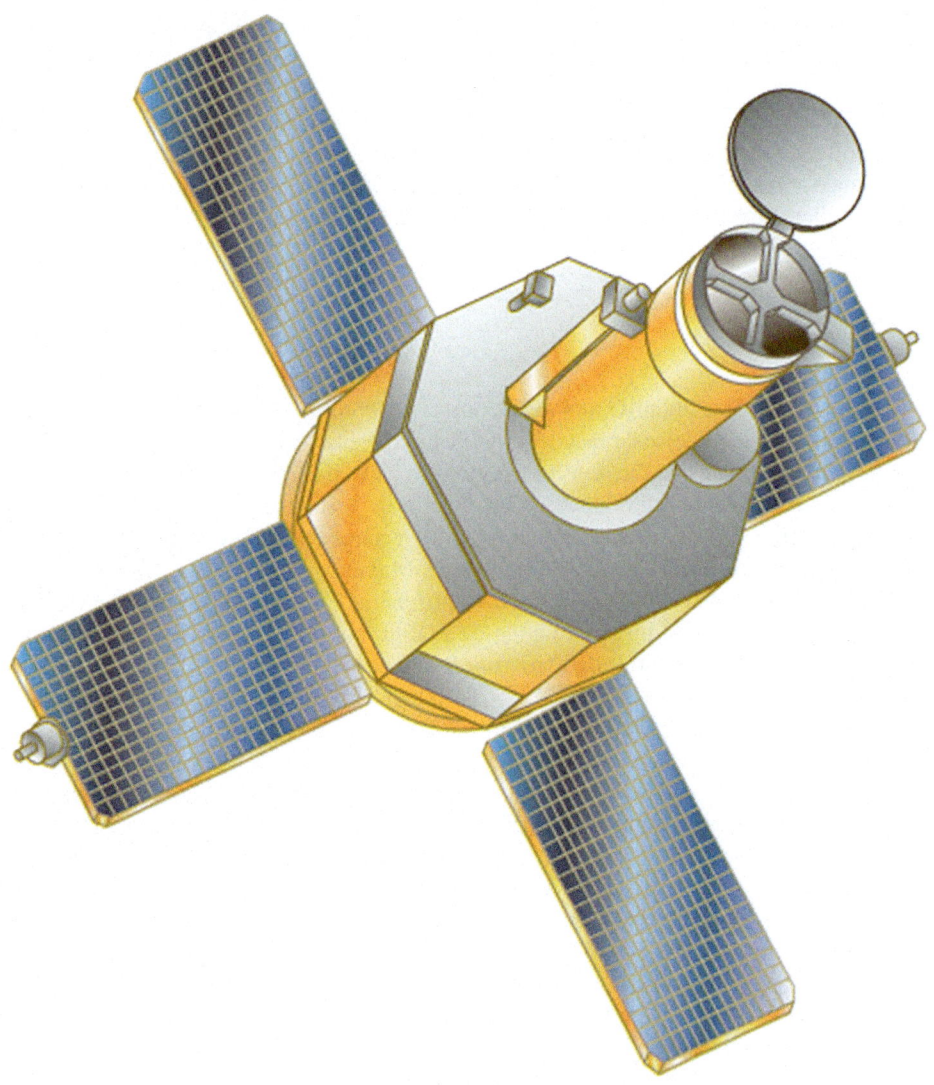

TRACE

TRACE made mainstream television, unusual for an Explorer mission, when Dan Rather of Columbia Broadcasting Service presented a program on the Sun. The program startled viewers out of their perception of the Sun as a 'nice, constant, yellow ball', instead showing it to them in ultraviolet light as a place of storms and violent energy with the potential to come and knock out power grids and mobile phones. After twelve years, TRACE had led to more than a thousand scientific papers and the creation of a unique data archive (trace.lmsal.com). TRACE took many images of the Sun in visible and ultraviolet light, including coronal loops, solar eruptions and 'solar moss'. It also found that the Sun's atmosphere was filled with ultrasound waves. From 2010, its role was largely taken over by the Solar Dynamics Observatory.

TRACE image of the Sun

SWAS (Explorer 74): A Watery Universe

SWAS, the Submillimetre Wave Astronomy Satellite, was launched on the Pegasus 160km off the coast of California on 6th December 1998. The 283kg satellite entered an orbit of 637km – 653km, 69.9°. The mission cost was $64m, including $13m for the launch. SWAS was a long-standing and much-delayed part of the Explorer program, originally planned for launch in 1995. Development to launch took nine years and it suffered badly from the early problems of the Pegasus XL. SWAS was to have taken the third Pegasus XL, but when the first two failed, it was moved back to 1999 while the problems were sorted out. When the next nine Pegasus XLs were successful, SWAS was moved up to December 1998 for launch on the 16th XL, the 25th of the Pegasus series. During its long period of grounding, SWAS was periodically taken out for testing.

JPL had begun to study a submillimetre Explorer in 1987 with a view to launch in the mid-1990s.[27] The submillimetre end of the wavelength is at 0.1mm to 1mm, beyond infrared and ideal for measuring interstellar gas and dust of temperatures 10 to 100K and the early formation of stars. The original proposal by Gary Melnick was for a large telescope with resolution of six arc seconds, placed in polar orbit to observe spectral lines emanating from dense molecular clouds and focussing on such targets such as Vega, Formalhaut and β Pectoris.[28] SWAS would map selected areas of the sky between 490 and 553GHz in four wavelengths to gain information on the chemical composition of interstellar galactic clouds, in order to determine star formation and find out how they cooled.

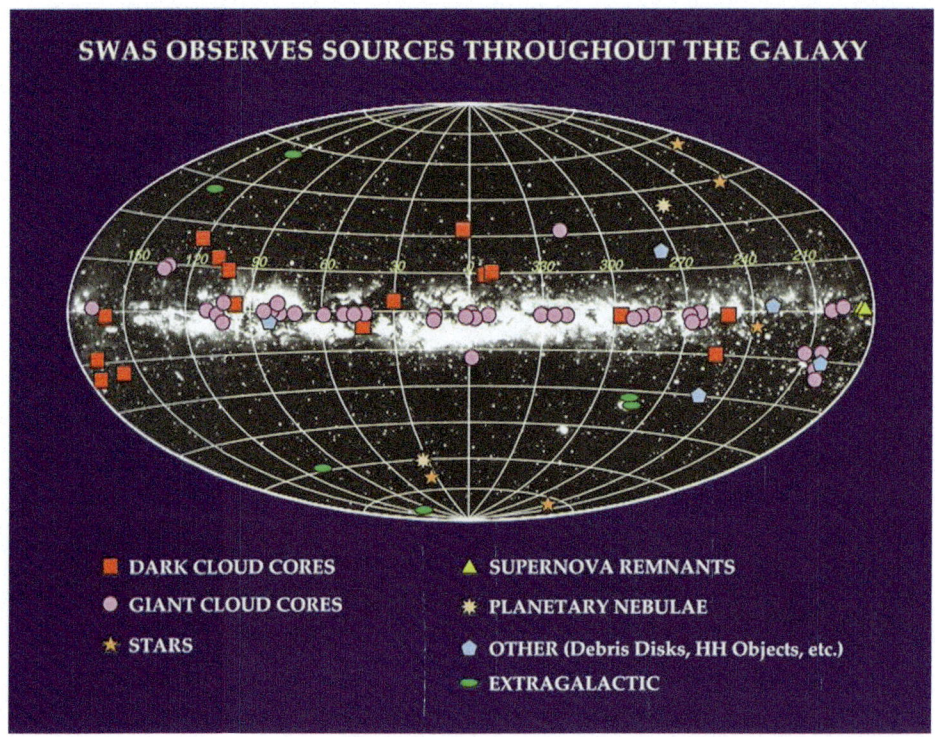

SWAS sources observed

Its main instrument was to study water, molecular oxygen, atomic carbon and isotopic carbon monoxide in submillimetre radiation which is blocked by Earth's atmosphere. The University of Cologne, Germany, provided a spectrometer to analyse radio signals. SWAS had an accuracy of 38arcsec, the intention being to aim at three to five targets per orbit. Data were dumped daily to NASA Wallops and Poker Flat, Alaska. SWAS had the highest frequency radio system yet put in space and included five solar panels generating 230w. Gary Melnick of the Harvard Smithsonian Center for Astrophysics, who became PI, described it as 'a sophisticated observatory in a very small package'.[29] The SWAS science team also comprised the University of Chicago, the University of Massachusetts Amherst, the National Air and Space Museum, Ames Research Center, Johns Hopkins University and the Smithsonian Astrophysical Observatory, which hosted the Science Operations Center.

SWAS had a 102kg, 102mm by 55cm by 71cm Cassegrain radio telescope with a beam width of four arc minutes. The submillimetre radiometers were a pair of passively cooled Schottky diode receivers, with receiver noise figures of 2,500–3,000K. The outputs of the two SWAS receivers had a final intermediate frequency from 1.4 to 2.8GHz.

In the event, SWAS made detailed 1° × 1° maps of numerous giant molecular and dark cloud cores. Closer to home, it observed the atmospheres of Mars, Jupiter, Saturn and several comets. SWAS made observations until 21st July 2004, far beyond its original planned lifetime and contributed important discoveries. It found that water vapour was present in almost all star-forming regions, including throughout the Milky Way. Water was detected in almost every dust cloud observed, with high amounts in warm gas, but low amounts in cold dense gas. Molecular clouds had an abundance of water ice in their interiors. Large amounts of water saturated the interstellar medium, but molecular oxygen was not found. A swarm of comets was found evaporating around an aging red giant star. Finally, SWAS was brought back into operation for three months to follow the crash of Deep Impact onto Comet P/Tempel 1, a comet which SWAS estimated was ejecting 328kg of water a second.

WIRE (Explorer 75): Back from the Dead

WIRE (Wide Field Infrared Explorer), the fifth SMEX, was launched on 5th March 1999 on a Pegasus XL but was lost soon after reaching orbit. Approved in 1994, WIRE was designed to observe star bursts 500 times fainter than those recorded by the Infrared Astronomical Satellite (IRAS) in 1983. The simple aim was to explain the history of star formation and from that how galaxies formed. It was also part of NASA's *Origins* program.

The project involved NASA JPL and the Space Dynamics Laboratory of Utah State University, with the spacecraft built at Goddard. The focal length of the infrared telescope was 1,050mm and its field of view was about that of Earth's Moon. WIRE was intended for an orbit of 97.2°, sun-synchronous at 540km. The mission was to be split between moderate depth surveys requiring 14-minute exposure times and deep surveys of up to eight-hour exposures. The aim was to focus on starburst galaxies billions of light years away, considered to be the main source of stars but blocked by dust in visible light. The mission cost $50m, including $15m for the Pegasus launcher.

The science goals of WIRE were to determine what fraction of the luminosity of the universe redshift was due to starburst galaxies; assess how fast and in what ways starburst galaxies evolved; and examine whether luminous protogalaxies were common. WIRE was designed to make a four-month, deep infrared extra-galactic study at 12 and 25 micrometres over ten to several hundred square degrees of the sky in star-forming regions. WIRE had a two-colour, solid hydrogen-cooled, infrared imaging telescope. The primary instrument was a 93kg, 30cm diameter Cassegrain telescope, with no moving parts and cooled by hydrogen, built by Utah State University. The science team came from JPL, Cornell University, California Institute of Technology and Ball Aerospace, with Perry Hacking of JPL as the PI. The spacecraft weighed 270kg.

The first launch attempt was made on 2nd March 1999, but was aborted a minute before drop because a locking pin on the Pegasus had not disengaged. Two days later, it was released 160km out over the Pacific. Pegasus fell for five seconds, ignited for nine minutes and delivered the satellite into its 540km orbit.

The first pass was over McMurdo tracking station in Antarctica 20 minutes later. The solar arrays had been deployed and everything appeared normal, but it was observed to be spinning when it came over Poker Flat ground station in Alaska a little later. WIRE appeared to be going out of control. Despite significant efforts to recover the spacecraft, it was declared a loss after four days. The investigation board determined that the telescope instrument cover was jettisoned prematurely, causing the solid hydrogen cryogenic supply to sublimate and the spacecraft to spin at one revolution a second. The supply became exhausted in 36 hours. The spacecraft was eventually brought under control, but with its cryogenic supply exhausted, it was now useless.

WIRE

A board of inquiry was convened at Goddard on 23rd March, with JPL forming its own review team.[30] The investigation determined that there was a digital logic design error in the pyro electronics box, which had fired explosive devices prematurely and thus ejected the telescope cover too early. It was not a component failure as such: simply turning on the electrical power in the box could activate the pyrotechnics, rather than later in the sequence as intended and that was what happened. This danger was not appreciated, the error was not detected and neither the box was peer-reviewed nor the system reviewed. NASA subsequently issued a Parts Advisory on the importance of testing for unwanted or premature outcomes from electrical start-ups and the board suggested the use of inhibitors for mission-critical pyrotechnics.

Some astronomers were not prepared to let the dead lie undisturbed. Although WIRE had been declared lost, there was nothing wrong with its attitude control system, nor its star tracker. The venting of the cryogen was complete after a week and it was possible to obtain full control and orientation of the spacecraft thereafter.[31] Recovery was declared on 11th March 1999. Operations were then transferred from Goddard Space Flight Center to Bowie State University's Satellite Operations and Control Center. Astro-seismologists were able to use the star tracker to make lengthy observations of bright stars (e.g. β Crucis and α Ursa Major), generally for 40 minutes at a time, especially to find evidence of starquakes. Astronomers used WIRE data on low-amplitude oscillations to probe the interior of a red giant, calculate its mass (ten times that of the Sun) and construct a model of its evolution.[32] Finally, WIRE re-entered the atmosphere on 10th May 2011. WISE took up where it left off.

TERRIERS (Explorer 76): Students Profiling the Ionosphere

TERRIERS was the Tomographic Experiment using Radiative Recombinative Ionospheric Extreme ultraviolet and Radio Sources, the second NASA Student Explorer Demonstration Initiative (STEDI) for small, low-cost ($4.4m), high-science spacecraft to be built in less than two years. It was launched on 18th May 1999 from Vandenberg, with 60 Boston students on standby to receive the first signals. Unfortunately, the 16w solar panels failed to find the Sun, it quickly lost battery power and no data were returned.

Mission control Vandenberg during Pegasus launch

TERRIERS was put together by the Center for Space Physics at Boston University, AeroAstro (providing the satellite and ground station), the University of Illinois Urbana – Champain, the Naval Research Laboratory, the Haystack Observatory at MIT, Phillips Laboratory and Cleveland Heights High School. The PI was Daniel Cotton of Boston University, where the ground station was located. Their objective was to make 3D latitude, altitude and thermospheric profiles of the electron density of the ionosphere in ultraviolet, radio and visible radiation, hence the term 'tomography'. The instrumentation comprised five Tomographic EUV spectrographs (TESS) for nightglow and dayglow observations (80–140nm, 1–2nm resolution), a Gas Ionization Solar Spectral Monitor (GISSMO), two photometers for 630nm night airglow and a dual-frequency radio beacon for electron content measurements. TERRIERS weighed 123kg. In two months, two small Explorers had now availed of Goldin's 'permission to fail'. It was hoped that the quota was now used up.

FUSE (Explorer 77): Empty Space? Not any more

FUSE, or Far Ultraviolet Spectroscopic Explorer, was originally proposed in 1981 by the National Academy of Sciences, but was not funded until 1989. The proposal responded to the request for missions that addressed questions about the fundamental nature of the universe, what NASA called its *Origins* program. When its budget passed the $300m mark and budget cuts hit in 1994, FUSE was scheduled for cancellation, but was rescued by PI Warren Moos, professor of physics and astronomy at Johns Hopkins, a specialist in space telescopes and a veteran of Apollo, Hubble and Voyager, together with project manager Dennis McCarthy. They proposed to save it by cutting the cost to $120m, moving management out of NASA to Johns Hopkins University (JHU), reducing the scientific objectives and using off-the-shelf commercial hardware. Now downsized, FUSE became the first MIDEX, with a funding cap of $226m. JHU offered to take over for $108m, so this became the first large-scale mission planned and run out of a university, taking over mission control at 100 minutes after launch. A ceremony was held during which control of the mission was formally passed from Goddard to Johns Hopkins. It may have been the most challenging astrophysics mission based in a university.

FUSE: attaching a solar panel

The aim of this 1,335kg satellite was to study star formation and the mass of the universe by examining the early relics of the big bang, principally hydrogen and deuterium. JHU worked with the universities of Colorado and Berkeley, with instruments also provided by Canada and France, for pointing and ultraviolet systems. FUSE was equipped with four 35cm diameter co-aligned telescopes to investigate trace gases of interstellar and intergalactic gases in the 90 to 120nm range, with telescopes of 10,000 times greater sensitivity than had been available in the 1970s. Its cameras, called the Fine Error Sensor, could see stars of 14th magnitude. The spacecraft bus was built by Orbital Sciences Corporation and the scientific instruments by Johns Hopkins.

The idea was that FUSE would build on the work of IUE in the far ultraviolet spectrum, its key challenge being to determine the ratio of deuterium (heavy hydrogen) to hydrogen in the interstellar medium between stars (it varied 50% in known samples). The initial focus would be on the Milky Way and our neighbouring galaxies, the Magellanic Clouds. Essentially, FUSE was to explore the tenuous regions of interstellar and intergalactic space hitherto considered to be empty and Goddard advertised the spacecraft as 'searching for the fossils of the big bang'. Its control centre was the Bloomberg Center for Physics and Astronomy in Johns Hopkins University in Baltimore, Maryland. The primary FUSE ground station was in the University of Puerto Rico, which it overflew six times a day, with another in Hawaii.

FUSE arrived at Cape Canaveral on 1st April 1999 and launched on 23rd June 1999 on a Delta II into a 775km circular orbit. After calibration and allowing spacecraft temperatures to subside, observations started in September 1999. On 12th January 2000, Warren Moos announced that thousands of exploding stars had created a halo of gas around the Milky Way, with evidence of hot gas around other galaxies. The hot gas halo around the Milky Way extended far above and below the galactic plane.[33] FUSE found molecular nitrogen in dense interstellar gas, while finding H_2 in the galactic interstellar medium and its halo. In effect, it challenged existing models of the chemical evolution of galaxies. Significant deuterium was found, which came from the early universe and which astronomers had been trying to discover for many years. Less deuterium had been destroyed since the big bang than had been previously modelled. FUSE also found 'missing baryons' (subatomic particles) in the local intergalactic medium.

The satellite studied η Carinae, a star a hundred times more massive than our Sun, to observe an eclipse of its two binary stars.[34] It found the sudden disappearance of far ultraviolet and then x-ray radiation, indicating that the smaller binary was hotter than the larger. The chief observer was Ted Gull at Goddard, a student of η Carinae since 1981, who had written over 200 papers on the star, made a 3D model of its structure based on these observations and is probably the world's leading expert on this bizarre star. FUSE also found a debris disk around β Picotris, which could be a place of planet formation. Closer to home, it detected molecular hydrogen in the atmosphere of Mars, important for an understanding of the role of water in the planet's evolution. It was possible to make calculations that the volume of water escaping from Mars could have filled an ocean 30m deep.

206 **Faster, Better, Cheaper?**

FUSE mission control, Johns Hopkins University

In December 2001, FUSE lost two of its four reaction wheels and, without an operational guidance system, was given up for dead. Engineers at Johns Hopkins devised a clever system of using electromagnets on the spacecraft to generate local magnetic fields and run electric current through its torque bars, so it was back in operation the following year. The software was then reprogrammed for the mission to resume after this near-death experience. This went so well that a guest observer program was introduced in 2002. FUSE kept this going until the last reaction wheel gave out. It was decommissioned October 2007, but will likely orbit for another twenty years. FUSE had observed 3,000 objects in 100,000 observing hours. Its data were archived at the Mikulski Archive for Space Telescopes (MAST) on an open, no-registration basis. FUSE was the basis of 400 scientific papers.

IMAGE (Explorer 78): Imaging the Magnetosphere

Three Explorers were launched in 1999, quite a contrast to the 1980s, but IMAGE was the first Explorer of the new century. Launched from Vandenberg on 25th March 2000 on a Delta II, its aim was to image Earth's magnetosphere and to see the 'invisible' plasma populations in the magnetosphere, the data being subsequently made available by the Southwest Research Institute in San Antonio on a non-proprietary basis. The mission coincided with solar maximum, so intense activity was expected. IMAGE had an education and outreach site called POETRY which encouraged students to question scientists about the mission.

IMAGE (Explorer 78): Imaging the Magnetosphere

IMAGE was 2.25m in diameter, 1.52m tall and weighed 494kg. Viewed from either end, it had the form of a regular octagon. Solar cells on its side provided power of 250w for a nickel-cadmium battery. The mission's objectives were to determine the dominant mechanisms for injecting plasma into the magnetosphere, how the magnetosphere responded to changes in the solar wind and how and where plasmas are energized, transported and lost in storms. IMAGE aimed to detect energetic neutral atom emissions from the ring current, inner plasma sheet and polar ionospheric outflows; make plasmaspheric imaging at extreme ultraviolet wavelengths; radio sound the magnetopause and other boundary layers; and image far-ultraviolet auroral emissions. It had four beryllium-copper 500m long antennae. The spacecraft was built by Lockheed in Sunnyvale, California and the equipment by Goddard and the Southwest Research Institute, San Antonio. The Science and Mission Operations Center was in Goddard and the PI was James Burch. By this stage, these spacecraft had become sufficiently automated or able to take care of themselves that ground control could be operated on a 40-hour week, rather than continuously around the clock, but there was a warning system to bring staff in should an alarm be sounded.

Observations were divided into duskside, dayside, dawnside and polar. IMAGE's orbit extended to R7.2, or 45,900km (perigee was just under 1,000km, so it was very elliptical). It was expected that the apogee would move from 40°N at the start to 90° (the pole) a year later and then return to 40°N two years after arrival in orbit. The rate of spin was 0.5/min. Communications were by three S-band antennae (two omni-directional), with continuous real-time data at 44kbps and a full downlink data dump every 14 hours at 2.28MBps.

IMAGE picture of the south pole

NASA subsequently reported that IMAGE had made the first global images of the geomagnetic storm ring current and proton aurora and the first remote measurements of plasmaspheric densities using radio sounding. IMAGE brought a wealth of new knowledge of the magnetosphere: determination of the spatial extent and location of the polar cusp as a function of the interplanetary magnetic field; observations of ionospheric ions after the arrival of a coronal mass ejection, indicating the existence of direct heating of the topside ionosphere; confirmation of the theory of plasmaspheric tails; discovery of several new and unpredicted features of the plasmasphere including 'shoulders', 'fingers', corotating 'voids' and isolated flux tubes; discovery of sub-auroral proton arcs; and a better understanding of the difference between magnetic storms and substorms. It identified plasmaspheric cavities as source regions for kilometric continuum radiation and made the first measurements of solar wind neutral atoms and interstellar neutral atoms from inside the magnetosphere. The National Oceanic and Atmospheric Administration used the real-time data for space weather forecasts and warnings.

HETE 2 (Explorer 79): The Mystery of Gamma Ray Bursts

The concept of a satellite capable of multi-wavelength observations of Gamma Ray Bursts (GRBs) was discussed at a meeting of astronomers in Santa Cruz, California in 1981. MIT worked up a proposal in 1986 for a multi-wavelength small satellite mission to solve the gamma ray burst mystery. In 1989, NASA approved funding for a new, low-cost Explorer to search for GRBs and funded what became HETE (High Energy Transient Explorer) in 1992. The original spacecraft contractor for HETE was AeroAstro and its instruments comprised four wide-field gamma ray detectors (CESR, Toulouse, France); a wide-field coded-aperture x-ray imager (Los Alamos National Laboratory and the Institute of Chemistry and Physics (RIKEN) of Tokyo, Japan) and four wide-field cameras (MIT). The aim was to detect gamma ray bursts, pinpoint them and transmit in real time, giving ample time for astronomers worldwide to co-investigate.

HETE was launched with Argentinian satellite SAC-B on 4th November 1996 from Wallops Island. Due to a battery failure in the Pegasus third stage, HETE was trapped within the Dual Payload Attachment Fitting, could not deploy its solar panels, lost power and died. Both payloads de-orbited on 7th April 2002. New Explorers were one-off projects and if they failed, there were no resources to start again and re-fly the mission. As it was explained, 'permission to fail' did not mean 'permission to try again'. There was one exception. NASA determined that gamma ray burst science was still a priority and a second mission, called HETE 2, was approved in July 1997. Some spare hardware was available and the replacement was built entirely at MIT. To put the mission in context, it was a precursor of the more capable Swift that followed (Explorer 84).

For this second attempt, an equatorial orbit was selected and there were changes in the instrumentation. The ultraviolet cameras were replaced by a CCD-based coded-aperture imager for soft x-rays (the Soft X-ray Camera, or SXC) and optical CCD cameras, which also served as star trackers. A new gamma-ray instrument, FREGATE, was added to provide more accurate burst triggers. The replacement satellite, weighing 124kg, was ready

HETE 2 (Explorer 79): The Mystery of Gamma Ray Bursts

by January 2000 and was set for a launch on 28th of that month from a new launch site, Kwajalein Atoll. Among the tracking stations would be Cayenne, French Guyana and the old British Singapore station.

At this stage, NASA intervened to delay the launch. Following the loss of three western satellites on Chinese rockets in the 1990s, the Congress had introduced the International Traffic in Arms Regulations (ITAR) aimed at preventing the transfer of technology to China. This was applied to all countries, including American allies, who had to undergo a process of certification that they were not aiding and abetting China. Neither the French tracking station in Cayenne nor Singapore station had been approved for ITAR certification by the State Department, so the mission was delayed until they were.

HETE on Pegasus

HETE 2 was eventually launched on 9th October 2000. Its task was again to carry out the first multi-wavelength study of gamma ray bursts with ultraviolet, x-ray and gamma ray instruments, with the ability to localize bursts with an accuracy of several arc seconds, in near real-time. Its goals were to determine the origin and nature of cosmic gamma ray bursts (GRBs) through simultaneous, broad-band observation in the soft x-ray, medium x-ray and gamma ray energy ranges, aiming at their precise localization and identification within several arc seconds in as near to real-time as possible. Ground-based observers would be alerted within seconds, so that they could also have a chance to observe.

HETE was one of the most international Explorer missions. Led by the Center for Space Research at the Massachusetts Institute of Technology, national participants included the University of Chicago; the University of California, Berkeley; the University of California, Santa Cruz; and the Los Alamos National Laboratory (LANL). International participants included the French Centre d'Etude Spatiale des Rayonnements (CESR), the Centre Nationale d'Etudes Spatiales (CNES), the Ecole Nationale Supérieure de l'Aéronautique et de l'Espace (Sup'Aero), the Italian Consiglio Nazionale delle Ricerche (CNR), the Instituto Nacional de Pesquisas Espaciais (INPE) in Brazil; the Institute for Chemistry and Physics (RIKEN) in Japan; and the Tata Institute of Fundamental Research (TIFR) in India.

HETE 2 duly got to work. Its first task was a 60% survey of the celestial sphere. Its orientation was called anti-solar, the instruments operating during the night, so that bursts would always be at least 120° from the Sun and at the best time for ground-based observers. HETE 2 was expected to find about 15 GRBs a year. In fact, it detected six GRBs in 2001, 19 in 2002, 25 in 2003, 19 in 2004, 12 in 2005 and three in 2006. HETE 2 declined in March 2007 as its batteries began to deteriorate.

Its landmark discovery, in October 2002, was that GRBs had their origin in star explosions that marked the end of a star and the beginning of a black hole. On 4th October 2002, HETE 2 spotted gamma burst GRB 021004 and was able to report on the death of a gigantic star and its replacement by a new spinning black hole. This is called a collapsar model event. That December, HETE 2 also detected dark gamma ray burst afterglow which lasted for two hours. HETE 2 in effect solved the mystery of short gamma ray bursts – they were explosive collisions of two neutron stars, or a neutron star and a black hole merging.

HETE 2 confirmed the connection between gamma ray bursts and supernovae (GRB 030329). It found light in what had been called dark bursts, which were not completely dark after all and it identified a new, less energetic form of GRBs called X-Ray Flashes (XRFs). HETE found a new class of burst, the short gamma ray burst, of milliseconds duration, rated as one of the top discoveries in the world in 2005. Three were found, with analysis suggesting that these were the disruption of a neutron star by a black hole, enough to tear the neutron star apart and prompt an emission of extraordinary violence.

WMAP (Explorer 80): Finder of Dark Matter

WMAP followed COBE to make more accurate measurements of the ripples of the big bang. The idea of a follow-up to COBE was proposed to NASA by astrophysicist Charles Bennett as the Microwave Anistropy Probe (MAP) in 1995. Its aim was to identify really minute differences in background radiation, to as little as 0.000001°, with 45 times greater sensitivity and 33 times the angular resolution of COBE. The idea won the strong endorsement of the National Research Council in 1996 and the SSB the following year, determining that the top scientific priority was to decide the early geometry of the universe. It was the second MIDEX.

WMAP (Explorer 80): Finder of Dark Matter

MAP was selected the following year for 2001 launch. It would later acquire the additional letter 'W' for David Wilkinson, the experimental physicist who had greatly contributed to it and COBE beforehand, but who died in 2003. Rather than use a Dewar cooling system like COBE, this time MAP had a single uncooled radiometer isolated from the rest of the spacecraft, assisted by insulation and passive cooling. MAP was big, at 3.8m tall, 5m wide and weighing 840kg (including 72kg propellant). It had 419w panels and housed two back-to-back Gregorian telescopes to survey the sky twice independently in five frequency bands from 22–90 GHz, as well as 20 differential radiometers covering five frequency bands, the system being cooled to 90K. The mission cost $95m for the spacecraft, $50m for the launcher and $14m for operations.

WMAP launch

WMAP was launched on 30th June 2001 aboard a Delta II 7425-10, the 286th Delta from pad 17B at Cape Canaveral and the first of the 7425-10 series, with a new 3m composite fairing. It followed a complex three-month phasing trajectory, first to 3,218–299,274km; then 4,827–350,762km; then passing 5,149km by the Moon on 30th July before settling in at L2 halo orbit 1.5 million km from Earth in the anti-Sun position, a trajectory that involved seven burns and then station-keeping firings. WMAP demonstrated the possibilities of operating a satellite at L2.

WMAP at L2

In February 2003, WMAP scientists presented the outcome of its 12-month survey of the big bang cosmic background, with a full-scale map of the universe followed by a series of oval sky maps of low and full-resolution as well as catalogues of individual sources.[35] They presented the hoped-for level of detail, showing cold spots and hot spots. WMAP marked a defining point in precision cosmology: a new age for the universe, 13.7 billion years (±0.5%) and its flatness (0.4%). Mean temperature of the microwave background was the same as measured by COBE, at 2.7K, with only small variations (5×10^{-5}). WMAP estimated its composition at 24% dark matter, 71.4% dark energy and 4.6% atomic matter. These are now the standard reference figures – and sparked off the hunt to identify and explain dark matter. WMAP confirmed the theory of 'inflation' for what happened to the universe after the big bang. The universe as we know it emerged 400 million years after what is now called the cosmic dark age and stars began to re-ionize it. WMAP suggested anomalies, such as a dark flow that tugged galaxies toward Centaurus.

WMAP's original mission was two years (to 2003), but extensions were given to August 2010. The spacecraft was then moved to a graveyard orbit. To take these studies a stage further, ESA launched Planck on Ariane 5 in 2009 to near L2, with ten times more sensitivity in nine bands. It operated until 2013, with its first star maps coming out in 2010 and a full set of results in 2013. Planck provided the greater subtlety of data hoped for, a new estimate of the age of the universe (13.798 billion years), a re-estimate of the composition of the universe (4.9% ordinary matter, 26.8% dark matter and 68.3% dark energy; and new types of neutrinos (muon and tau). This raised a further question, for the universe was much more simple, flat and uniform that it should be, given the violence that first brought

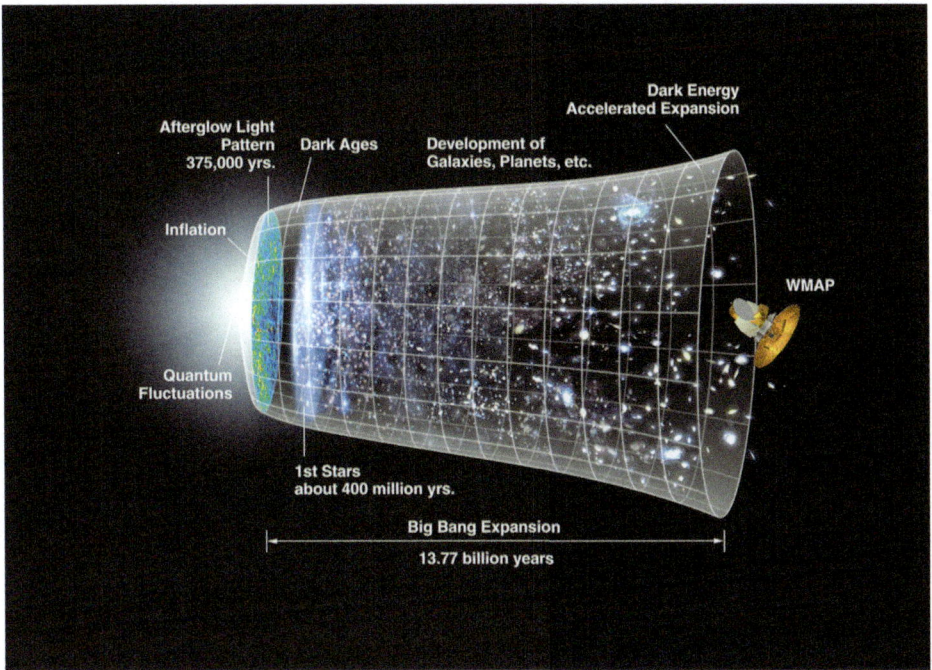

WMAP new timeline

it into existence. Between them, COBE, WMAP and Planck, with their 20 years of background radiation studies, provided the baseline for a revolution in cosmology that continues to the present. They paved the way for the James Webb Space Telescope to follow.

RHESSI (Explorer 81): The Legacy of Reuven Ramaty

HESSI (High Energy Solar Spectroscopic Imager), like Rossi, was an Explorer that acquired an additional name, for it later became RHESSI. The 'R' this time came from Reuven Ramaty, one of the fathers of solar gamma ray astronomy, who died just before the launch. Hard x-ray imaging dated back to the Solar Maximum Mission ('Solar Max') in 1980, the Japanese *Hinotori* and a proposal made to develop this research with a Solar High-Energy Astrophysical Plasmas Explorer (SHAPE) in 1986. By the time it was adjudicated several years later, there was not enough time to launch it for the solar maximum of 1990, so it was turned down. It was re-proposed as a MIDEX High Energy Solar Imager (HESI) in 1995 with the second 'S' later added to acknowledge the importance of spectroscopy. Now called HESSI, it was eventually approved in October 1997 as the sixth SMEX, at the same time as GALEX (Explorer 83). The original cost was $67m and it was set for a Pegasus launch in 2000. Robert Li of University of California Berkeley was the PI.

All was coming along nicely for the 2000 launch until the shake table went out of control during a vibration test at JPL in December 1999, breaking the solar panels, two of the three telescope mounts and other instrument parts. Then a Pegasus launcher failed on 7th June 2001 (not an Explorer mission), forcing a further delay into the following year. The launch eventually came 19 months after the original scheduled date that had been designed to take advantage of solar maximum in July 2000, so it missed about a thousand solar flare events. Even with a late launch, it was still expected to record about a thousand, but over a longer period.

RHESSI was finally launched on 5th February 2002 into a 600km circular orbit at 38°, falling to 554km by 575km after six years. Even that launch was nerve-wracking, as the aircraft lost radio contact with base two minutes before launch and had to fly back on a racetrack course while the problem was fixed.

The specific purpose of RHESSI was to take high resolution x-ray and gamma ray images of solar flares as they happened, to learn about particle acceleration and energy release in solar flares. RHESSI was designed to provide high-resolution imaging and spectroscopy on the x-rays to gamma rays continuum, using an imager made of nine rotating modulation collimators with germanium detectors cooled to 90K. The system was to combine high resolution, precision, wide energy range and continuous data. Its only instrument was an imaging spectrometer to construct flare images from the patterns of light and shadows coming from high energy radiation passing through the telescope's grids. The spectrometer separated the light electronically into its component colours, while x-ray and gamma ray detectors counted the number of photons and measured their energy. From this data, scientists could reconstruct high-resolution colour pictures of solar flares and loops using computers.

RHESSI was 2.16m tall, 5.76m wide including its solar panels and weighed 293kg. Data were transmitted at a rate of 1.8GB a day and by 2008 had accumulated 4TB. The spacecraft was sun-pointed and revolved at 15rpm. The University of California, Berkeley was responsible for the mission, while international cooperation came from Genoa, Italy; Glasgow, Scotland; Graz, Austria; Switzerland and France. RHESSI was operated from a multi-mission facility at the same Space Sciences Laboratory in Berkeley that also controlled FAST, CHIPS and THEMIS.

RHESSI

Once in orbit, RHESSI detected its first flare on 12th February. Although the solar maximum had passed, there was thankfully abundant solar activity left in the cycle and the RHESSI catalogue built up to over 40,000 flares. By solar minimum though, both the cryogenic cooler and detectors had begun to degrade. Still, by 2008 RHESSI had detected over 11,000 flares above 12keV, 950 above 25keV and 30 above 300keV with 18 showing gamma ray lines, as well as 25,000 micro-flares above 6keV from active regions of the Sun, with details of frequency and energy. RHESSI found that the Sun was not perfectly spherical: it was only 0.001% perfect, fatter at the equator than at the pole, the difference being about 6km on a radius of 700,000km.

In addition to its solar observations, RHESSI observed 420 gamma ray bursts but did not have the facility to locate them precisely. It picked up a huge flare from the soft gamma ray repeater SGR 1806-20 which happened to be in a line just 5.25° from the Sun at the time. It is thought to be a magnetar and the first 0.2 seconds of this flare released more energy than the Sun in a quarter of a million years. RHESSI followed the accreting x-ray binary A0535+26 in 2005, eleven years after its previous outburst which was at times brighter than the Crab nebula. Closer to home, Terrestrial Gamma Ray Flashes (TGFs) had only been previously observed by the Compton Gamma Ray Observatory, which had identified 75, but RHESSI found over 100 a year. These were millisecond bursts, generally associated with cloud-to-cloud lightning strikes at 13km to 15km altitude with beams of up to 30MeV that could be seen from orbit and sometimes entered the magnetosphere. Altogether, RHESSI found a thousand TGFs and scientists concluded that they came from intra-cloud lightning, rather than cloud-to-ground sprites.

RHESSI led to 670 publications, 970 citations and was the basis of 30 PhDs. RHESSI workshops were held in Paris, France and in California (Santa Cruz and Berkeley) and a 500-page Space Science Review book called *High energy aspects of solar flares* was published.[36] NASA's mission summary identified its legacy as the discovery of gamma ray ion structures in solar flares and loops; determination of the energy content and spectrum of flare energetic electrons, the measurement of non-thermal emissions from the corona; the detection of twin coronal x-ray sources; the initial downward motion of x-ray sources in flares; detailed hard x-ray and white light imaging of flare ribbons; the location of a super-hot x-ray source (30MK, the flare of 23rd July 2002); and the most precise measurement of the Sun, showing that it is flattened to make it oblate, which may affect the solar cycle.

CHIPS (Explorer 82): From the Local Bubble to Galactic Tertulia

CHIPS (Cosmic Hot Interstellar Plasma Satellite) was a Berkeley University Explorer (UNEX) launched together with ICESAT on 13th January 2003 aboard a Delta II from Vandenberg, to an orbit just over 575km. The aim was to obtain spectra of the diffuse extreme ultraviolet background. CHIPS was selected on 11th September 1998 as the first UNEX mission, following the STEDI missions SNOE and TERRIERS. Selected at the same time was Inner Magnetosphere Explorer (IMEX) to study how the radiation belts responded to the solar wind. Both were set to fly in 2001, but CHIPS was the only UNEX to get airborne. The budget for CHIPS was $13m, bringing a new meaning to the term 'low cost'. It was a suitcase-sized satellite of only 60kg.

CHIPS was originally to launch as a secondary payload with the FAISAT commercial communication satellite on a Cosmos rocket in mid-2001, with Russia now offering commercial launches to generate cash in hard economic times.[37] In 1994, the U.S. government forbade the launch of government payloads on foreign rockets, but both the proposers and NASA had assumed that this would not present a problem, given the small size of the university-based project. They were wrong. In October 1998, the Office of Science and Technology Policy (OSTP) required a review of the project, but eight months later the project was still stalled, undermining the whole purpose of fast turnaround in the UNEX series. In May 1999, the CHIPS team had to abandon the Russian launcher.

This was not as easy as it sounded, as it was challenging to find a primary payload that would have space for CHIPS and happened to be heading into the right kind of orbit. NASA offered the Shuttle, but its orbit was so low that CHIPS would have required its own propulsion system for the final lift to the desired orbit, which was too complicated and expensive a proposition. Eventually, an opportunity arose to fly CHIPS as a secondary payload to 600km on a Delta II mission whose primary payload was ICESAT. The Delta II, though, was not ideal, because to fit CHIPS into the shroud with the right centre of gravity requirement meant downscaling the satellite, reducing its instrument throughput by 50% and working on six rather than nine channels. In the end, CHIPS was a small, three-axis stabilized spacecraft using four momentum wheels, three torque coils, a Sun sensor, a rate sensor and 14v batteries. Solar arrays were mounted on three sides, feeding nickel-cadmium batteries, with some small keep-alive arrays. The PI was Dr. Mark Hurwitz of UC Berkeley.

CHIPS satellite

Both CHIPS and ICESAT were headed for 94° polar orbits, which required a second stage shutdown as the Delta II headed southbound out of Vandenberg, a long coast, then a second stage re-ignition. ICESAT was released first after 64 minutes and then CHIPS at 83 minutes. Its only instrument was a multi-channel nebular spectrograph, focussing diffuse EUV radiation onto a photon-counting micro-channel plate detector. Its purpose was to study the local bubble of hot, million-degree but low-density plasma around the solar system, believed to have come from a supernova between two billion and ten billion years ago. It would also be used to learn how plasma cooled, applying this knowledge to other galaxies beyond our Milky Way. High temperature plasma was everywhere in the universe, but had never been seen in the ultraviolet range. The specific task of CHIPS was to provide spectral maps of the extreme ultraviolet sky in the 90Å to 260Å range. It was intended that a full-sky survey would determine electron temperatures, ionization conditions and cooling mechanisms of the interstellar bubble's million-degree plasma. This was a relatively unexplored part of the spectrum and scientists were uncertain about how hot interstellar plasma shed its thermal energy reservoir.

CHIPS launch

The first CHIPS scientific results were presented not at a mere tertulia, but a galactic tertulia, in Granada, Andalucia, Spain in June 2003.[38] Arriving English-language speakers were perplexed: what was a tertulia? Richard de Grijs did his homework, with the help of Wikipedia and found that tertulias were originally informal, social literary or artistic get-togethers, particularly in Latin cultural contexts, often but not always held in public places. This particular galactic tertulia was defined as 'an open discussion on outstanding questions in contemporary star cluster-related astrophysics'. CHIPS was never able to find extreme ultraviolet radiation from the hot interstellar gas, so it was redirected to look at the Sun. Either the radiation expected was hotter or cooler than the range surveyed, or it did not exist in the first place.

CHIPS was switched off on 11th April 2008 after five years of operation. It would have cost another $100,000 to keep it going for the rest of the year and that was not available, to the annoyance of some of the scientists involved. According to Mark Hurwitz, the whole mission had been run on a shoestring. The Cosmos fiasco had not helped but despite that CHIPS had achieved good results.

GALEX (Explorer 83): How Galaxies Evolved

GALEX, or GALaxy Evolution Explorer, was launched on 28th April 2003 from a Pegasus XL off Cape Canaveral. It was a 280 kg Small Explorer (SMEX) – although counter-trend it came in 52.8% over budget. The launch was delayed from March when a fastener and clip were found loose, which, if they had fallen into the inside of the spacecraft could well have wrecked the mission. The mission was formally concluded ten years later in June 2013, having presented a wealth of data on stars, black holes, galaxies and dark energy.

GALEX was selected at the same time as HESSI, October 1997, as a two-year SMEX mission to explore the origin and evolution of distant stars and galaxies with an ultraviolet telescope. The cost was set at $65m with a Pegasus launch set for 2001. An alternate mission was identified should either GALEX or HESSI not reach launch – the Broadband Observatory for the Localization of Transients (BOLT), which aimed to pinpoint the location of gamma ray bursts.

It was estimated that the universe had about ten billion galaxies, most arranged in clusters. The Milky Way, for example, belongs to a cluster of 40 of the Local Group and in turn is part of the Local Supercluster, with voids in between. The task of GALEX was to make the first all-sky survey of galaxies, compile a map of them and help inform our understanding of star formation. It was to observe up to a million near and distant galaxies in ultraviolet, focussing on the early stages of star-forming galaxies across ten billion years of cosmic history. It would sweep the skies for 28 months with ultraviolet detectors, looking for galaxies dominated by young, hot, short-life stars, white dwarfs and quasars. With the age of the universe then calculated at 13.7 billion years, the main period of star formation was calculated at eight to ten billion years ago, but new precision was hoped for. Specific scientific objectives were to learn what triggered star formation in galaxies and determine how fast they formed; follow star formation, evolution and death, estimating their speed of evolution; track the formation of heavy chemical elements (i.e. more than hydrogen or helium) in stars; create an ultraviolet map of the universe; and study ultraviolet bright quasars. GALEX was designed to make eight surveys, grouped into all-sky, 150 nearby galaxies, wide field, medium spectroscopic, deep imaging, deep spectroscopic, ultra-deep imaging and medium imaging.

The spacecraft was a 1m wide, 2.5m tall aluminium structure with 290w power from two solar wings of $3m^2$. It was designed to survey in near ultraviolet in the 180 to 300nm range and far ultraviolet in the 130 to 180nm range. GALEX had a 128MB computer and a 24GB solid state recorder. Four reaction wheels were used for stabilization, but it had no thrusters. It had two transmitters: S-band at 2MB/sec and X-band at 24MB/sec.

Galex

The principal instrument on GALEX was a 50cm ultraviolet telescope with two ultraviolet detectors (near and far ultraviolet) and a spectrometer to break down the light into its constituent colours and chemical elements (e.g. carbon, oxygen). It was designed to operate only in Earth's shadow, recharging during daylight but never pointed at the Sun or Earth. The telescope was a Cassegrain design – named after French sculptor Guillaume Cassegrain who invented it in 1672 – and was not that different in concept from small amateur telescopes. Its field of view was 1.25°, about three times the diameter of the Moon. Other countries involved were the Republic of Korea (Yonsei University) and France (CNES). Data went to CalTech, Pasadena, being archived under a system developed by Johns Hopkins University in Baltimore. The PI was Chris Martin at Caltech, the location of the mission science centre.

GALEX belonged to a class of wide-area surveyors like Swift and WISE and in time surveyed millions of galaxies in ultraviolet across their ten billion years, fulfilling its mission brief by accumulating data to suggest the ways in which they evolved. Specifically, GALEX found a huge comet-like tail behind a speeding star (Mira); the remnant of a star shredded by a supermassive black hole; giant rings of new stars around old, hitherto presumed dead galaxies; a black hole devouring a star; and confirmed the concept of dark energy suggested earlier by WMAP.

GALEX image Messier 84

The spacecraft was handed over to CalTech on lease in 2012 so that it could continue to be operated on private funds. It was eventually decommissioned on 28th June that year, but will orbit another 65 years.

Swift (Explorer 84): The Far end of the Universe

Swift, a collaborative mission with Britain and Italy, was launched on a Delta II from Cape Canaveral's pad 17 on 20th November 2004, its purpose being to detect and analyse gamma ray bursts. Though it was a MIDEX medium explorer, Swift was the size of the one of the 1980s Explorers at 5.6m tall by 5.3m wide and 1,470kg in weight. It came in 68.8% over budget, but the huge scientific reward suggests it was well worth it.

Swift started out as the 'International Gamma Ray Burst mission', its aim being to tell whether a GRB meant the birth of a black hole or the collision of two neutron stars. It was expected to detect about 200 bursts. Swift was among 35 proposals submitted to NASA in August 1998 and was the only one selected in February 1999 according to the criterion of 'best science value'. The concept was then awarded $350,000 for a four-month implementation study to examine cost, management, technical aspects, educational outreach and small business involvement.

The spacecraft cost $250m (including British and Italian contributions) and had a set lifetime of two years. Its mission control was at Penn State University, Pennsylvania, with tracking by the Luigi Broglio Space Centre, Malindi, Kenya, which would receive 6GB of transmissions a day. British participation came from the University of Leicester, which provided parts of the x-ray telescope, as well as the UK Swift Science Data Centre. Much of the optical telescope was built by University College London's Mullard Space Science Laboratory. Italian participants were the Italian Space Agency and Brera University, Milan. The PI was Neil Gehrels at Goddard. Swift followed HETE, with mission objectives to determine the origin of gamma ray bursts; classify them and search for new types; follow the development of GRB explosions; use GRBs to study the early universe; and perform the first sensitive hard x-ray (also called soft gamma ray) survey of the sky. Swift had three instruments:

- Burst Alert Telescope (BAT), covering 1/6th of the sky at a time, to detect and locate a GRB;
- X-ray Telescope (XRT), to capture the spectrum of the GRB;
- Ultra Violet/Optical Telescope (UVOT), to capture the ultraviolet and optical properties of the afterglow (first detected in 1997 by the small Italian satellite BeppoSAX). This was a 30 cm telescope with a precision of one arc second, similar to that used on ESA's XMM Newton mission.

Swift's breakthrough was its ability to swivel its telescope quickly to follow a GRB. Bursts could be as short as milliseconds to as long as 100 seconds. After a burst, the position would be determined in 20 seconds to within 1–4 arc minutes and the telescope would repoint toward the object within 75 seconds. Swift was able to react within an average 52 seconds to a newly-discovered GRB: 'no satellite turns faster' said the promo.[39] It followed what was called a 'lights-out monitoring' regime, which kept routine operations cost at a minimum while still enabling it to respond to astronomical discoveries on a 24/7 basis. Its detector was three times more sensitive than that loaded on the Compton Gamma Ray Observatory (GRO).

Swift arrived at Cape Canaveral by truck on 29th July 2004 and had a perfect launch into an orbit of 584km – 604km, 20°, once four hurricanes had finally dissipated. Control was handed over to the Mission Operations Center in State College, Pennsylvania some 80 minutes after launch. The first four months were spent turning on the instruments and calibrating them. First light was on 5th January 2005 and Swift caught its first GRB later that day, although 5th April 2005 was the official first day of operations. From early on, Swift began detecting GRBs at the rate of two a week, yet there would also be gaps of up to a month when none might be detected, followed by rapid activity: 94 new GRBs were found in the first year, so the previous discovery rate had effectively doubled. This was a great start and Swift won the 'best of what's new' award in *Popular Science* for 2005.

Swift was followed by a duty team in Britain, Italy and the United States, each of whom would get a text message by mobile phone when there was a burst. By August 2015, the team had received 977 GRB texts and it was expected that they would pass the 1,000 mark by 2016. It was estimated that there was a GRB in the universe every day, so with an average of 90 a year, Swift was picking up a quarter of them. They came in irregularly – there might be three a week and then a quiet period.[40]

Swift in orbit

Swift scientists issued a lively newsletter which conveyed some of the excitement and enthusiasm around the mission. It carried many pictures of galaxies of all kinds of wonderful shapes, sizes and colours, the purple whirlpool ones being the most beautiful. It also kept a diary of key dates and events, such as those noted here.

In May 2005, Swift found its first short burst, defined as less than two seconds. This enabled the mystery of such short bursts to be solved – typically they happen when neutron stars merge – which *Science* magazine rated as one of the 'top five science breakthroughs of the year'. On 4th September, Swift found a long, 200-second burst at the edge of the visible universe, between 500 million and one billion years after the big bang in Pisces, called GRB 050904. This was 12.8 billion light years distant, a new record for distance. Not long after, another was calculated at a distance of ten billion light years. Swift found a burst that continued to generate an afterglow for 125 days, which must have required some kind of on-going injection of energy. On 25th April 2008, Swift found the brightest ever flare, coming from EV Lacerta, a close (16 light years away), young and dim star. It blazed so brightly that Swift's instrument closed down as a safety precaution. It was still shining eight hours later. On 10th June 2007, Swift picked up a five-second burst in Vulpecula (J195509+261406), about 15,000 light years away which then flashed 40 times over the next three days, then stopped for eleven days, flashed once more and disappeared. Swift scientists determined that this was a hibernating magnetar 18km across, one of about twelve known, that experienced a starquake and briefly came to life.

One GRB was as short as a twentieth of a second (GRB 050509B). On 9th January 2009, Swift happened to be observing supernova NGC 2770 when, by fortunate coincidence, an x-ray burst came from the same galaxy, the start of another supernova. GRB 080319B was the brightest object ever seen in the sky, 2.5 million times brighter than anything before, remarkably coinciding with the death of writer Arthur C. Clarke (19th March 2008). GRB 080913 then broke the distance record at 12.8 billion light years distant in September 2008, to be followed by GRB 090423 at 13.035 billion light years in April 2009, when the universe would have been only 630 million years old. Six days later, GRB 090429B beat this, at 13.14 billion light years when the universe was 520 million years old. In March 2010, Swift logged a record four bursts in a single day and it passed the 500 mark of GRBs the following month.

Swift image of Andromeda

Some GRBs were ultra-long, with emissions for hours, while others were as short as two seconds and some even less.[41] Scientists had the impression that short gamma ray bursts were likely to come from colliding pairs of orbiting neutron stars, while longer bursts, more than two seconds, were likely caused by massive stars collapsing and then exploding. These events were so violent that any planet within 100 light years would likely have its atmosphere blown off. Gamma ray bursts emitted carbon, nitrogen, iron and gold into space.[42]

The GRB universe kept on growing. On 18th February 2006, Swift picked up a long GRB of more than half an hour some 440 million light years away, associated with an adjacent supernova explosion at the same time and visible to astronomers around the world. On 14th June 2006, Swift found a new, hybrid class of GRB, one which exhibited the characteristics of a short burst but lasted for 100 seconds – more typical of longer ones – only 1.6 billion light years away in the constellation Indus, which then disappeared without a trace. A burst detected on 29th July 2006 left an afterglow that could still be detected a record four months later, suggesting a continued source of energy there.

Swift also turned its attention to comets. It imaged the break-up of comet Schwassman-Wachmann 3 – astronomers counted 66 fragments – and the imaging revealed hydroxyl molecules, evidence of sunlight making water dissolve. In May 2014, Swift picked up a fresh comet, Siding Spring (Comet C 2013 A1), as it hurtled into the solar system and passed 138,000km from Mars in October 2014. It found Siding Spring 2.46AU away in the constellation Eridiani as it first began to sublimate and generate water at the rate of 49l a second, modest compared to some comets. The previous closest cometary encounter was Lexell which came within 2.3 million km of Earth on 1st July 1770.[43] Swift followed Comet Garradd as it emitted large amounts of dust and gas. On 4th July 2005, Swift observed NASA's Deep Impact probe hit Comet Tempel 1 and the aftermath of the impact in the days that followed. Amazingly, the brightness of the comet grew for two days, as the debris filled the comet's coma and then subsided.

In summer 2008, Swift turned to the Andromeda galaxy, one of the few just faintly visible to the naked eye from Earth, taking 330 images of 24-hour exposure time amounting to 85GB. Andromeda (M31) is the largest, closest spiral galaxy, 2.5 million light years away and 220,000 light years across. Swift made a new portrait of Andromeda possible, finding 20,000 ultraviolet sources there, typically hot young stars. Next, Swift identified a model middle-weight black hole, NGC 5408-X-1 some 15.8 million light years away in Centaurus. This was important because most of those found to date were either heavyweight or lightweight. The April 2011 issue of *Nature* reported how a distant galaxy's dormant black hole (J1644 + 57) had awakened and consumed a star 3.9 billion years ago. All galaxies have a black hole at their core, but this one was twice as big as the one in the Milky Way. It was exceptionally bright, largely because the jet of the flare was pointed at Earth (a 'beamed emission'). In 2011, Swift swivelled to Cygnus OB2, a collection of 3,000 hot and massive stars 4,700 light years away and in particular the binary OB219, one of which was 50 times the Sun's mass, the other 45, separated at their closest by only three astronomical units. X-rays flared up between them fourfold at their closest approach.

Individual examples of its discoveries are given. Swift detected magnetar SGR 1806-20 some 50,000 light years away, so powerful that it could erase a credit card half the distance from the Moon. In April 2013, Swift found a new magnetar in the galactic centre. A year later, Swift recorded a blast from a nearby red dwarf 10,000 times more powerful than the greatest solar flare. In 2008, Swift spotted small, dim binary red dwarves in DG Canum Venaticorum, each about a third of the mass of our Sun. Over two weeks, seven massive solar flares were observed erupting from one, the hottest being 10,000 times more powerful than anything from Earth's Sun. This binary system was unusually young, only 30 million years old. Traditionally, such episodes lasted only about a day.[44]

At 2.32pm on 15th June 2015, Swift was the first satellite to detect a burst of high energy rays on V404 Cygni, which quickly became the brightest object in the sky, fifty times more so than the Crab nebula. V404 Cygni had not been this bright since 1989, when it had been observed by Japan's *Ginga* x-ray satellite and the *Mir* space station. Swift alerted other orbiting spacecraft, such as MAXI (Monitor of All-sky X-ray Imaging) on the *Kibo* module of the International Space Station and the European-Russian *Integral*. Called a low-mass x-ray binary, V404 Cygni began emitting repeated bright flashes from a few minutes to several hours. Searches through archival data suggested two-decadal gaps between outbursts, with those prior to 1989 being in 1938 and 1956 and with quiet periods in between. It appeared that over time, the black hole would attract in material from its orbiting star, gather it in a disc and reach a tipping point when it was heated up and exploded in powerful jets of particles. These outbursts were rare, once or twice in a lifetime events, but were now being emitted at 70 flares a week. The V404 Cygni binary comprised a black hole and star orbiting one another in our Milky Way 8,000 light years distant. Later images from Swift showed rings of x-ray light, like ripples from a stone in a pond, with the concentric rings expanding and then contracting, all resulting from a large flare on 26th June.[45]

Swift image of Triangulum

Next, Swift, with colleagues operating the Japanese *Suzaku* observatory, classified a new form of obscured Active Galactic Nuclei – those obscured by so much gas and dust that no optical light can escape. Two were found, 80 million and 350 million light years from Earth. In June 2007, they then found an eight-millisecond pulsar, a binary pair of a neutron star or pulsar with a low-mass companion white dwarf orbiting each other every 54.7 minutes across a distance similar to the Earth and Moon. The white dwarf had been the size of the Sun, but lost its mass to its companion and was now ten times the size of Jupiter. After two weeks, it faded and disappeared.

On 19th March 2008, Swift observed GRB afterglow so bright that it could also be seen in visible light, some 100,000 times brighter than the norm, coming from a distance of 7.5 billion years in the direction of constellation Bootes (though no professional or amateur astronomer observed it, the detections being by telescopes operating automatically). At once, automated telescopes in Chile and Texas swivelled automatically to capture it. The light came in a very narrow jet that happened to be pointing right at Earth. It was bright enough to be seen by the naked eye (magnitude 5 or 6) and it was the brightest GRB ever detected. Swift was then turned on η Carinae, where it found x-ray fluxes coming and going every five years, possibly a function of the orbit of the two binaries passing one another or one of them eclipsing an x-ray region.[46] On 9th January 2008, its telescope then caught two supernovae exploding at the same time, the new one for the first time. Under the direction of Professor Julian Osborne of the University of Leicester in England, Swift was used in 2016 to look for the x-ray and ultraviolet emissions associated with recently-discovered gravitational waves, but did not find them, indicating to scientists the challenge of how to narrow down the methods for searching for them.[47]

In 2014, Swift was rated one of NASA's most successful scientific missions and its operations were extended to 2016. It led to 2,000 papers and 40,000 citations, about half in the GRB field. In its first ten years, Swift made 315,000 observations of 26,000 targets, including 6,200 Targets Of Opportunity (TOO) requests from 1,500 scientists. It studied 300 supernovae and also picked up comets and asteroids. The telescopes were also used to survey more than 700 active galaxies and the monster black holes within them. Some 2,200 snapshots were taken to map the Large Magellanic Cloud, 163,000 light years away, in ultraviolet. Swift catalogued 2,000 hard x-ray sources. Swift science has been outstanding and addressed the heart of the *Origins* question raised in the 1980s. It was fair to claim that Swift revolutionized the study of GRBs.

THEMIS (Explorers 85–9): Five Satellites Find Magnetic Reconnection

THEMIS (Time History of Events and Microscale Interactions during Substorms) was a wordy title for an unusual mission (Themis was the goddess of justice in Greek mythology). The purpose of THEMIS was to investigate the Earth's aurora, the solar wind, the lunar wake and the distant magnetotail with five satellites. It was the largest number of satellites ever launched together by NASA (the USSR had begun an 8-in-1 series as far back as 1970). The $200m mission involved Germany, France, Austria and Canada. Frank Snow was the project manager, Vassilis Angelopoulos was the PI and Craig Pollock the program scientist.

The aim of the mission was to pinpoint the origin of auroral substorms, using more spacecraft than ever before, each with identical instrumentation. Although most were harmless, highly energized substorms had the potential to cause electrical disruption on Earth. Substorms occurred both in larger storms and in isolation, but it was not known exactly what sparked them off. Substorms generally developed so quickly that satellites had not been precisely able to find their starting point. They were frequent – every four hours, more so in spring – the main storms being associated with the solar rotation (27 days), solar maximum and the decline to solar minimum (11 years). According to NASA, there were two competing theories to explain the triggering of magnetic

THEMIS (Explorers 85–9): Five Satellites Find Magnetic Reconnection

substorms: magnetic reconnection and current disruption theories. The specific mission objectives for THEMIS were to establish when and where substorms began, determine how the individual components of the substorm interacted, find out how substorms powered the aurora and identify how local current disruption mechanisms coupled to the global substorms. The idea was that the probes would be released simultaneously and would use their own thrusters to reach orbits of one, two and four days, from a sixth to half-way to the Moon, on the night side of Earth in its magnetic tail.

In these orbits, THEMIS was to map the north American continent magnetically every four days for 15 hours, giving a 3D picture. It was intended that the probes align perfectly over northern Canada and Alaska during these mapping operations. Eleven schools in the northern U.S. installed magnetometers to make simultaneous measurements, whilst ground-based observers would also document the storms from 20 locations in Canada and Alaska with cameras and magnetometers. It was expected that they would observe 30 storms over two years.

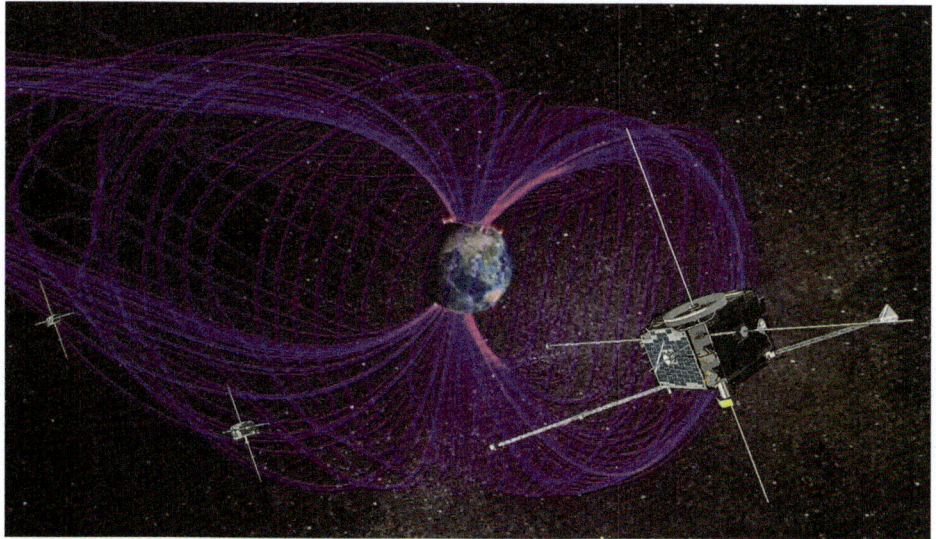

THEMIS formation

The spacecraft were box-shaped, housing two fuel tanks and with four sides of solar panels feeding a lithium ion battery. Their electronics had to be especially hardened, but the power demands were only 36w, less than a domestic 40w light bulb. Each spacecraft weighed 126kg, including 49kg of fuel for four hydrazine thrusters. Downlink was transmitted at 1 Mbps on S-band. The tape recorder had five days of storage and kept a data volume of 400MB a day. The spacecraft were spun at 20 rpm. Each had two 5m booms, four 20m booms, one 1m and one 2m, their overall purpose being to measure electric fields in different directions. The spacecraft were designed to change orbit to be able observe magnetic storms from different heights: 1 from 210,000km (R34); 2 from 120,500km (R19); 3 and 4 from 67,600km (R11.6); and 5 from 77,200km (R13.1), as well as from different inclinations: 1 from 3.9°, 2 from 9.8°, 3 and 4 from 7° and 5 from 12°. The five instruments were:

- Flux Gate Magnetometer, from the Technical University of Braunschweig, Germany;
- Search Coil Magnetometer, from the French Centre des Environnements Terrestre et Planetaire, to measure magnetic field fluctuations;
- Electrostatic analyser, to measure thermal electrons and ions, tracking their flow;
- Sold State Telescope, to sense the expansion of the heated plasma sheet; and
- Electric Field Instrument, all from University of California, Berkeley.

THEMIS five probes together

THEMIS was launched by Delta II 7925 from Cape Canaveral on 17th February 2007. The THEMIS satellites duly circled Earth in what might be called a string-of-pearls formation. Observations and discoveries came quite quickly. THEMIS observed a two-hour storm on 23rd March 2007, determining the level of energy released in a storm to be in the order of 500,000 billion joules, equivalent to a scale 5.5 Earthquake, suggesting the existence of magnetic ropes connecting the magnetic field to the solar wind. On 26th February 2008, THEMIS probes were able to determine, definitively, the triggering event for the onset of magnetospheric substorms – the purpose of the mission. Two probes, a third of the

THEMIS (Explorers 85–9): Five Satellites Find Magnetic Reconnection

way out to the Moon, found what appeared to be a magnetic reconnection event 96 seconds before auroral intensification, so 'magnetic reconnection *is* the trigger', concluded Vassilis Angelopoulis. THEMIS had discovered a flux rope pumping a 650,000 amp current into the Arctic. Magnetic ropes connecting Earth's upper atmosphere directly to the Sun confirmed the theory of solar-terrestrial electrical interaction postulated by Kristian Birkeland in 1908 (Birkeland currents).

In August 2008, the THEMIS team published its initial findings from the mission, based on observations of the storm of 26th February that year: auroral storms were caused by magnetic reconnection, when the Earth's magnetic field lines storing energy from the solar wind were stretched far out into space and snapped, like a rubber band that has been pulled too far. The electrical currents were then recreated, a process that triggers the substorms. Further THEMIS results were presented to a plasma workshop in Huntsville, Alabama in October 2008, where debate centred on Flux Transfer Events (FTEs) and examined the processes of reconnection in these magnetic cylinders stretching into the sky. The principal outcome was that the solar wind, far from trickling steadily down toward Earth, arrived in the course of dynamic, short and sudden events. THEMIS also discovered that Earth's magnetic field often developed two holes that allowed leaks of solar particles.

A second stage of the mission began three years after launch, when two of the THEMIS spacecraft were re-routed to the Moon to observe the solar wind from far outside the magnetosphere. These were renamed ARTEMIS (Acceleration, Reconnection, Turbulence and Electrodynamics of the Moon's Interaction with the Sun) P1 and P2. This involved a complicated series of 90 manoeuvres, the first use of the Earth-Moon libration point and risking the spacecraft in darkness for eight hours when their batteries had been designed for only two. Direct manoeuvres to lunar orbit would have used up all the fuel available very quickly, so subtle, fuel-efficient gravity assists were required. The first step was to increase the apogees of the orbit, which already reached half-way out to the Moon. The second was to reach the Lagrangian or L point, which P1 did to the farside of the Moon on 10th August 2010 and which its P2 companion did to the nearside point on 22nd October. There, both were well placed to study the cavity that the Moon made in the solar wind, although P1 lost the end of its 30m boom which either fell off or was knocked off. P1 then joined P2 on the nearside L point on 9th January 2011. Eventually, ARTEMIS P1 (THEMIS B) entered a 3,543km by 27,000km lunar orbit on 2nd July 2011 by firing its engine to slow it by 50m/sec. ARTEMIS P2 (THEMIS C) entered lunar orbit two weeks later on 17th July 2011.

Results continued to come in. When solar particles, including Coronal Mass Ejections (CMEs) reach Earth, they are fended off by the protective magnetosphere, although some get through and descend to Earth. On 17th January 2013, Goddard identified a moderate solar storm when there was the possibility that the plasmasphere might send up a protective plume to protect Earth's environment. Thankfully, three THEMIS spacecraft happened to be flying 45 minutes apart through this phenomenon and were able to see the way in which cold dense plasmaspheric material stretched up to the magnetic reconnection point where the CME met the magnetopause, providing Earth with an additional layer of protection for several hours. First results were published by Brian Walsh in *Science Express* on 6th March 2014.[48]

The THEMIS spacecraft detected a range of electromagnetic waves, such as chorus waves, hiss, Ultra Low Frequency waves and EMIC (Electromagnetic Ion Cyclotron)

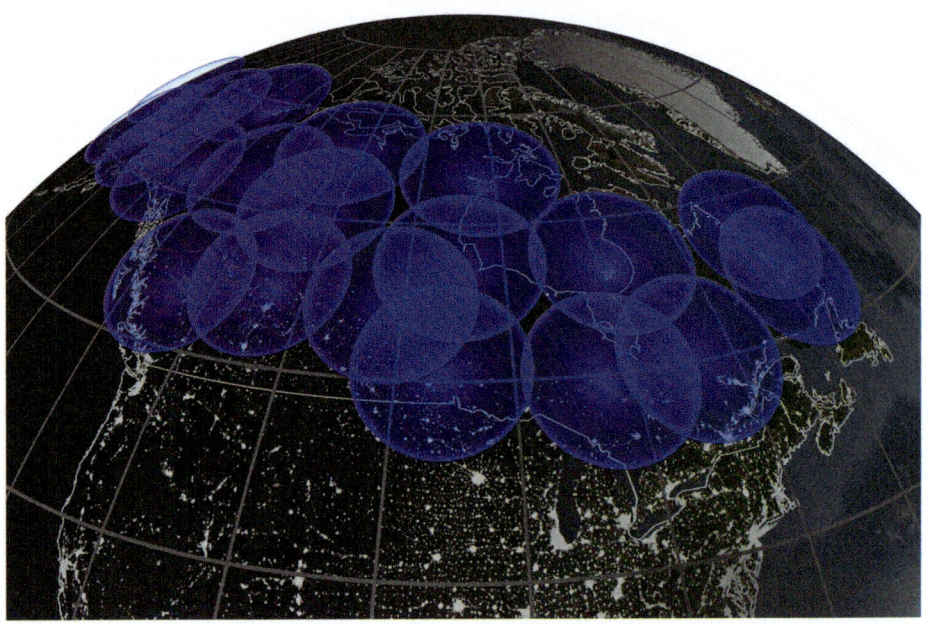

THEMIS network of ground stations

waves, as well as during solar storms, particle drop outs and sudden losses throughout the magnetic system. In the view of PI Vassilis Angelopoulis, it was evident that the solar wind was powering the magnetosphere 'even by seemingly innocuous rotations in the magnetic field'. THEMIS data were subject to detailed analysis by Galina Korotova at Russia's long-standing research centre for the magnetosphere, IZMIRAN in Troitsk. She found that even small perturbations in the solar wind could have large effects near Earth, with wave-particle interactions in the solar wind in the turbulent region upstream from the bow shock acting as a gate valve, causing undulations throughout the magnetopause and energizing particles in the Van Allen radiation belts.

THEMIS travelled through more than 50 solar storms in its first six years. When the THEMIS spring science working group members gathered to analyse results in Fairbanks, Alaska, in March 2013, they were rewarded with a dazzling night-time display of aurorae at an outdoor temperature of -40 °C and a 74-page *Results* report was published.

In November 2016, Lynn Wilson and David Sibeck of Goddard published important findings in *Physical Review Letters*, reporting that the foreshock – the protective barrier at the outer edge of Earth's magnetic field – was accelerating electrons up to the speed of light for short periods, up to a minute. Hitherto, their behaviour had been explained by collisions making them bounce around the bow shock, but there was no evidence that this was happening. THEMIS indicated that there was an unknown source within the fore-shock itself to make this happen.

Aim (Explorer 90): Polar Ice

The purpose of the Aeronomy of Ice in the Metosphere (AIM) mission was to investigate polar ice clouds, especially noctilucent clouds, scientifically known as Polar Mesospheric Clouds (PMCs). The ninth SMEX mission, AIM was approved in 2002, along with SPIDR, for a 2006 launch. The cost was $140m, including launch.

AIM in fairing

AIM was the first mission dedicated to investigating the mysterious ice clouds at the edge of space in Earth's polar regions, generally 80km high in the mesosphere. Such clouds had first been reported in the aftermath of the eruption of Krakatoa, east of Java in Indonesia, in 1885. First satellite-borne observations dated to 1969 and in the Explorer program to the Solar Mesosphere Explorer. Little was known about what formed them, nor why they were becoming more common, frequent and visible at lower latitudes. Generally, they appeared from mid-May to mid-August in the northern hemisphere and from mid-November to mid-March in the southern. They were noctilucent, meaning that they were illuminated by sunlight over the horizon and could only be seen at night. They seemed to be composed largely of water ice or forms of frozen ice crystals, although how they could form at such an altitude was a mystery. Explanations ranged from cosmic dust, to gravity waves, to a combination of cold temperatures and water vapour. Their increasing frequency led to theories of a connection to global warming: they could act as a warning 'canary in the mine'. There was even speculation that they were sparked off by the exhaust of the Space Shuttle (though no pattern had been established to connect them to Shuttle launches). Noctilucent clouds had been observed on Mars, but were made of carbon dioxide rather than water ice. The mission's duration was expected to enable AIM to follow two entire seasons at both poles.

AIM was a LEOStar model spacecraft weighing 195kg, of which 64kg was the instrumentation. The intended orbit was sun-synchronous at 600km, 97.7° and the spacecraft was due to begin its mission immediately after entering orbit, when the transmitter would activate and send its first messages via the TDRS system. The six solar panels of its single array would deploy soon afterwards. Data transmission was 1.3GB a day on S-band. The Mission Operations Center was the Laboratory for Atmospheric and Space Physics (LASP) at the University of Colorado, Boulder. The University of Hampton, Virginia would house the data centre, which was the location of the PI, James Russell. The Center for Atmospheric Sciences was founded there in 1986 and he and his colleagues were considered the most knowledgeable experts on the phenomenon. James Russell had a background at NASA Langley, was involved in the Gemini and Apollo programs, worked on the Nimbus 7 weather satellite and had published more than 350 papers by the time of this mission. Hampton had an unusual history, being founded as a black college in 1868 for freed slaves and there was a high level of student involvement in the mission. There was also an international partner in the form of the British Antarctic Survey. AIM was to measure air pressure and temperature, moisture content and the dimensions of these clouds and had three instruments:

- Solar Occultation for Ice Experiment (SOFIE) to estimate the temperature, gases and chemicals of the clouds;
- Cloud Imaging and Particle Size Experiment (CIPS), which had four cameras to make daily panoramic images;
- Cosmic Dust Experiment (CDE), to record dust entering the atmosphere from the cosmos, to see if there were any connections to polar clouds.

AIM was launched on a Pegasus from Vandenberg on 25th April 2007 and made its first cloud observations on 25th May, such clouds being seen by ground observers over northern Europe on 6th June. AIM went on to provide the first global view of the clouds in the northern hemisphere – for the whole 2007 season – to a resolution of 5km by 5km. AIM found that the clouds appeared every day, were widespread and were highly variable by hour and by day. They were ten times brighter than previously measured. Their main layer was 83km to 90km, with a continuous layer of mesospheric ice. They had complex features typical of normal tropospheric clouds.

The mission was so promising that it was extended in May 2008 until September 2012. According to analysts in the University of Colorado, these clouds were the highest in the atmosphere and were made of tiny ice crystals. Improbably, there were long-distance connections between those over the north pole and those over the south pole, later called teleconnections. Northern stratospheric winds caused the southern mesosphere to become

AIM image of Arctic clouds

warmer and drier, producing fewer such clouds, but when they picked up again, the southern mesosphere became colder and wetter and the clouds came back. Going through the records, they found a statistical link between cold weather over north America and the clouds over Antarctica.

Turning to the southern hemisphere, AIM was able to observe the first clouds of the 2013–4 Antarctic summer appearing on 20th November, expanding in bluish colours to cover the continent in five to ten days. According to James Russell, the trigger was methane greenhouse gas rising from the lower atmosphere, oxidized to form water vapour and then mixed with the smoke of meteors burning up in the upper atmosphere. AIM may indeed have proven to be an effective 'canary in the mine' for greenhouse gases.

But was there a Shuttle connection? As it headed into orbit on its last mission on 8th July 2011, the Shuttle released 350 tonnes of water vapour exhaust, as it normally did. Its plume blew across into the Arctic remarkably quickly – within 21 hours – where it formed noctilucent clouds full of ice particles detected by AIM. The plume spread over 3,200 to 4,000km, with the clouds forming at 88km altitude. The outcome was reported in the *Journal of Geophysical Research* on 27th August 2012. This episode had the advantage of measuring a known release of known composition at a known time with subsequent observations, showing that noctilucent clouds could be human-formed as well as illustrating the speed of high-altitude winds.

IBEX (Explorer 91): Where the Solar System Meets the Stars

IBEX (Interstellar Boundary Explorer) was an unusual mission, for although it was stationed in Earth orbit, its purpose was to explore the boundary of the solar system with interstellar space. IBEX was launched by Lockheed L-1011 from Kwajalein Atoll on 19th October 2008, with a Pegasus XL putting it into a highly elliptical initial orbit of 219–250,281km, 11°. Its target orbit was so distant that it required IBEX to have its own ATK Star 27 solid rocket motor for the final stage. The motor raised its orbit to one of 7,000km out to 320,000km, nearly out to the Moon. Later, it moved into a more stable, lunar-resonant orbit called P/3.

The purpose of IBEX was to make the first comprehensive map of the boundary between the solar system and interplanetary space, beginning at twice the distance of Pluto, between 84 and 94 times the distance of the Earth to the Sun. The place where the solar wind begins to slow down and interact with the interstellar medium is called the heliosheath, which has three parts: the termination shock (the innermost part of the boundary), which was the principal focus of IBEX; the heliopause (the outermost part of the boundary); and the part in between the inner and outer boundary. Voyager 1 crossed into the heliosheath at 94AU but Voyager 2 did so at 84AU three years later, suggesting that it expanded and contracted.[49] It was hoped that IBEX would operate for eleven years, a complete solar cycle. The mission manager was Gregory Frazier, a veteran of COBE, Swift and XTE/Rossi.

IBEX was developed by the company that built the Pegasus itself, Orbital Sciences Corporation (OSC), since renamed Orbital ATK (and now often referred to simply as 'Orbital'). Orbital was contracted to develop IBEX in 2005 for the Southwest Research Institute at San Antonio, Texas, with PI David McComas. The spacecraft was the hexagonal drum LEOStar which had already flown 45 times. The satellite weighed 110kg, including 30kg of hydrazine fuel and 26kg for instruments. Power was 85w and downlink 320kbps. The mission control was at OSC in Dulles, Virginia, while science operations were conducted from the Southwest Research Institute, San Antonio and Boston University. IBEX carried two narrow-angle neutral atom-detecting imaging sensors, IBEX lo (10eV to 2keV) and IBEX hi (500eV to 6keV), working with a Combined Electronics Unit to make a full sky map every six months. On orbit checkout took 45 days.

Within four months, David McComas was able to say that IBEX was going to produce great science. That autumn, the Chicago planetarium presented the show *IBEX - search for the edge of the solar system*. IBEX compiled the first global maps of the heliosphere in Energetic Neutral Atoms (ENAs), along with measurements of low energy oxygen, hydrogen and helium atoms coming from interstellar regions. It picked up the charged particles deriving from electrons and ions coming in from the galaxy that were captured on arrival in the heliosphere and sent inward into the solar system and Earth. The surprising discovery, which puzzled scientists for many years, was of a distinct ribbon of ENAs, henceforth called 'the ribbon', a narrow 20° arc that stretched across much of the sky that protected the solar system from the rest of the galaxy. These were possibly interstellar magnetic field lines marking the solar system's direction of travel through the Milky Way. The new IBEX model of the interstellar magnetic field showed it to warp around the heliosphere, even in a perpendicular. Some of the cosmic rays coming in from the galaxies were 12 times more energetic than photons of visible light from the Sun.[50] The heliosphere appeared to be egg-shaped, with interstellar magnetic field lines wrapping themselves around it.

IBEX

IBEX on right, motor on left

By summer 2014, IBEX had completed ten full sky maps of the heliosheath. The maps focussed on ENAs, neutral hydrogen atoms formed in the heliosheath by the charge exchange of solar wind protons with interstellar hydrogen. ENAs declined by 35% over the five years and showed considerable variability with energy and location. The decline was probably connected to a 40% drop in solar wind dynamic pressure in the course of the solar minimum at the time.[51] Moreover, IBEX's instruments saw how, when solar wind arrived at Earth's magnetosphere 56,315km out, the interaction created ENAs which then spun away in all directions.

By this stage, IBEX had generated 169 scientific papers.[52] Results so far were presented at the COSPAR 40 meeting in Moscow in August 2014, with IBEX complementing the data sent back by Voyager as it headed into interstellar space. IBEX was able to provide information on the speed of the solar system and its heliosphere through the local interstellar cloud and the flow of the local interstellar medium (18 to 23 km/sec). More IBEX results were published in October 2015, in a supplement to the *Astrophysical Journal Supplement*.[53] The main outcomes focussed on the heat and direction of the interstellar flow.

Explaining the ribbon continued to be a challenge for mission scientists Nathan Schwadron and David McComas.[54] In March 2016, scientists speculated that the 'IBEX ribbon' which streamed particles back into the solar system were solar particles that had reached the outer heliosphere, undergone complex changes and been sent back, a process that took three to six years. In effect, solar wind protons reached the boundary region, gained an electron which made them neutral, then passed through the heliopause but lost the electron again. This made them gyrate, pick up another electron and travel back inward.

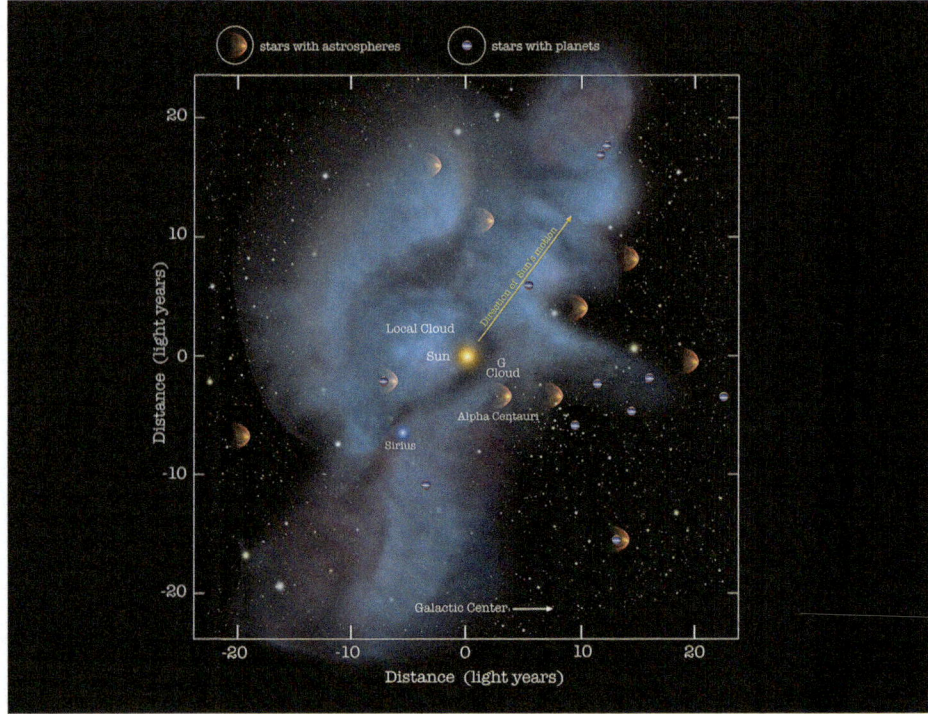

IBEX model of cosmic neighbourhood

The mission was subsequently extended to 2017. The most recent decadal survey had already recommended a successor probe to build on the achievements of IBEX, called the Interstellar Mapping and Acceleration Probe (IMAP) – a $500m mission with higher sensitivity and resolution – and a tribute to the new science opened up by IBEX.

WISE and NEOWISE (Explorer 92): Asteroid Hunter

WISE was NASA's Wide-field Infrared Survey Explorer, launched on a Delta II 7320 from Vandenberg into a 525km sun-synchronous orbit, 95 minutes at 97.5°, on 14th December 2009. It was originally proposed as a mid-infrared all-sky survey in the 1990s, was selected in 2002 and eventually acquired the WISE name in 2003. WISE survived a 50% budget cut in 2006, which led to a 17-month launch delay and the final mission cost was $320m. This was the last MIDEX mission in a series that had comprised FUSE, IMAGE, WMAP, Swift and THEMIS, with a cost cap of $180m, one which WISE clearly exceeded. Like the others though, it was worth it. WISE was also one of the last launches in the Delta II program: the military moved on to the more capable Delta 4 and Atlas V and then the cheaper Falcon 9, making the low production runs for the Delta II no longer profitable. This left the Explorer program with Pegasus, which necessarily limited its experimental payload to 450kg, for which WISE, at 661kg, would have been too big.[55]

WISE, later NEOWISE

WISE was the first dedicated survey of the mid-infrared region, its chief attribute being its ability to use infrared to see through dust. It was intended to find brown dwarfs, sometimes also called failed dim stars, some of which might lie beyond the edge of the solar system, in the Oort cloud far from the Sun. So far, a few hundred brown dwarfs had been discovered, but they were very difficult to see. WISE was expected to double or treble the number and possibly reveal whether they hosted planetary systems. It would also survey active galactic nuclei, or galaxies with active black holes at their core and try to obtain new information on dark energy. Additional discoveries were expected with new stars, planetary debris, the structure of the Milky Way and galaxy clusters. IRAS had discovered Ultra Luminous Infrared Galaxies (ULIRGs), where new stars were created at a rapid rate in the younger universe when it was three billion years old. These were likewise difficult to see in optical telescopes and were obscured by dust, so it was hoped that WISE would find more.

WISE was also expected to detect asteroids as small as 3km across, being visible by their movement against the stellar background. The pictures would be sent to the Minor Planet Center in Cambridge, Massachusetts, for refinement of their orbits. Infrared imaging provided a more consistent idea of the size of an asteroid, because visible imaging had difficulty distinguishing between large, dark objects and small but bright ones, each of which could give out the same level of light. WISE was also expected to improve our understanding of how the combination of rotation and sunlight gradually changed an asteroid's orbit, called the Yarkovsky effect.

WISE was 2.85m tall, 2m wide at its eight-sided base and had the appearance of a thermos flask. For orientation, it had two star trackers, four reaction wheels and 14 Sun sensors. There was a fixed high-gain antenna and one solar panel, 2m by 1.6m, with 684 cells generating 551w. Its main instrument was a 40cm telescope with infrared cameras in four wavelengths. This weighed 347kg, the camera being frozen by a cryostat with 15.7kg hydrogen at 8K. WISE had the capacity to take 7,500 images every day and cover each position in the sky eight times. The strategy for WISE was to take images every 11 seconds as the satellite precessed around the celestial sphere over six months, compiling 50GB of raw data a day leading to an all-sky catalogue. Its aim was a thousand times better sensitivity than IRAS, made possible by advances in detector technology. The infrared cameras had a sensitivity of a million pixels each (1,032,256 to be precise) compared to 62 on IRAS. The aim was to collect a million images and catalogue half a billion objects.[56] The telescope was to obtain images of 2.75 arcsec resolution and operated in four mid-infrared bands – 3.4, 4.6, 12 and 22 microns, compared with 3.5 to 4.9 microns (COBE) and 12–25 microns (IRAS). This was a wide-angle survey mission, in contrast to Spitzer and Herschel, which focussed on only 1% of the sky at a time. WISE was intended to provide a source catalogue for the James Webb Space Telescope to come.

The mission duration was for one and a half sky surveys over ten months. It was divided into checkout (one month); whole sky survey (six months); and a second but partial scan to uncover more objects and determine changes (three months). At that stage, the hydrogen cryostat was expected to warm. WISE was managed by JPL, with Edward Wright as the PI at UCLA. Transmissions would go via the TDRS system, at the rate of 100 MB/sec four times a day to its ground station in White Sands, New Mexico for onward relay to JPL. CalTech established a WISE Science Data Center for cataloguing and subsequent on-line archiving and distribution.

WISE separated after 55 minutes, deployed its solar panel after 150 minutes, powered up its instruments five days later and opened its telescope after 16 days. First light was 3,000 stars in the constellation Carina. WISE began with a six-month, four-channel, all-sky baseline survey, the beginning of a successful career. Its orbit was designed in such a way that its telescope should never be pointed toward the Sun, but its orbit was likely to evolve by 2017 into one that would put it in too much sunlight to function effectively.

WISE and NEOWISE (Explorer 92): Asteroid Hunter

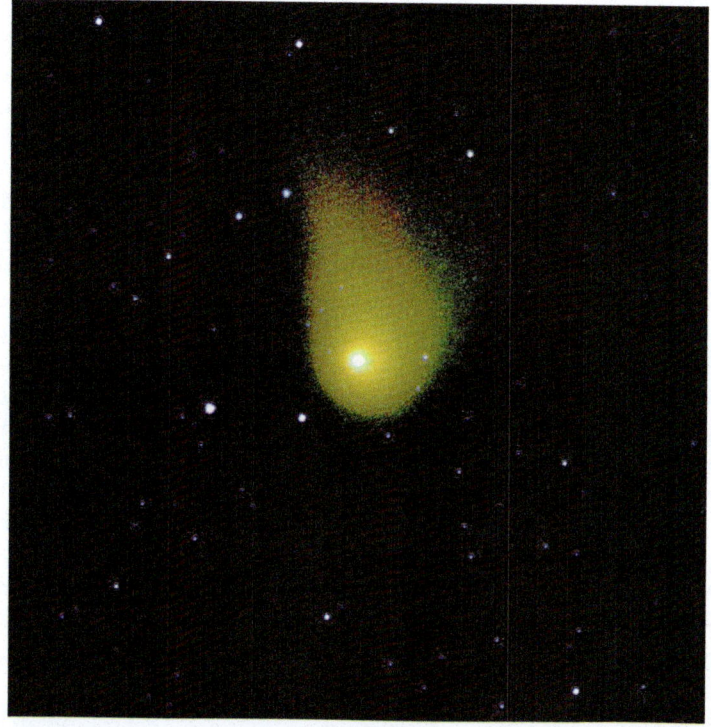

WISE Comet Kristensen

WISE duly made 7,500 images a day during its primary mission from January 2010 to February 2011, a total of 2.7 million images. It observed 158,000 rocky bodies, discovered 21 comets, observed 34,000 main belt asteroids and discovered 133 near-Earth objects (defined as asteroids coming within 45 million km). WISE scanned the sky twice as planned, but its coolant ran out as expected in October 2010. Fortunately, two of the four infrared detectors were still viable at the warmer temperatures. Its main mission concluded in February 2011 when it was decommissioned ('finally' turned off, said the announcement). The first batch of WISE data, covering slightly more than half the sky, was released to the astronomical community in spring 2011, the rest in 2012 (it was called ALLWISE). The catalogue had data on over 747 million objects drawn from the 2.7 million images and was designed to be open to anyone, not just professionals or astronomers. This database became the basis of future research. For example, in 2016, astronomers matched WISE data against blazars (compact quasars associated with supermassive black holes) investigated by the Fermi GLAST gamma ray telescope.[57]

As an example of its infrared outcomes, red giant WISE J180956.27−330500.2, discovered in 2010, stood out because of its infrared glow. Its most interesting feature, though, was that it was now 100 times brighter than photographs taken from the ground in 1988 and had not been visible to IRAS at all. The glow came from when it exploded, leaving behind large amounts of dust which the star was still heating, causing it to glow. Writing in *Astrophysical Journal Letters*, Poshak Gandhi and Issei Yamamura of the Japan Aerospace Exploration Agency (JAXA) described these dust eruptions as rare, once in 10,000 years and that only one other was known, called Sakurai's Object.

Normally, this stage would have marked the end of the mission, but WISE was now reinvented as NEOWISE with asteroids becoming the second main focus of the mission. There were two reasons. First, Near Earth Objects (NEOs) were increasingly perceived as a threat to Earth. In 2013, one had slammed into the atmosphere over Chelyabinsk, Russia and some debris had reached the ground. Although there were no fatalities, the worrying fact was that it had not been spotted on the way in. Second, NASA was trying to identify targets such as an asteroid for a manned Orion spacecraft mission in the late 2020s. With its coolant at 75K, WISE's two infrared channels were still sufficient for asteroid hunting. Asteroids are difficult for amateur astronomers to spot because they are so dark, but they would glow in WISE's telescopes. Unlike light, which is a poor indicator of size (small bright asteroids can appear to be as big as large dark ones), infrared signatures correspond accurately to size. As a result, WISE was renamed NEOWISE in autumn 2013 and charged with hunting for NEOs that might come close to Earth, characterizing 150 known NEOs and finding up to 2,000 more.[58] The PI for NEOWISE was Amy Mainzer of JPL.

The first new NEOWISE discovery was asteroid 2013YP139 on 29th December 2013, which it spotted several times before it moved out of field of view. YP139 was 43 million km from Earth at the time, about 650 m in diameter and dark as coal. Initial observations of its orbit suggested that it was tilted to the plane of the solar system, but with the potential to come within 500,000km of Earth, bringing it into the class of 'hazardous'. Its discovery was immediately relayed to the Minor Planet Center, which passed it on to the community of amateurs and professionals that track asteroids.

WISE asteroid Eurphrosyne

In its first 25 days of operations, NEOWISE found 857 minor bodies and 22 NEOs, an average of almost one a day. At that time, there were 10,500 NEOs, but only 10% had been characterized, a number NEOWISE was expected to double. On 14th February 2014, NEOWISE found a new comet, one in a retrograde orbit at a distance of 230 million km, coming in from the outer depths of the solar system. In its primary mission, WISE had already found 21 comets, but this was the first after coming out of hibernation. In its first full year as NEOWISE, it discovered and characterized 40 NEOs, 245 previously known near-Earth objects and three new comets, as well as observing 32 other comets.

WISE found the 300m diameter asteroid 2010TK7 in the L4 point ahead of Earth's orbit, the first – and so far only – Earth Trojan to be found, orbiting the Sun along with the Earth. Trojans are asteroids in stable orbits similar to those of the terrestrial planets (L4 in the case of Earth, L5 in the case of Mars). The first of seven L5 Trojans, asteroid 5261 Eureka, was discovered in 1990.[59] WISE data suggested to scientists that there might be as many as 28 distinct new asteroid families. Analysis suggested that centaurs, small objects orbiting between Jupiter and Neptune, were old comets rather than asteroids. WISE was unable, though, to compute which asteroid class might have been responsible for the extinction of the dinosaurs 65 million years ago.

By the end of 2015, NASA reckoned that it had found 98% of the 13,000 NEOs then estimated to orbit in our vicinity. NEOWISE had estimated the size of 8,000 asteroids, including 201 near-Earth ones, by that time. Now NEOWISE scientists were able to make an assessment of the 163 comets that the spacecraft had observed, the largest infrared survey of comets.[60] Although observations of recent cometary visits to the inner solar system had detected high levels of water sublimation as they headed through perihelion passage, NEOWISE suggested that the predominant form of sublimation further out and in lower temperature was carbon dioxide (CO_2) and carbon monoxide (CO), both difficult to detect from Earth because of their abundance here. The two carbons were especially observed by NEOWISE in comets further than 600 million km out, four times the distance of Earth to Sun and beyond the orbit of Neptune, as well as in long-period comets that took more than 200 years to circle the Sun. In another discovery, Comet Hartley 2 was found to be leaving a trail of golf-ball-size grains behind as it orbited the Sun. By mid-2017, NEOWISE had taken 7.7 million images to characterize 693 Near Earth Objects (NEOs), of which 114 were new and some potentially hazardous. One discovery was of a comet (C/2010 L5 WISE) which made sporadic tails.

Although its main mission was over, NEOWISE was still able to return to some of its original WISE assignment. WISE made a huge range of individual discoveries. One remarkable picture was that of the Helix nebula in a field of view crossed by the track of at least one asteroid. The Helix had an orange core (with a white dwarf at its centre) set in a turquoise disk. The expelled material from the dwarf will eventually fade away, a dying system. In summer 2014, WISE found a new star only 7.2 light years away, the fourth closest from the Sun (α Centauri, the nearest, is four light years). It was a dim brown dwarf, the coolest ever found, its temperature like the north pole, as cool as -48 °C. Called WISE JO85510.83-071442.5, it was too cool to be detected by visible light telescopes because brown dwarves lack the mass to burn nuclear fuel and radiate starlight, but had sufficient glow to be picked up by WISE. The discovery of such a star so close to our Sun drew attention to the fact that the inventory of Earth's closest stellar neighbours was still quite incomplete.[61] WISE also found a new class of cold stars, whose temperatures were as cool as the human body.

At the other extreme of light, in May 2015, WISE found the most luminous galaxy ever, with the light of more than 300 trillion suns, a member of the new class of Extremely Luminous Infrared Galaxies (ELIRGs). It was catalogued as J224607.57-052635.0.[62] The discovery was published by lead author Chao-Wei Tao in *Astrophysical Journal*. The scientists speculated that at its heart was an immense black hole 12.5 billion years old, dating to the birth of the universe at 13.8 billion years, sucking in gas, heating it and then blasting out light at enormous temperatures. WISE observations provided fresh data on star formation, early galaxies in the local universe, massive galaxies and obscured quasars. Some 32.9% of bright galaxies had recent star formation.[63]

Finally, NEOWISE was used to search for planet X, considered to be a large object in the Edgeworth Kuiper belt, possibly a brown dwarf three to ten times the mass of Jupiter. Just as Uranus and Neptune had been found by 18th and nineteenth century astronomers trying to explain orbital perturbations in the solar system, so too had their modern equivalents noted the clumping of comets in particular regions, most likely the outcome of the presence of a large, dark body somewhere between 40,000 and 100,000 AU distant. Its existence had been re-proposed by CalTech astronomers Mike Brown and Konstantin Batygin. Irish astronomer Eamonn Ansboro believed that there might be two such objects there, one in the inner Oort cloud and one in the outer. NEOWISE offered the best opportunity for discovery and from 2015 onward, NEOWISE data were being used to try search for this elusive, dim object. No trace was found, initially at least.[64]

NuSTAR (Explorer 93): X-rays on a Mast

NuSTAR (Nuclear Spectroscopic Telescopic Array) was launched by Pegasus XL from Kwajalein Atoll on 13th June 2012. This was a JPL mission to search through cosmic dust for high-energy x-rays (like hospital x-rays) and black holes with unprecedented detail in the medium to hard x-ray band (2–80keV), 200 times more sensitive than RHESSI. Its objectives were to census black holes; map radioactive material in young supernovae; study the birth of stellar elements and how stars exploded; and observe their jets and cosmic accelerators. Although designed for astrophysics generally, it could also be turned toward our Sun. NuSTAR had an uneven start. Although its optics and detector technologies had already been tested on the High Energy Focussing Telescope on a balloon in 2005, it was abruptly cancelled in February 2006 because of funding shortfalls. It was restarted in September 2007 when it became clear that the James Webb space telescope would not fly any time soon. It was the eleventh SMEX and cost $170m, including launch.

The delay was actually fortuitous, as NuSTAR needed a mirror of 3,000 segments and the reinstatement came just as Will Zhang had perfected the technique of making super-thin curved mirrors. NuSTAR was an unusual configuration, because its main instrument was a 10m girder-like mast which took 25 minutes to deploy. It was based on the experience of the Shuttle Radar Topography mission, STS-99, which had successfully deployed a mast in Earth orbit. The telescope of NuSTAR was a light bucket that looked like a polished artillery canister, as fragile as an eggshell, with 133 parallel cylinders made of paper-thin glass. Instead of being reflected toward a focal point, light exited through these cylinders onto the nuclear telescopic array 10m back, which would collect and focus the hard x-rays. The most difficult part of its fabrication was the manufacturing of the cylinders, which required high-precision computers, layers of material a micron thick with epoxies and glue, all done in a dust-free room that was once home to a cyclotron particle collider. It was 'not for the faint-hearted.'[65] The 9,000 mirrors had to be nested at slight, almost parallel angles of shells as small as 0.2032mm each, inside each other in order to concentrate on the same focal plane. They were coated with platinum, carbon, tungsten and silicon at the Danish Technical University Space Centre. The girder had to extend perfectly: in 50 pre-flight tests, it never failed once, but space history is full of examples of equipment that worked perfectly on Earth but not up there when it mattered. A lot was riding on the extension, as NuSTAR had a 100-member astronomy team ready to go.

NuSTAR mast

In collecting the x-rays, the team would get a picture of neutron stars, clouds of space dust and black holes, especially those at the centre of our galaxy where only a small number had been discovered so far. NuSTAR held out the promise of images ten times crisper than anything before, able to penetrate dim, supermassive black holes as well as ones hitherto undetected. It would also be able to locate supernova remnants and collapsed stars. The only comparator was the Chandra observatory, which likewise had a 10m focal length telescope. According to PI Fiona Harrison, who had championed it for over ten years, NuSTAR offered a 50 to 100 times improvement on previous telescopes and was the first satellite to focus hard x-rays in the 5 to 80keV band.[66] NuSTAR was designed to search well above the range of the big Chandra (<10keV).

The basic spacecraft measured 1.2m by 2.2m (but 10.9m when extended), had a total weight of 350kg and used a LEOStar 2 bus built by Orbital in Dulles. The solar panel was 2.7m^2 in area, generating 729w to feed two lithium ion batteries. The Mission Operations Center was in the University of California Berkeley, but the principal ground station was

Malindi in Kenya, operated by the Italian Space Agency, because of its location close to the equator. JPL was responsible for mission management and Goddard for program management. NuSTAR's orbit was 575 by 600km, but it hugged a 6° inclination in order to avoid the South Atlantic Magnetic Anomaly, although it would still skim it from time to time.

NuSTAR Pegasus launch

Once in orbit, the optics were calibrated against the Crab supernova remnant and bright quasar 3C273. NuSTAR quickly generated a scientific return. In October 2014, NuSTAR found the brightest pulsar ever, a dead pulsating star with the energy of ten billion suns. Originally categorized as an Ultra Luminous X-ray (ULX) source, it turned out to be a pulsar 12 million light years distant in the Messier 82 galaxy, emitting signals every 1.37 seconds.[67] Early in 2015, NuSTAR spotted the collision of two galaxies 134 million light years away, together called Arp 299 and each with a supermassive black hole in the middle.[68] The right pair was devouring gas while the left pair was dormant or hidden behind dust, a process of accretion, when a black hole drags in gas and builds up its mass. NuSTAR was ideal for this discovery, for high-energy x-rays can penetrate thick gas. The two scientists involved were Andrew Ptak and Ann Hornschemeier. The possible presence of these black holes had been postulated by NASA's Chandra and Europe's XMM Newton, but they operated in lower x-ray energy bands.

In May 2015, Steve Boggs of the University of California Berkeley published the findings of observations of the earlier supernova 1987a, which had been a dramatic supernova explosion 166,000 light years distant. Then, telescopes on Earth and in space (*Chandra, Kvant*) found it emitting cobalt 56. In 2012, Europe's *Integral* had found titanium 44. The most important finding from NuSTAR was that the emissions were asymmetrical, sending out material in a lopsided direction, leading to the categorization of 1987a as a type II core-collapse supernova. Titanium 44 was ejecting at 2.6 million km/hr. in one direction, with the core of the supernova, a neutron star, in the other. NuSTAR also made observations of Cassiopeia A, finding it similar.[69] In summer 2015, NuSTAR was pointed at nine galaxies where supermassive black holes were suspected but obscured by gas, detecting five of them.[70] In March 2016, NuSTAR x-ray imaged spiral galaxy NGC 1068 some 47 million light years distant, finding a rough torus of thick gas and dust clouds surrounding the supermassive black hole at its centre.[71]

Although other space-borne deep space instruments might have been damaged by looking at our Sun, the high x-ray emissions of the Sun were not considered bright enough to pose a risk to NuSTAR and held out the promise of detecting x-ray flashes.[72] NuSTAR's initial images were able to pick out sunspots and it was hoped that later images would identify nano-flares (charged particles and high-energy radiation smaller than solar flares and dark matter particles called axions) to enable better understanding of the nature of the hot temperatures of coronas.[73]

NuSTAR pulsar (centre)

In summer 2016, NuSTAR, together with ESA's X-ray observatory XMM Newton, contributed to the confirmation of a gravitational vortex around a black hole, in line with Einstein's theories of gravity.[74] The discovery was attributed to Adam Ingram, then writing a PhD at the University of Amsterdam while observing Quasi Periodic Oscillations

(QPOs) in black holes. The target was black hole 1743-322, which was observed for 260,000 seconds with XMM and 70,000 seconds with NuSTAR. In 2016, Fiona Harrison received the COSPAR Harrie Massey award for her achievements with NuSTAR, with some of her cited achievements including mapping the Cassiopeia A remnant, finding the first pulsar in our central black hole and determining that the ultra-luminous x-ray source in galaxy M82 was a hundred times greater than its theoretical limit.

In January 2017, Rafaella Margutti published an article in *The Astrophysical Journal* based on NuSTAR observations of supernova 2014C, in a spiral galaxy between 36 and 46 million light years distant, which appeared to defy all the well-established rules of how supernovae should behave by transforming itself from a hydrogen-scarce type I supernova into a hydrogen-rich type II supernova. In March 2017, NuSTAR solved the mystery of the dominant source of high x-ray emissions in the Milky Way's nearest galaxy, Andromeda, determining it to be a pulsar.

IRIS (Explorer 94): Into the Chromosphere

IRIS (Interface Region Imaging Spectrograph) was a SMEX mission launched just over a year after NuSTAR, on 27th June 2013. Its purpose was to study the solar atmosphere in the interface region between the photosphere and the corona, with a telescope to enable continuous observations of the Sun for two years. The Pegasus XL was dropped from its carrier 140km NW of Vandenberg off the California coast near Big Sur, at an altitude of 8,000m. IRIS was set for a 620km – 670km, 97.9° polar orbit for two years, designed to complement the work of the Solar Dynamics Observatory and Japan's *Hinode*. It began work after 30 days of engineering checks and 30 days of calibration.

IRIS being loaded

IRIS (Explorer 94): Into the Chromosphere

The purpose of IRIS was to focus on the Sun's lower atmosphere to see how solar material moved there and generated solar wind and ultraviolet radiation. It was the first mission designed to observe the range of temperatures of the chromosphere and transition region simultaneously, not only at very high resolution, but rapidly, every few seconds. IRIS was intended to enable better understanding of how energy and plasma moved from the Sun's surface (the photosphere) through the chromosphere layer of the Sun's atmosphere to its outer corona, observing the physical forces in the largely unexplored area near the its surface, a fundamental challenge in solar and heliospheric science. The task of IRIS was to provide significant new information to increase our understanding of energy transport into the corona and solar wind and provide an archetype for all stellar atmospheres. It was hoped that IRIS would provide key insights into the solar drivers of space weather from the corona to the far heliosphere, resolving in space, time and wavelength the dynamic geometry from the chromosphere to the low-temperature corona and the physics of this magnetic interface region. IRIS aimed to understand how internal convective flows powered atmospheric activity, what types of non-thermal energy dominated the chromosphere and beyond, how the chromosphere regulated mass and energy supply to corona and heliosphere, how magnetic flux and matter rise through the lower atmosphere and the role of flux emergence in flares and mass ejections.

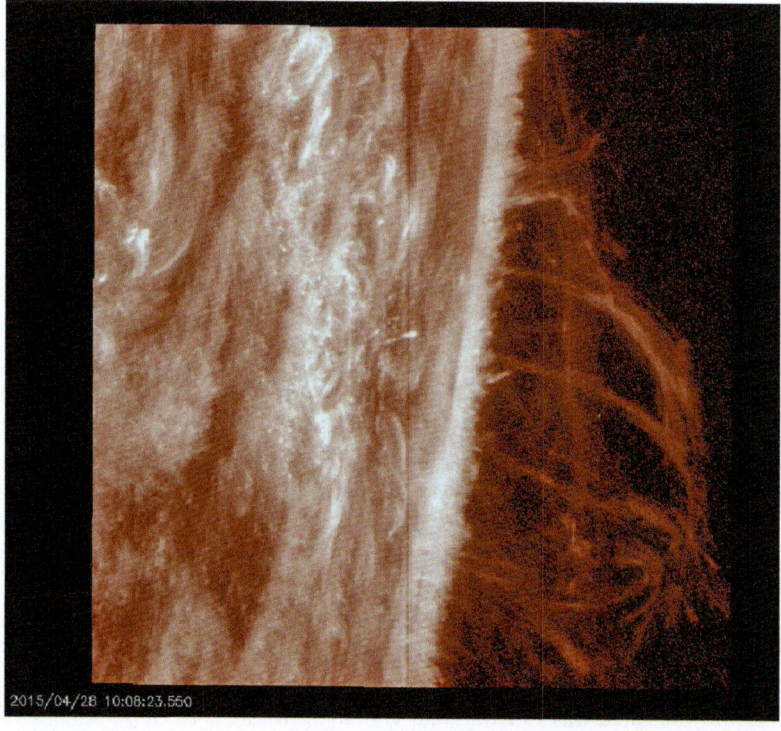

IRIS solar prominences

The mission was led by NASA's Ames Research Center, Moffett Field, California, with science operations guided by Lockheed Martin Solar and Atmospheric Laboratory in Palo Alto, California and processing and archiving by Stanford University in California. The ground stations were in Svalbard, Norway; Fairbanks, Alaska; McMurdo, Antarctica; and Wallops Island, Virginia. The satellite was 2m tall and weighed 183kg. It was shaped like a telescope with a base and two solar panels 3m across. Data were transmitted by x-band at 15 Mbps over 15 downlink passes, or a total of 60GB a day. IRIS carried one instrument, a multichannel imaging spectrograph with a 20cm ultraviolet telescope able to follow temperatures from 6,000K to 1mK. It operated over 1,332Å to 1,406Å and in the near ultraviolet in the range 2785Å to 2835Å. The Harvard Smithsonian Center for Astrophysics in Cambridge, Massachusetts built the telescope, while Montana State University designed the spectrograph.

IRIS quickly captured dozens of solar flares, one x-class flare and a Coronal Mass Ejection (CME). Its imaging took both photographs and spectrographs which enabled scientists to measure temperature, velocity, density and composition. It found that CMEs sped away from the Sun at over two million km/hr. and followed sunspot-ejected material that accelerated outward and then rebounded at much faster speeds. On 29th March 2014, IRIS observed an x-class flare on the right-hand side of the Sun, a first. IRIS tackled what was called the coronal heating problem, of how the solar corona reached temperatures as high as one million °C. With Japan's *Hinode*, IRIS scientists concluded that the corona was heated not by solar flares but by magnetically driven waves, a process called resonant absorption.[75] Interpretations of IRIS data by this stage suggested that coronal heating was not even, but was instead the outcome of bomb-like events in which the magnetic fields of the corona clashed and explosively realigned. In 2015, the IRIS mission was extended to September 2018, with a view to two possible further extensions.

Conclusions

The 'faster, better, cheaper' period covered the first SMEX (Explorer 68) to IRIS (Explorer 94), some 26 missions altogether over 1992-2016. This was a much improved rate of launching ('faster') than the thin period of the mid-1970s to the end of the 1980s, although one of these missions belonged to that earlier period, the delayed XTE/Rossi. While some missions exceeded their budget caps during this period, they were balanced by those that came in under cost and the many that were on target ('cheaper'). Explorer funding rose in this period, from a starting point of $82.1m in 1989 to $109.1m in the year of Goldin's accession (1992). Although NASA's budget contracted, Explorer funding held up, $115.8m the following year to $117.5m in 1997. The cost of individual missions fell substantially, the prime objective. Goldin had also set down the principle of 'permission to fail' (although not permission to try again, except HETE) and two missions were unsuccessful (TERRIERS and WIRE). Of the three new classes of Explorer, most were SMEX or MIDEX, with only two UNEX flown. There was an additional system of Missions of Opportunity (MO). Not only that, but an old launcher had been phased out (Scout, with SAMPEX) and a new one introduced (Pegasus).

Were they 'better', though? Certainly, the scientific results were substantial and vindicated the choice of missions, some of which were speculative and adventurous. SAMPEX located a new radiation belt. SWAS found the universe to be watery, FUSE found that it was far from empty and WMAP that it comprised large amounts of dark matter. Fresh knowledge was collected on galaxies (GALEX), the infrared universe (WISE) and the x-ray sky (NuSTAR). There were significant gains in our knowledge of the Sun (ACE, TRACE, RHESSI, IRIS), the magnetosphere (IMAGE, THEMIS) and noctilucent clouds (AIM). Two missions pushed back the barriers of what was known of the local bubble (CHIPS) and the boundary between the solar system and the heliosphere (IBEX). There were two gamma burst missions, HETE 2 and then the spectacularly successful Swift. The program showed a high degree of adaptability, with two THEMIS satellites sent into lunar orbit and WISE converted into the NEOWISE asteroid hunter as the Chelyabinsk meteor laid bare the deficiencies in Earth's defences. Some missions were technically complex, notably NuSTAR. Compared to the state of depression in the space science community in the late 1980s or early 1990s, these outcomes suggest that the series passed the 'better' test too.

Table 4.1 summarizes the rate of approvals in the 'faster, better, cheaper' period. Table 4.2 summarizes them by category. Note that in both cases, missions were selected, deselected, categorized and re-categorized, so they should be treated with caution. As may be seen, there was a rapid set of approvals in the late 1990s, but no missions were selected for 2003–9.

Table 4.1 'Faster, cheaper, better' Explorer mission selections

Date	Type	Proposals	Selections
1986			SAMPEX, SWAS, FAST, TOMS
			Later TRACE, ACE, IMAGE
1997	SMEX	40	RHESSI, GALEX
1998	UNEX	23	CHIPS, *IMEX*
1998	MIDEX	27	Swift, *FAME*
1999	SMEX	21	*SPIDR*, AIM
2001	MIDEX	21	THEMIS, WISE
2003	SMEX	22	IBEX, NuSTAR
2008	SMEX	32	*GEMS, JANUS, TESS, CPEX,* IRIS, *NICE*
2010	EX		TESS

Subsequent cancellations in italics

Table 4.2 'Faster, cheaper, better' missions summary

Type	Missions	Cost
SMEX	SAMPEX, FAST, SWAS, TRACE, WIRE, RHESSI, GALEX, AIM, IBEX, NuSTAR, IRIS, *ICON, TESS*	<$120m
MIDEX	ACE, FUSE, IMAGE, WMAP, Swift, THEMIS, WISE	$180-200m
UNEX	SNOE, TERRIERS, HETE, CHIPS, TWINS, *GOLD*, NICER	<$15m
MO	HETE 2, CINDI, *Suzaku* (Astro E), TWINS, *Astro H, GOLD, NICER,* Integral	<$55m

Source: National Research Council; Goddard current listing. Italics for in development (except Astro)

The formal Explorer program includes a number of Missions of Opportunity (MO) which may incorporate missions that arose in particular circumstances from the main program (e.g. HETE failure), instrumentation (CINDI) or cooperation with other countries (Japan's *Suzaku*). Some of these missions are reviewed in the next chapter. Overall, the period reviewed in *Chapter 4: Faster, better, cheaper?* may reasonably be considered a period of recovery.

American space science fared better than its original competitor, the Soviet and then Russian Cosmos program which began in 1962. Whereas in the 1990s the United States revived the Explorer program after a low launch rate in the 1980s, the old Cosmos program effectively came to an end during the Russian period, with space science being one of the main casualties of the post-Soviet transition. Russia continued to launch small scientific satellites (e.g. *Sergei Vernov*) but no longer in a programmatic framework. Its only large-scale space science mission was the highly successful Spektr radio observatory, launched in 2011.

A final point concerned the home of Explorer, the Goddard Space Flight Center. Having been through 'refocussing', downsizing, reductions in force and more than one threat of closure, those clouds have lifted and there is no more talk of closing Goddard now. Early in the new century, Goddard opened a Science and Exploration Building, constructed on green principles. The local dump was closed and sealed, but Goddard inserted a gas line into it to extract methane, from which it draws its energy. New trees were planted and no fertilizer is used on the grass – either it grows or it does not. Goddard now has 1200 engineers and 400 PhD scientists. It still has the cradle-to-grave capacity to conceive, design, build, launch and complete missions on its own, within its own gates. This is a capacity shared only by JPL and the Johns Hopkins Applied Physics Laboratory and 'something the country needs to hold onto.'[76]

References

1. Ride, Sally: *Leadership and America's future in space*. 53pp, 1987.
2. *Report of the advisory committee on the future of the U.S. space program.* U.S. Government Printing Office, 1990.
3. Congress of the U.S. Congressional Budget Office: *The NASA program in the 1990s and beyond*. Washington DC, author, 1988.
4. Office of Technology Assessment: *Background paper - NASA Office of Space Science and Applications - process, priorities and goals*. Washington DC, Congress of the United States, 1992.
5. Congressional Budget Office: *The NASA program in the 1990s and beyond*. Washington DC, Congress of the United, States, 1988.
6. Roy, Stephanie: *The origin of the smaller, faster, cheaper approach in NASA's solar system exploration program*. Space policy, 1998.
7. John M. Logsdon (ed): *Exploring the unknown - selected documents in the history of the U.S. civil space program, vol. V, Exploring the cosmos*. NASA, SP 4407, 2001.
8. *Goddard News*, vol. 44, §1 December 1996.
9. *Goddard News*, vol. 2, §4 January 1998.

10. Anselmo, Joseph: *NASA howls over massive budget cut*. Aviation Week & Space Technology, 2nd August 1999.
11. Oberg, Jim: *NASA faster, cheaper but not better*. Posting 7th December 1999, Friends and Partners in space.
12. *Goddard News*, vol. 41, §11 November 1994.
13. From *Exploring the unknown*, vol. VI, chapter 7, there is a section *The new Explorers*.
14. *Goddard News*, vol. 34, §9, September 1988; vol. 35, §5, May 1989.
15. *Cinq nouveaux Midex*, Air & Cosmos, 5th February 1999.
16. Furniss, Tim: *Pegasus - winged workhorse*. Flight International, 13th August 1988.
17. McKenna, James: *NTSB cites confusion in Pegasus launch room*. Aviation Week & Space Technology, 21st June 1993.
18. Asker, James: *Speedy fix eyed for Pegasus XL*. Aviation Week & Space Technology, 3rd July 1995.
19. Fisk, L. *et al*: *An interpretation of the observed oxygen and nitrogen enhancements in low-energy cosmic rays*, Journal of Astrophysics, 1974.
20. *NASA's SAMPEX satellite locates new radiation belt*. Aviation Week & Space Technology, 7th June 1993.
21. Markowitz, A.G; Krume, M; Nikutta, R: *First x-ray-based statistical test for clumpy torus models: eclipse events from 230 years of monitoring Seyfert AGN*. Monthly notices of the Royal Astronomical Society, 12th February 2014; and Reddy, Francis: *RXTE reveals the cloudy cores of active galaxies*. Space Daily, 21st February 2014.
22. Stone, E.C. *et al: The Advanced Composition Explorer*. Netherlands, Kluwer, 1998, 24pp.
23. Strope, David: *Advanced Composition Explorer - lessons learned and final report*. NASA, GSFC, 1998.
24. Matson, John: *Sun down - high energy cosmic rays reach a space age peak*. Scientific American, 6th October 2009.
25. *Microscopic 'clocks' time distance to source of galactic cosmic rays*. Space Daily, 22nd April 2016.
26. Bailey, Scott: *Science instrumentation for the Student Nitric Oxide Explorer*. 11pp, undated.
27. McLaughlin, William: *Submillimetre astronomy*. Spaceflight, vol. 30, §2, February 1988.
28. *A Mission operations and data analysis plan* was presented in 1990 (Melnick, Gary: *Mission operations and data analysis plan*. Cambridge, Smithsonian Astrophysical Observatory, 1990).
29. Smith, Bruce: *SWAS launched on Pegasus XL*. Aviation Week & Space Technology, 14th December 1998.
30. WIRE Mishap Investigation Board: *Report*. GSFC, 8th June 1999.
31. Laher, Russ *et al: Attitude control system and star tracker performance of the WIRE spacecraft*. AAS 00–145.
32. Guenther D.B. *et al: Evolutionary model and oscillation frequencies for α Ursa Majoris - a comparison with observations*, American Astrophysical Journal, 530, 10th February 2000.
33. Powell, Joel: *FUSE*. Spaceflight, vol. 42, no 6, June 2000.
34. DeVorkin, David: *Mystery star*. Air & Space, January 2016.

35. WMAP Science Working Group: *The Wilkinson Microwave Anistropy Probe (WMAP): Five year explanatory statement.* 10th June 2008 (editor, M. Limon).
36. Lin, Robert; Dennis, Brian; Bester, Manfred & Mendez, Bryan: *The Reuven Ramaty High Energy Solar Spectroscopic Imager.* Heliophysics Senior Review, 29th February 2008.
37. Marchant, Will & Taylor, Ellen Riddle: *Status of CHIPS NASA university Explorer mission.* Paper presented at the 14th AIAA/USU conference on small satellites.
38. Hurwitz, Mark; Sasseen, Timothy; Sirk, Martin: *Ongoing results from the CHIPS mission.* Presentation at Galactic Tertulia, Granada, June 1983.
39. Cominsky, Lynn & Wanjek, Christopher: *Swift - International Gamma Ray Burst mission.* NASA, undated.
40. BBC Four: *Sky at Night*, 9th August 2015.
41. *Swift marks ten years of game-changing astrophysics.* Space Daily, 24th November 2014.
42. *Success for burst mission.* Space UK, §42, winter 2015.
43. Readdy, Francis: *NASA's Swift spacecraft tallies water production of Mars-bound comet.* Space Daily, 20th June 2014.
44. Hays, Brooks: *NASA's Swift satellite sees small star ejecting 'super flares'.* Space Daily, 30th September 2014.
45. *NASA missions monitor a waking black hole.* Space Daily. 22nd June 2015.
46. DeVorkin, David: *Mystery star.* Air & Space, January 2016.
47. *Swift satellites used to chase colliding black holes.* Space Daily, 21st June 2016.
48. Fox, Karen: *THEMIS discovers new process that protects Earth from space weather.* Space Daily, 13th March 2014.
49. *NASA's IBEX observations pin down interstellar magnetic field.* Space Daily, 29th February 2016.
50. Fox, Karen: *IBEX helps paint picture of magnetic system beyond the solar wind.* Space Daily, 17th February 2014.
51. Reisenfeld, Dan: *Interstellar Boundary Explorer completes first five years of operation.* Space Research Today, §190, August 2014.
52. *NASA's IBEX and Voyager spacecraft drive advances in outer heliosphere research.* Space Daily, 6th August 2014.
53. *IBEX sheds new light on solar system boundary; IBEX sets standard for understanding galactic material around solar system.* Space Daily, 21st October 2015.
54. Schwadron, N.A. & McComas, D.J: *Spatial retention of ions producing the IBEX ribbon.* The Astrophysical Journal, 764:92, 10th February 2013; and Schwadron, N. et al: *Global anisotropies in TeV cosmic rays related to the Sun's local galactic environment from IBEX.* Science Express Report, 22nd February 2014.
55. Corneille, Philip: *More wisdom with WISE.* Spaceflight, vol. 52, no 2, February 2010.
56. Mainzer, Amanda et al: *Preliminary design of the Wide Field Infrared Survey Explorer.* NASA, JPL, 2004.
57. *WISE, Fermi missions reveal a surprising blazar connection.* Space Daily, 25th August 2016.
58. *Resurrecting WISE for asteroid hunt.* Spaceflight, vol. 55, §11, November 2013.
59. Corneille. Philip: *Asteroids galore!* Spaceflight, vol. 56, §5, May 2014.

60. *NEOWISE observes carbon gases in comets.* Space Daily, 25th November 2015.
61. Dunne, Ann: *Star found only seven light years from Sun.* Astronomy Ireland, June 2014.
62. *WISE discovers most luminous galaxy in universe.* Space Daily, 25th May 2015.
63. Korean National Committee: *Report to COSPAR.* Seoul, author, 2014.
64. Ansboro, Eamonn: *Evidence for a ninth planet in our solar system.* Astronomy Ireland, May 2016; Duffy, Adam: *Planet Nine's properties revealed.* Astronomy Ireland, June 2016
65. Craig, David: *X-ray specs.* Columbia, spring 2007.
66. Harrison, Fiona *et al: The Nuclear Spectroscopic Telescope Array.* 9pp, undated.
67. *NuSTAR telescope discovers shockingly bright dead star.* Space Daily, 11th October 2014.
68. *Will the real monster black hole please stand up?* Space Daily, 9th January 2015.
69. *Star explosion is lopsided, finds NASA's NuSTAR.* Space Daily, 9th May 2015.
70. George Lansbury in the Astrophysical Journal (*Five monster black holes discovered.* Astronomy Ireland, August 2015).
71. *The heart of galaxy NGC 1068,* Astronomy Ireland, April 2016.
72. *Sun sizzles in high-energy x-rays.* Space Daily, 26th December 2014.
73. BBC Four: *Sky at Night,* 9th August 2015.
74. *Gravitational vortex provides new way to study matter close to a black hole.* Space Daily, 19th July 2016.
75. *Hinode, IRIS and ATERUI cooperate on 7-year-old solar mystery.* Space Daily, 25th August 2015.
76. Wright, Rebecca; Johnson, Sandra; Dick, Steven: *NASA at 50 - interviews with NASA's senior leadership.* Washington DC, NASA, 2008, notably interview with director Ed Weiler.

5

Future Missions and Conclusions

This chapter looks at future Explorer missions; instrumented missions; missions that might have happened or were cancelled; and comes to conclusions.

Future Missions: ICON, TESS, IXPE

There are three Explorer missions currently in the pipeline: ICON (2017), planet-hunter TESS (2018) and IXPE (2020). In addition, there is another instrumented mission, GOLD. NASA's $19bn budget request for 2017 included two Explorer missions for later in the decade.[1]

ICON: The Ionosphere from Below

Ionospheric CONnection (ICON) was a $200m Explorer class mission approved in April 2013 to fly in 2017. It will orbit at 550km, 27°. The mission was originally proposed in 2008, but was repurposed as a second proposal three years later.

An underlying assumption of the mission was that hitherto we have tried to explain the nature and movement of the ionosphere from *above*, but gave little attention to how it was influenced from *below*.[2] The source of daytime changes in the ionosphere, both in motion and density, has never been determined, making it the longest-standing puzzle in ionospheric physics. There are extreme changes at low and middle latitudes during even moderate magnetic storms. The ionosphere can be volatile on an hour-by-hour basis and as plasma grows it can even cause Global Positioning System (GPS) errors and electrical outages. The general aim of the mission was to understand the coupling of the planetary ionosphere and the upper atmosphere, as mediated by strong ion-neutral interactions, specifically searching for the source of strong ionospheric variability; the coupling of energy and momentum; and how solar wind and magnetospheric effects modify the internally driven atmosphere-spatial system. The aim was to answer questions regarding the physics behind the unpredictability of the space environment; how the space environment reflects the tropical rainfall season; how plasma grows so dense during magnetic storms; and how space weather seasons differ from Earthly seasons. ICON's payloads, listing the scientists responsible, are:

- Michelson Interferometer (named after the physicist Albert Michelson) for Global High resolution Thermospheric Imaging (MIGHTI) (Christoph Englert);
- Far Ultra Violet (FUV) spectrographic imager (Stephen Mende);
- EUV (EUV) limb profiler for daytime ionospheric height and density (Jerry Edelstein); and
- Ion Velocity Meter (IVM) to measure electrical fields (Rod Heelis).

ICON

ICON uses an Orbital ATK LEOStar 2 spacecraft bus, with the Pegasus XL launcher. It will have a one-month checkout, followed by a two-year lifetime. The instruments require 80w of power, provided by two solar panels and the downlink is 3.3GB/day to Berkeley, Santiago and Wallops. The Mission and Space Operations Control is at the University of California, Berkeley, with backup stations in Wallops Island and Santiago, Chile and ICON also includes an education and outreach program with events in California, Texas and Hawaii. The leader and PI of the project is Thomas Immel of the Space Sciences Laboratory at the University of California, Berkeley. Immel has stated that ICON would 'change the way we think about the Earth-space boundary,' the forces at play there and the dangers of disruption to communications and GPS. In the past ten years, he said, there had been changes in the ionosphere which no existing model could explain. ICON will make stereoscopic samples of the limb from three vantage points every seven minutes.

Scientists from the University of Berkeley Space Sciences Laboratory shipped the four instruments to Utah State University in Logan in March 2016, for integration onto the Payload Integration Plate, connection to the Instrument Control Package and then vibration and thermal testing.

TESS: Planet Hunter

Proposed in 2012 and due to fly in 2018, the Transiting Exoplanet Survey Satellite (TESS) is intended to be the first space-borne all-sky transit survey of exo-planets, be they Earth-size or giants, but with the particular ability to detect smaller – and presumably more Earth-like – planets beside bright stars. Looking for the tell-tale temporary, tiny diminution in brightness as a planet transits, it is expected that TESS will cover 400 times more sky than previously, examining more than 200,000 stars within 200 light years, starting with nearby, bright stars. It is hoped to find about 1,600 exo-planets. To date, most exo-planets have tended to be the larger ones – Jupiter-type gas planets – circling dim stars, but TESS should have between 30 to 100 times more capacity than existing observatories like Kepler. The holy grail of exoplanetology, though, is to find rocky worlds with atmospheres – 50 was the target for TESS – for which closer, brighter stars hold out the best prospects.

The TESS mission was originally researched privately by the MIT's Kavli Institute of Technology for Astrophysics (MKI) Foundation, Google and MIT in 2006. MIT proposed TESS as a SMEX mission in 2008. It was not selected, but was re-proposed in 2010 and selected in 2013. The head of the science team for TESS is Dave Latham, an astrophysicist who attended Carl Sagan's first class and who famously found the first exoplanet, HD 114762 in 1989, together with Israeli astronomer Tsevi Mazek.[3] By the mid-2010s, over 2,300 had been found by the Kepler space telescope launched in 2009 alone. The field of exoplanetology had grown to the point that a conference in Switzerland attracted 400 participants. The PI for TESS is George Ricker, who comes from x-ray and gamma ray astronomy and had been PI on HETE 2.

TESS

TESS will divide the sky into 27 observation sectors, covering the northern hemisphere in the first year and the southern in the second. Its main interest will be rocky, icy planets comparable to Earth in what is called the habitable, 'Goldilocks' zones (not too hot and not too cold). The key breakthrough intended for TESS is not just to see that exoplanets are there – that has already been done – but to get a picture of what they are like. TESS should be sufficiently sensitive to make it possible to compile data on the masses, densities and orbits of exoplanets. The instruments should be sensitive enough to provide information on their atmospheres, clouds, winds, chemical compositions and moons. Later, targets will be sent for further study for the James Webb Space Telescope. Selection will be by the Space Telescope Science Institute, Baltimore, Maryland and the TESS Science Center, while archiving will be the responsibility of the Mikulski Archive for Space Telescopes.

TESS is the eighth Orbital ATK LEOStar 2, weighing 350kg, with two solar arrays generating 400w and is to be launched from Cape Canaveral. Its instruments are four wide field 24° × 24° view CCD cameras, each able to take 16.8Mpixels. TESS will be the first Explorer to use the Space X Falcon 9 commercially developed rocket, which has been used since 2012 to re-supply the International Space Station with the Dragon cabin. TESS will transmit its data for three hours every two weeks at perigee on a high-speed downlink.

Much attention went into choosing the best vantage point for the mission. Ricker was searching for a 'Goldilocks orbit' in Earth orbit – not too near, not too far – to search for exoplanets. The orbit selected (R17 to R59, or 107,000km – 371,000km) was designed for reasonable launch opportunities and to be long-term; stable (many high altitude orbits are vulnerable to perturbations and waltzing around the sky); give maximum unobstructed sky coverage; avoid stray light; and be both good for downlink and benign from a temperature and radiation point of view, as well as avoiding the Van Allen belts. Technically, it is called P/2 lunar-resonant orbit, with a period half that of the Moon's orbit, 13.7 days.

Such an orbit has never been used before and will involve a lunar fly-by. Previous Explorers, such as the IMP series (see chapter 2) had found their Earth-Moon orbits greatly perturbed, but three computer engineers, Daniel McGiffin, Michael Matthews and Steven Cooley, published a paper in 2001 solving this problem. Lunar resonant orbits, set at 90° to the Moon, are very stable, unlikely to be perturbed and require little station-keeping adjustment. The TESS team came across the paper nine years later and was aware that IBEX was already working in a different type of lunar resonant orbit, called P/3. Several years were spent refining the orbit, which can meet all the criteria, tolerate an inclination from 60° to 90° and has the additional benefit of multiple launch opportunities.

IXPE

IXPE was the first of two selections from a request for astrophysics Explorers in September 2014. Fourteen were received and three subjected to additional review, from which the Imaging X-ray Polarimetry Explorer (IXPE) was selected with a view to a November 2020 launch. The $188m mission – which covers everything from design to launch to scientific analysis – will involve the Italian space agency, which will contribute x-ray detectors. Martin Weisskopf of the Marshall Space Flight Center in Huntsville, Alabama, is the Principal Investigator.

IXPE will carry three telescopes which will detect high energy cosmic x-ray radiation vibrating or polarized in the environment of black holes, to learn about their turbulent, extreme, gravitational, electrical and magnetic environments. Because black holes and neutron stars cannot be observed directly, it is all the more important to make instruments that can measure their effects on their immediate environments. It is hoped that IXPE will improve knowledge of black holes, supermassive black holes, cosmic x-rays, neutron stars, pulsars and other exotic objects.

GUSTO Gondola

GUSTO was selected in mid-2017 from this group. However, it is not a satellite, nor an instrument (like NICER), but a balloon gondola, to be launched from the McMurdo station in Antarctica in 2021 for a 100–170 day flight. Costing $140m, GUSTO stands for Galactic/extragalactic Ultra Long Duration Balloon Spectroscopic Terahertz Observatory. It will be equipped with a telescope with carbon, oxygen and nitrogen detectors to map the interstellar medium – the clouds of dust between stars. The PI is Christopher Walker of the University of Arizona, Tucson, with other partners drawn from Johns Hopkins APL, JPL, MIT and the Netherlands Institute for Space Research, SRON. The Russian radio observatory, Spektr R, in high Earth orbit, has already provided maps of interstellar dust clouds and it is hoped that GUSTO will help scientists understand how they form and evolve.

Instrumented Missions, Past and Future

The Explorer program includes a number of instruments fitted to other spacecraft. This can be confusing, for a quick reading can mislead one into thinking that these were stand-alone satellite missions in the Explorer program, which they are not. Past missions are CINDI, TWINS and *Hitomi*, while future missions are two *instruments* flying on other spacecraft under the Missions of Opportunity program, GOLD and NICER.

CINDI

The Coupled Ion-Neutral Dynamics Investigations (CINDI), was an air force mission to understand the dynamics of the Earth's ionosphere and the interaction between electrically neutral and electrically charged gases in the upper atmosphere, along with its influence on the structure of the ionosphere. In particular, CINDI was designed to research giant bubbles of plasma that would rise above the night-time equator and damage radio communications, finding out where, when, how they arose and their potential impact. The PI was Rod Heelis of University of Texas, Dallas, with Robert Pfaff as the project scientist. CINDI comprised two instruments provided for the U.S. Air Force Communication Navigation Outage Forecast System (CNOFS) satellite, whose overall purpose was to predict the behaviour of equatorial ionospheric irregularities which can cause major problems for communications and navigation systems. The aim of CINDI was to investigate the physical connections between the ion and neutral gases that led to and promoted the growth of equatorial plasma structure, in particular the behaviour of F-region neutral winds, neutral atmosphere wind velocity and the charged particle drift velocity in the equatorial upper atmosphere at altitudes between 400km – 860 km. CNOFS was launched on 16th April 2008 by Pegasus. The CINDI mission has concluded, but its data are still under investigation.

CINDI

TWINS

Two Wide-Angle Imaging Neutral-Atom Spectrometers (TWINS) was another project linked to the air force, being launched on the National Reconnaissance Satellite (NRS) spacecraft USA 184 on 28th June 2006 and USA 200 on 13th March 2008 into high orbits. There were two parts, TWINS A and TWINS B, intended to provide new capacity for stereoscopically imaging the magnetosphere. TWINS was based on the MENA instrument on the IMAGE mission and comprised a neutral atom imager in the 1keV – 100 keV range with 4°x4° angular resolution and one-minute time resolution, as well as a Lyman-α imager to monitor the geocorona. The concept was that they would compile 3D images of the structures and dynamics of the magnetosphere from their altitude at two locations.

Astro-H/*Hitomi* SXS

For this venture, a Soft X-Ray Spectrometer (SXS) was fitted to the Japanese astrophysics mission, Astro-H. The Japanese Astro series was focussed on x-ray observations, apart from the infrared Astro-F and the cancelled Astro-G. Jointly developed by NASA Goddard, the University of Wisconsin and JAXA, the Japanese space agency, the first task for SXS was to measure the velocity field of x-ray-emitting gas and winds in clusters of galaxies and their energy output. Second, it was intended to measure metal abundances and observe matter in extreme gravitational fields, such as the event horizon of black holes. Third, SXS was to determine the chemical abundances and velocity structure of supernova remnants to try to understand what made them explode. Astro-H was launched into clear afternoon skies from Tanegashima, Japan on 17th February 2016, on a H-IIA rocket into a 580km orbit, along with three micro-satellites. As is the norm in Japan, it acquired a new name once in orbit, *Hitomi*, Japanese for 'eye pupil'.

Only a month later, disaster struck. On 26th March, communications were lost and ground observers, notably the expert Arizonan satellite watcher Paul Maley, spotted debris – originally five and then eleven objects – with the satellite itself spinning. Its beacon was picked up over Tanegashima on the 28th and what was thought to be its beacon by Santiago, Chile, the following day, but these signals may have been spurious. Although it was possible that the satellite was hit by space debris or even a meteorite, an on-board explosion, decompression or leak of fuel or coolant was the more likely culprit. JAXA announced the news on 5th April and expressed the hope that the spinning would slow, telemetry could be recovered and at least some of the mission resumed. Later in April, Sergei Bilardi, an undergraduate physics engineering student at Daytona Beach's Aeronautic University in Florida, used a 1m telescope able to take 100 samples a second to observe passes of its main bus and fragments, which were typically tumbling and flashing every 2.6 seconds. On the 28th, JAXA came to the conclusion that the solar panels had separated from the spacecraft and finally gave up on it. The mood in the Japanese space science community was gloomy at the loss of such a promising satellite so soon. Three senior executives subsequently volunteered to take a 10% pay cut for four months by way of atonement.

Before it failed, *Hitomi* provided new observations of the Perseus galaxy 250 million light years away with its soft x-ray spectrometer. It took images of slowly moving gas – 150 km/sec – 60 kiloparsecs across the central nucleus of the galaxy. Previously, it had been thought that the core of the galaxy would have a hot, fast moving engine injecting shocks and bubbles into the galaxy and thereby creating turbulence, but instead it was remarkably quiet. Findings were presented in *Nature* by Andrew Szymkowiak, Meg Urry and Paolo Coppi of Yale University, with PI Tadayuki Takahashi of JAXA, in an article called *The quiescent intracluster medium in the core of the Perseus cluster*. So there was at least one scientific outcome.

NICER: Aboard the ISS

The Neutron star Interior Composition ExploreR (NICER) was a new departure for the Explorer program, for it was not a satellite, but an experiment installed on the International Space Station (ISS). Technically, it was an astrophysics Mission of Opportunity within the Explorer program. NICER was an 18-month mission dedicated to the study of the gravitational, electromagnetic and nuclear-physics environments of neutron stars. The idea was to use x-ray timing and spectroscopy, as developed earlier by RXTE, to explore the exotic states of matter inside neutron stars where density and pressure are higher than in atomic nuclei; compare the predictions of theoretical models with the thermal and non-thermal emissions of neutron stars in the 0.2-12keV x-ray energy band; and probe their interior structure, the source of their dynamic phenomena and their particle acceleration mechanisms. NICER's principal instrument was the Station Explorer for X-Ray Timing and Navigation (SEXTANT), designed to view neutron stars in soft x-ray light, with 56 x-ray detector enclosures and as many sunshades. There was a *Star Trek* twist to the mission, because by measuring and timing the arrival of the x-ray photons from neutron stars, it could ultimately devise systems whereby spaceships could explore deep space by using pulsars as beacons. It was beaten to the line by China, however, which launched the first spacecraft for this purpose, XPNav, in November 2016. The principal investigator was Goddard's Keith Gendreau, with other partners being MIT and the Technical University of Denmark. NICER was launched to the space station on the SpaceX CRS-11 Dragon unmanned resupply craft on 3rd June 2017. The Dragon took two days to reach the ISS, where astronaut Peggy Whitson used the station's remote arm to grab and berth the cargo ship. NICER was deployed on 14th June and by mid-July had been calibrated against 40 targets.

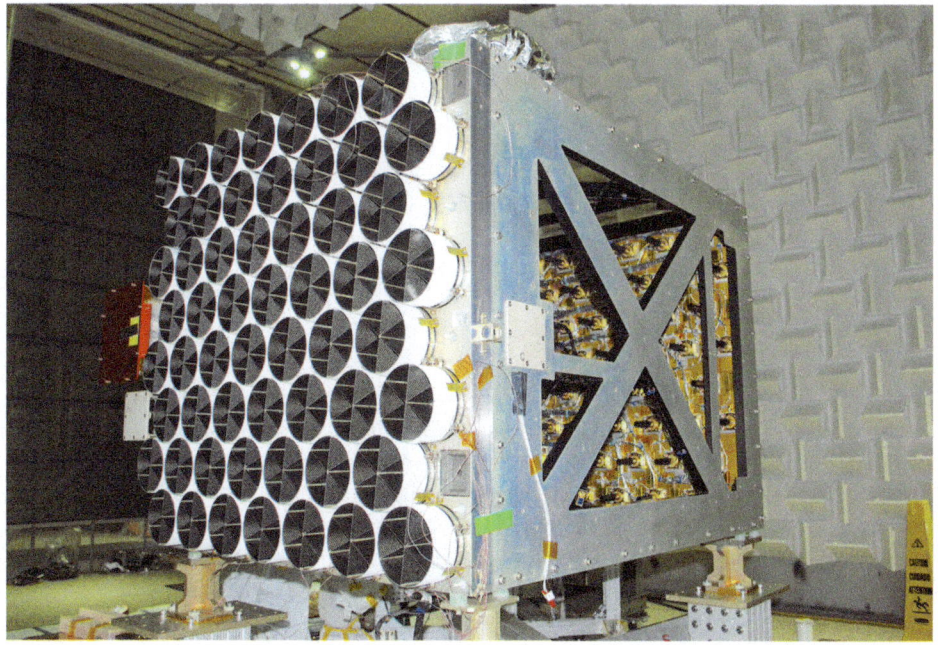

NICER

GOLD

Global-scale Observations of the Limb and Disk (GOLD) is a 28kg instrument due to fly on a 24-hour commercial communications satellite in late 2017. The mission, which cost $55m, was originally selected in 2008 and was reselected in 2013. From its vantage point in 24-hour orbit, it will image the full disk of Earth every 30 minutes day and night to study geomagnetic storms, atmospheric waves and tides, vertical ion drifts and the relationship between the thermosphere and solar extreme ultra violet. It will give readouts on every scan of the oxygen/N_2 density ratio. The imager was built at the Laboratory of Atmospheric and Space Physics (LASP) at the University of Colorado, while the Science Data Center will be at the University of Central Florida (UCF). This is UCF's first mission. GOLD will transmit downlink at 1.6MB/sec.

According to PI Richard Eastes at UCF, global ultraviolet imaging is now the frontier of thermosphere-ionosphere science. Earth's thermosphere and ionosphere have many processes that are still not well understood, affected ('forcing' is the word used) from above (e.g. solar ultraviolet and x-rays, solar wind and magnetosphere, geomagnetic storms, magnetospheric substorms, electric field processes, solar cycle, solar rotation) and below (e.g. turbulence, convection, tides, planetary waves).[4] GOLD is intended to address fundamental questions about how the thermosphere-ionosphere (T-I) system responds to geomagnetic storms, solar radiation and upward-propagating tides, by making temperature and composition measurements on a global scale, rather like weather satellites. Like ICON, it will assist in the understanding of events that cause GPS errors and blackouts.

Missions that Never Flew

The Explorer program also had a number of missions that, for one reason or another, never got to fly.[5]

Owl (1966)

In 1965, Rice University in Texas was given approval to design and build its own satellites as a precursor University Explorer program. Two almost identical satellites were to be launched by Scout rockets from Vandenberg into high-inclination 89° orbits of 930km – 1,100km to study the aurora and Van Allen radiation by day and by night, examining the wavelengths of oxygen (5577Å), nitrogen (3914Å) and hydrogen (4861Å). The satellites weighed 70kg, were 76cm in diameter, 84cm high and had 8,000 solar cells for electrical power. They were cancelled in 1966 following both financial and developmental problems.

MSS (1970)

The MSS (Magnetic Storm Satellite) was a 36kg spacecraft developed between the air force and NASA. It was designed to study magnetic storm and auroral mechanisms and related phenomena by direct measurements in the inner magnetosphere of the low-energy proton and electron flux, the vector magnetic field and the electric field. It was scheduled to be launched in late 1970 on a Scout booster from Wallops Island, but was cancelled.

1998 MIDEX: NGSS, AMM, ASCE, FAME

In the 1998 MIDEX selection, five went for further study. Swift was the first selected and would be the only one to fly. The others were:

- Full Sky Astrometric Mapping Explorer (FAME), which was selected but cancelled;
- Next Generation Sky Survey, (NGSS), a four-channel infrared sky survey;
- Auroral Multiscale Mission (AMM), four spacecraft in near-polar elliptical orbit; and
- Advanced Solar Coronal Explorer (ASCE), a powerful space telescope to study solar processes and explosive coronal mass ejections.

The latter three concluded after the study stage. FAME got the furthest. It was a 1,123kg satellite due to fly on a Delta in 2004 to obtain precision and brightness measurements of stars with a CCD telescope. It was a space astrometry mission intended to measure the positions, proper motions, parallaxes and photometry of no less than 40 million stars brighter than 15th magnitude, with between 50 and 500 micro arc seconds of accuracy. NASA became increasingly doubtful of its feasibility and it was cancelled. The U.S. Naval Observatory and the USAF Space Text program (STP) took it over, but cancelled it in turn.

SPIDR (2003)

SPIDR (Spectroscopy and Photometry of the Intergalactic medium's Diffuse Radiation) was to measure the amount of hot gas (105–106K) found between galaxies in what is called the inter galactic medium (IGM). Its task was to analyse the quantum, distance, spectrum and intensity of these gases (expected to be combinations of ionized oxygen and carbon) to better understand dark matter. Its launcher was to be a Pegasus XL out of Vandenberg with a Star 26C kick motor, but it was cancelled in July 2003 due to instrumentation problems.

GEMS (2012)

GEMS (Gravity and Extreme Magnetism SMEX) was a small x-ray observatory whose purpose was to detect and measure the polarization of x-rays coming from energetic and enigmatic objects, such as ultra-dense neutron stars, stellar-mass black holes, ultra-massive black holes at the centres of distant galaxies, accretion disks, magnetars, dead stars and the remnants of supernovae. In more simple terms, the idea behind GEMS was to discover the shape of space around black holes and magnetars. After two years, the project was approved in June 2009, with Jean Swank as PI and Sandra Cauffman as project manager. GEMS would have the angular resolution and capacity to determine the shape of magnetars, something not done before. Polarized X-ray emission would probe the bending of space, the curvature of light in regions of extreme gravity, the effects of super-strong magnetic fields and how cosmic rays are accelerated by supernova remains. The instrument, an x-ray telescope on a boom to keep it at a distance from its detectors, was to be built by

Goddard and the spacecraft by Orbital. GEMS was 288kg, in the shape of a long cylinder and was due to take a Pegasus XL out of Cape Canaveral in 2014 into an orbit at 575km, 28.5°. GEMS failed its 10th May 2012 confirmation review: it had become too expensive and was taking too much time.

CATSAT

CATSAT (Cooperative Astrophysics & Technology Satellite) was the third STEDI mission and a project of the University of New Hampshire in Durham and Leicester University in Britain, both of which would form satellite operations centres. It was a 168kg satellite that would have used a Delta to fly to a high 500–700km orbit. It was scheduled to carry three instruments to measure Gamma Ray Bursts (GRBs) over 500eV to 5MeV, into the soft x-ray range:

- Soft X-Ray spectrometer (SXR) to measure bursts and their density in the soft x-ray region;
- Hard X-Ray spectrometer (HXR) to measure the hard X-ray region and search for cyclotron features;
- Directional Gamma ray Spectrometer (DGS) for directional information and to search for annihilation features.

Because the SXR had to be kept below -30°C, CATSAT was selected for what was called a solar terminator orbit, which would keep the SXR always pointed away from the Sun. The SXR had a protective door to act as a Sun shield, with an alarm if it came within 5° of exposure to the Sun. Although the satellite was almost completed, it was cancelled due to cost overruns.

IMEX (2003)

The Inner Magnetosphere Explorer (IMEX) was set to fly into a highly elliptical, ten-hour-long, 350km – 35,000km orbit into the Van Allen radiation belts, to measure its populations of protons, electrons and other charged particles, aiming at the 2001 solar maximum. The mission cost was set at $12.8m and it used an adapted form of the SNOE design. The Principal Investigator was John Wygant of the University of Minnesota, Minneapolis with co-investigators Daniel Baker and Xinlin Li in the University of Colorado's Laboratory for Atmospheric and Space Physics (LASP). LASP was responsible for design, construction, testing and operations of the spacecraft. IMEX was to be a secondary payload on a USAF Titan 4B military mission. Its six instruments were:

- FIELDS, focussed on 20Hz electric fields and using a Langmuir probe to measure plasma density and temperature with five-minute resolution (University of Minnesota);
- Fluxgate Magnetometer for 20Hz magnetic field measurements from 0.016-64000nT (Goddard);
- High Energy Particle Experiment (HEPEX) with Ring Current Ion Sensor (RIS) and magnetic spectrograph to measure the distribution of hydrogen (30–500keV), oxygen (50–500 keV) and helium (Aerospace Corporation);

- Energetic Electron Spectrograph (EES), a magnetic spectrograph to measure electrons (10 keV – 1.5 MeV);
- Relativistic Electron Detectors (RED), five threshold sensors to measure the omnidirectional flux of electrons (2MeV – 15 MeV) and protons (20MeV – 100MeV);
- Electrostatic analysers to measure distributions of electrons and ions (3eV – 30keV) (University of California Berkeley).

IMEX was cancelled in 2003 for two connected reasons: exceeding its cost cap and difficulties in getting a slot on an air force mission.

2008 Selection

Three Explorer mission were selected in 2008, but were not subsequently manifested:

- Coronal Physics Explorer (CPEX), a mission to use a solar coronograph to study the processes which accelerate the solar wind and generate the coronal mass ejections, the PI being Dennis Socker of the Naval Research Laboratory;
- Joint Astrophysics Nascent Universe Satellite (JANUS), a gamma ray burst monitor directed at the most distant galaxies to study star formation, the PI being Peter Roming of Pennsylvania State University;
- Neutral Ion Coupling Explorer (NICE), to use instruments to discover how winds and the upper atmosphere drive electrical fields and chemical reactions in Earth's ionosphere, the PI being Stephen Mende of the University of California, Berkeley.

This listing is certainly incomplete. In a story common to space programs the world over, cancelled missions tend to disappear from the record. Rarely is there an official announcement that a mission is cancelled or de-selected and, unless it is controversial, it is not the stuff of a gripping press story. For very human reasons, people whose missions are cancelled rarely draw attention to their disappointment or sense of professional loss. The cancelled missions do illustrate the competitive nature of the Explorer program in the era of 'faster, better, cheaper', revealing the rigorous nature of selection stages and the financial ruthlessness applied to missions that suffered cost overruns. Nevertheless, they raise some intriguing 'might-have-beens' of the space science that could have been learned had they gone ahead.

Conclusions

There is no doubt that Explorer was not only the first but has been one of the most important of the space programs of the United States. Although much less visible in the public mind than manned, lunar, or interplanetary missions; or less impactful on daily lives than applications programs, for example in telecommunications, it vindicated the concept of scientific exploration by small spacecraft. The volume of information collected by the Explorers has been enormous, as the previous chapters illustrated, covering a wide range of broad areas, from astrophysics to the magnetosphere; a huge field, from the edge of the solar system to the poles of Earth; a range of imaging, from gamma rays to infrared; and from mainstream science to the exotic. Novel techniques have been developed, from

Earth-Moon orbits to atmospheric diving. Fundamental discoveries were made in cosmic background radiation and the evolution of stars.

Many of the spacecraft proved to be durable. At time of writing, 13 missions were operational, the oldest being ACE, currently (2017) in its twentieth year. Some had operated for long periods of time, like IMP 8 (Explorer 50, 26 years) and IUE (19 years). Mission extensions were the norm.

Explorer's trajectory breaks down into four parts. The first (chapter 1) was as America's first satellite, especially after the imperative to get *any* satellite into orbit as a response to Sputnik, a period that shaped the first five, army Explorers. The second period (chapter 2) was a scientific program of small satellites, which took us from NASA's Explorer 6 to the end of the numerical program, Explorer 55 in the early 1970s. The third period (chapter 3) was the most difficult, with only a limited number of larger, albeit successful missions, coinciding with the crisis in space science in the post-Nixon period when the resource-hungry Shuttle was under construction. In the turbulent period of 'faster, better, cheaper' (chapter 4), Explorer returned to its vocation as a program of small scientific spacecraft, albeit with nothing like the frequency of the early years. The rules of engagement, though, were quite different from the earlier, prolific period. Instead of being a program generated in-house, missions went through the cycle of ideas-proposals-selection-execution outside NASA, generally to the universities, with NASA applying arm's length supervision. This trajectory meant that the pace of the Explorer program was uneven, as shown in Table 5.1: 35 in the first ten years (1958-67); 20 in the second (1968-77); only nine in the third decade (1978-87); and seven in the fourth (1988-97). 'Faster, better, cheaper' brought

Table 5.1 Explorer launches 1958-2018

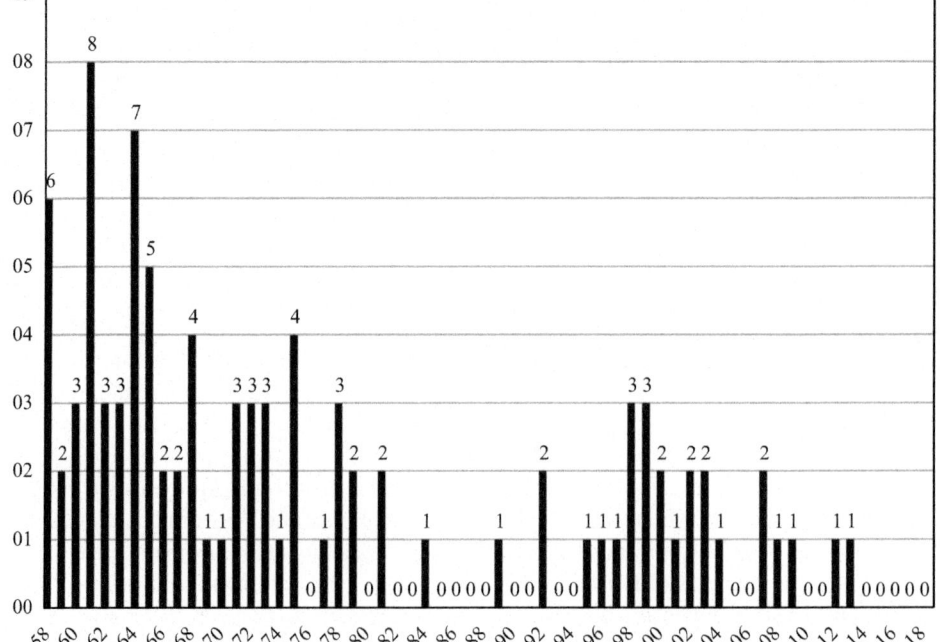

about a burst of launches from 1995-2004, but the rate since then has been slow, reflecting sharply constrained NASA budgets in the late 2000s.

As may be seen, Explorer launches were at their highest during their early years, with up to eight launches. The first year in which there were no launches was 1976, but this was followed by 1980, 1982 and 1983, the start of the thin period for the program. There were no launches from 1985 to 1988, nor in 1990–1, nor 1993–4. A consistent but low launch rate resumed at this point, coinciding with the introduction of 'faster, better, cheaper', but there were still blank years in 2005–6, 2010–11 and 2014–16. The projected future launch rate is also quite thin, even if the missions are promising.

What was the view of Explorer within the scientific and political community during this period? What may we anticipate for the future? The 1990 report on the future of the space program, chaired by Norman Augustine, acknowledged the on-going debate between 'big projects' vs 'little projects'.[6] It had been asserted, the report said, sometimes justifiably, that cost overruns in large projects were often to the direct detriment of small research and technology undertakings. Some large projects, it argued, were unavoidable and there was a natural tendency toward large scale, but there was no single answer to the issue. It acknowledged the level of 'discontent and unease' about the lack of priority given to space science, documented in the 1986 *Crisis in space and Earth science*, especially program stretch-outs, delays and cancellations that wasted the time of creative researchers, squandered resources and decreased flight opportunities. It recommended that space science should continue to be funded at or above a minimum level of 20% of the program, with small and fast-paced projects and the increasing use of universities as prime contractors (rather than NASA field centres). Both the SSB and the Office of Space Science and Applications *Strategic plan* recommended an increase in the Explorer launch rate the following year. In reality, the move to 'faster, better, cheaper' was already under way, as formally articulated by Goldin two years later and this did pave the way for a resumption of pace in the Explorer program.

Just as the loss of the *Challenger* (1986) proved to be a catalytic event, leading to a recasting of the space program, so too did the loss of the *Columbia* (2003), prompting another restatement of U.S. space goals. This took the form of President George W. Bush's *Vision for space exploration* (2004), which in a short text, re-cast the space program, committing the United States to retire the Space Shuttle, support the International Space Station and return to the Moon with a view to a first voyage to Mars. Apart from a commitment to robotic exploration and the search for life beyond Earth, however, he did not outline specific goals for space science.[7] NASA issued a more detailed iteration the following month, but this also focussed on the high-visibility goals of manned flight, robotic planetary exploration and exoplanets.[8] That same year, the President's commission on the implementation of United States space exploration policy, chaired by Peter Aldridge, had a section entitled *Exploration and the science agenda*, but it did not point NASA toward any particular program or method, such as Explorer. Instead, it iterated a set of scientific themes, *Origins, Evolution* and *Fate*, inviting a broad-ranging investigation of the universe in conjunction with the science community.[9] The U.S. *National Space Policy, 2006*, authorized by the President on 31st August 2006, referred only to the importance of space science, but gave no indication as to how this might be realized.

Was space science in general – and the type of program represented by Explorer – once again slipping down the priority list? Was there a danger of a replay of the 'crisis in space science' of the 1980s? Some in the scientific community felt moved to re-state the case.

In the American Geophysical Union, Baker and Worden referred to the significant benefits of small satellite missions, defined as less than 750kg.[10] They reduced cost at a time of restrained budgets, yet still allowed the time to obtain scientific results, with valuable hands-on experience for scientists and their students. New technologies could be used, for which the scope was more limited on large-scale projects with long lead-in times. Specifically, they focussed on Explorer, which they described as a 'continuous, important element' of the nation's space program. Originally, it fulfilled a role somewhere between sounding rockets and observatory-class missions. It was reasonably inexpensive, quickly developed and with frequent opportunities for experiments. It could not undertake this role, they said, when the missions became ambitious projects with long development times and larger budgets. They pointed out that in 1984, the SSB had argued for scientists to have frequent access to space in new, small missions, a view reiterated by the National Research Council in 2006. They were critical of the SMEX launch rate for being half that originally projected. The authors spoke of how the United States had a broad spectrum of 'small, relatively unpublicized scientific space programs', suggesting that their value was under-appreciated.

In 2009 George Abbey, former head of the Johnson Space Center (JSC) and a legend within NASA, bemoaned the lack of interest in space science, with his co-author, Neal Lane, arguing that 'any truly visionary plan for NASA's future should specify science as one of NASA's principal goals, otherwise the unique contributions that NASA can make to astronomy and to planetary, Earth and space science will be lost' and America would cede its traditional leadership role in these frontier areas of science to other parts of the world. The last four years, during which science continued to lose ground in NASA's budget decisions, had proven the validity of those concerns, he said.[11] The slow Explorer launch rate since then appeared to confirm his fears, although there was no sign of a loss of American pre-eminence in space science.

As part of the *Astro 2010* decadal review of astronomy and astrophysics, *A vigorous Explorer program* was proposed by Martin Elvis and almost fifty co-authors.[12] According to them, Explorers did cutting edge science that no one else did, either before or since the program restructure in 1988, be that in precision cosmology, compact star mergers or the metallicity of redshifts. Its current budget was $150m a year for the next five years. The main attributes of the program, they argued, were addressing critical science gaps, rapid implementation, innovation and the engagement and training of science and engineering students over the lifetime of a project. The science return, they said, was much broader than just the new knowledge about space science enabled by a specific mission. The PI-led programs played a particularly important role in training the next generation of scientists and engineers; strengthened the scientific and technical infrastructure, including instrument and spacecraft developers, launch services and the managing institutions; and generated excitement in the science and larger communities through each team's enthusiastic promotion of the mission. They proposed an annual astrophysics launch, with maximum duration from approval to launch of four to five years. The fact that they had to make a case for Explorer at all was an important statement in itself, but at least it ensured that the case would not be lost by default or by taking the program for granted.

The space program at this stage took an abrupt change of course with the arrival of Democratic President Barack Obama in 2009. He ordered a review of the space program, which concluded that the *Vision for space exploration*, enunciated by his predecessor, was no longer affordable and cancelled plans for returning to the Moon and speeding onward toward Mars. The President's *National space policy of the United States of America* (2010) expressed the need for a 'strong program of space science for observations, research and analysis of our Sun, solar system and universe to enhance knowledge of the cosmos, further our understanding of fundamental natural and physical sciences, understand the conditions that may support the development of life and search for planetary bodies and Earth-like planets in orbit around other stars'. The NASA 2011 *Strategic plan* has a *Strategic goal 2: Expand scientific understanding of the Earth and the universe in which we live*, which is subdivided into Earth science, climate and environmental change; the Sun; the origin of the solar system; and life in the universe, as guided by the National Academy's decadal surveys.

None of this, though, gives us much sense of the level of priority to be given to programs like Explorer. Others did have views. In *Solar and space physics - science for a technological society* (2014), the National Research Council followed the *Vigorous Explorer program* by specifically recommending an expansion to Explorer in a section called *Enhance the Explorer program*.[13] It drew attention to Explorer's achievements in generating Nobel prizes and applauded the management system whereby a PI had the responsibility and authority to make critical decisions on cost and schedule. The strength of the program was its ability to respond rapidly to new concepts and developments in science, as well as its synergistic relationships with larger, strategic science missions. Investigations were competitively selected to address the highest priority science that could be accomplished with small class missions. Explorer had proved to be one of the most cost-effective and best cost-controlled avenues for space science missions. The council applauded the achievements at the edge of the heliosphere (IBEX), reconnection physics on the Sun (RHESSI), explosive releases of energy in Earth's magnetosphere (THEMIS) and ice clouds over Earth's poles (AIM), as well as SNOE, CINDI and TWINS, which had all proved that a relatively modest investment could address many scientific challenges. The council recommended increasing Explorer heliospheric missions to one every two to three years and specifically endorsed an Interstellar Mapping and Acceleration Probe (IMAP), as an extension to IBEX, to study the processes between the heliosphere and the local interstellar medium, but with 20 times higher resolution.

Once again, these exchanges illustrated that the debate of the role of space science in general and the Explorer program in particular was alive, if – in the view of some of protagonists – not as well as it should be. Priorities within space programs are the outcome of a multitude of factors at work: public opinion, the mobilization of enthusiasts (like the Planetary Society), the interplay of institutional actors (such as the Academy of Sciences), the political system (the Congress and the President) and key financial players (e.g. the Office for the Management of the Budget (OMB)). John Logsdon explained how space science always had to compete for its share against manned flight, military, intelligence, applications and political or prestige interests.[14] This will probably always be the case. President Kennedy, though, set down a minimum requirement, when he enunciated the principle that the U.S. should be pre-eminent in *all* areas of space activity – what is called across-the-board supremacy – which included space science. It is of course possible to set a prominent role for space science without necessarily having an Explorer-type program, but such a program gave space science salience. Homer Newell articulated the role of the Explorer program well:

> *Explorer-class satellites permitted scientists to perform a wide range of space science experiments. The lower costs meant that more of the funds available could*

be put into the scientific research itself. These advantages account for the unvarying insistence of the scientific community that NASA continue to provide sounding rockets and small satellites. Whenever larger projects appeared to threaten the funding of the smaller ones, the scientific community rose in defense of the smaller. Over this issue, the scientists came as near to unanimity as they ever did.[15]

Explorer will certainly continue to compete for its place in space science, while space science will continue to look for its place in the overall space budgets, as it should be in a democracy. Newell took the view that the Explorer program achieved what it set out to do and made important discoveries in every field to which it was applied. More than that, thousands of scientists were attracted to work in these fields and a strong constituency built up for the support of space science. The Explorer program now passes its 60th year, the longest, continuously running program in the United States space program as a whole. Whatever the ups-and-downs of its evolution, there is every reason to expect that it will go on – and that it will continue to bring the world more ground-breaking science.

References

1. Morring, Frank: *Fallback position*. Aviation Week & Space Technology, 15-28 February 2016.
2. *ICON - a pioneering mission for space physics and aeronomy.* Presentation to AGU, 2-7th December 2012, San Francisco.
3. Reichhardt, Tony: *Kepler's children - the scientists searching for other Earths.* Air & Space, August 2016.
4. Eastes, Richard: GOLD presentation at COSPAR, 11pp.
5. Useful sources for unflown missions are space.skyrocket.de (Gunter's space pages) and www.nasaspaceflight.com.
6. Committee on the future of the U.S. space program: *Report of the advisory committee.* Washington DC, 1990.
7. Statement by President George W. Bush, 14th January 2004.
8. NASA: *The Vision for Space Exploration.* Washington DC, author, 2004.
9. *A journey to inspire, innovate and discover.* Report of the President's Commission on the implementation of United States space exploration policy. Washington DC, 2004.
10. Baker, D.N. and Worden, S.P: *The large benefits of small-satellite missions.* EOS, vol. 89, §33, August 2008.
11. Abbey, George & Lane, Neal: *United States space policy - challenges and opportunities gone astray.* Washington DC, American Academy of Arts & Sciences, 2009.
12. Elvis, Martin: *A vigorous Explorer program.* Harvard, Smithsonian Astrophysical Observatory, undated.
13. National Research Council of the National Academies: *Solar and space physics - science for a technological society.* Washington DC, author, 2014.
14. Logsdon, John: *Opportunities for policy historians - the evolution of the U.S. civilian space program*, from Roland, Alex: Spacefaring people. NASA, SP-4405, 1985.
15. Newell, Homer: *Beyond the atmosphere - early years of space science.* Dover, New York, 2010.

Bibliographical Note

For a study of the Explorer program, the best single-point source is NASA in general and the Goddard Space Flight Center in particular. Most missions have dedicated sites, the volume available being variable depending on the state of development of the mission and its scientific outcomes. One comment is that few individual mission histories or summaries appear to be available, the IMP series being an exception. Goddard also published *Goddard News*, its in-house newsletter, most of which are available from 1960. This research drew substantially on NASA's historical data and documents series, which were especially useful for the early stages of the program. Many periodicals have given useful accounts of Explorer missions and their outcomes, notably *Aviation Week & Space Technology, Flight International, Space Daily* and *Spaceflight*.

Brzezinsky, Matthew: *Red moon rising - Sputnik and the rivalries that ignited the space age.* London, Bloomsbury, 2007.

Butler, Paul: *Interplanetary Monitoring Platform - engineering history and achievements.* Washington DC, NASA, 1980

Donald Le Galley (ed): *Space physics.* New York, Wiley, 1964.

Explorers: searching the universe 40 years later. NASA facts series. Greenbelt, Md, Goddard Space Flight Center. 1998.

Ezell, Linda Neuman: *NASA historical data book, vol. II, programs and projects 1958-68.* Washington DC, NASA, SP-4012; *vol. III, programs and projects 1969-78,* do.

Foerstner, Abigail: *James Van Allen - the first eight billion miles.* Iowa City, University of Iowa Press, 2007.

Gatland, Kenneth: *Missile and rockets.* London, Blandford, 1975.

King-Hele, D.G. *et al: Table of Earth satellites, 1957-89.* Farnborough, Royal Aerospace Establishment, 1990.

Morgan, George D: *Rocket girl- the story of Mary Sherman Morgan, America's first female rocket scientist.* New York, Prometheus, 2013.

Newell, Homer: *Beyond the atmosphere - early days of space science.* Washington DC, NASA, SP-421.

Naugle, John: *First among equals - the selection of NASA space science experiments.* Washington DC, NASA, 1991.

Portree, David: *NASA's origins and the dawn of the space age.* NASA, Monographs, §10.

Roland, Alex: *A spacefaring people - perspectives on early spaceflight.* Washington DC, NASA, 1985, SP-4405.

Rosenthal, Alfred: *Venture into space - early years of the Goddard Space Flight Center.* NASA, US Government Printing Office, 1968.

Rosholt, Robert: *An administrative history of NASA, 1958-1963.* Washington DC, NASA, 1966, SP-4101.

Rumerman, Judy: *NASA historical data book, vol. VII, 1989-98.* Washington DC, NASA, 2009.

Turnill, Reginald: *Observer's book of unmanned spacecraft.* London, Frederick Warne & co, 1974.

Journals and Magazines

Air & Cosmos
Aviation Week & Space Technology
Flight International
Goddard News
Space Daily
Spaceflight
Space Research Today

Websites

Mark Wade: Encyclopedia Astronautica
Gunter: space.skyrocket.de

Appendix 1

List of Explorer Missions

			Weight	Launcher	Site	Orbit km	Incl.	Per.
1	1958 Jan 31	Explorer 1	5kg	Jupiter C	KSC	347-1,849	33.2	107.2
2	1958 Mar 5	Explorer 2	Failed to reach orbit					
3	1958 Mar 26	Explorer 3	5kg	Jupiter C	KSC	186-2,800	33.4	115.7
4	1958 Jul 26	Explorer 4	8kg	Jupiter C	KSC	257-1,352	50.2	100.9
5	1958 Aug 24	Explorer 5	Failed to achieve orbit		KSC			
	1958 Oct 22	(Explorer 6)	Failed to reach orbit		KSC			
	1959 Jul 16	Explorer S-1	Failed to reach orbit		KSC			
6	1959 Aug 7	Explorer 6 S-2	64kg	Thor Able	KSC	245-42,400	47	765
7	1959 Oct 13	Explorer 7 S1A	42kg	Juno II	KSC	523-857	50.3	98.6
	1960 Mar 23	Explorer S-46	Failed to reach orbit		KSC			
8	1960 Nov 3	Explorer 8 S-30	41kg	Juno II	KSC	394-1,331	49.9	102.2
	1960 Dec 4	Explorer S-56	Failed to reach orbit					
9	1961 Feb 16	Explorer 9 S-56A	7kg	Scout X-1	W	757-2,433	38.8	118
	1961 Feb 24	Explorer S-45	Failed to reach orbit					
10	1961 Mar 25	Explorer 10 P14	35kg	Thor Delta	KSC	221-181,100	33	5,013
11	1961 Apr 27	Explorer 11 S-15	37kg	Juno II	KSC	480-1,458	28.8	104.5
	1961 May 24	Explorer S-45A	Failed to reach orbit					
	1961 Jun 30	Explorer S-55	Failed to reach orbit					
12	1961 Aug 16	Explorer 12 EPE-A, S-3	38kg	Thor Delta	KSC	790-76,620	33.4	1,587
13	1961 Aug 25	Explorer 13 S-55A	86kg	Scout	W	125-1,164	37.7	97.5
14	1962 Oct 2	Explorer 14 EPE-B, S-3	40kg	Thor Delta	KSC	2,558-96,229	42.3	2,184
15	1962 Oct 27	Explorer 15 EPE-C, S-3	45kg	Thor Delta	KSC	306-17,610	17.9	314.7

(continued)

274 List of Explorer Missions

#	Date	Name	Weight	Launcher	Site	Orbit km	Incl.	Per.
16	1962 Dec 16	Explorer 16 S-55B	100kg	Scout X-3	W	744-1,159	52	104.1
17	1963 Apr 3	Explorer 17 AE-A, S-6	185kg	Thor Delta	KSC	254-891	57.6	96.1
18	1963 Nov 26	Explorer 18 IMP-A, S-74	62kg	Thor Delta	KSC	192-197,616	33.3	5,666
19	1963 Dec 19	Explorer 19 AD-A	7kg	Scout X-4	WTR	597-2,391	78.6	115.9
	1964 Mar 19	BE-A	Failed to reach orbit					
20	1964 Aug 25	Explorer 20 IE-A, S-48	44kg	Scout X-4	WTR	857-999	79.9	103.6
21	1964 Oct 4	Explorer 21 IMP-B, S-7A	62kg	Thor Delta	KSC	191-95,590	33.5	2,097
22	1964 Oct 9	Explorer 22 BE-B, S-66A	52kg	Scout X-4	WTR	872-1053	79.7	104.3
23	1964 Nov 6	Explorer 23 S-55C	134kg	Scout X-4	W	463-980	51.9	99.2
24	1964 Nov 21	Explorer 24 AD-B	9kg	Scout X-4	VAFB	530-2,498	81.4	116.3
25		Explorer 25 IE-B Injun 4	40kg			526-2,319	81.3	114.3
26	1964 Dec 21	Explorer 26 EPE-D, S-3	46kg	Thor Delta	KSC	284-10,043	19.8	205.7
27	1965 Apr 29	Explorer 27 BE-C	60kg	Scout X-4	W	932-1309	41.2	107.7
28	1965 May 29	Explorer 28 IMP, S-74b	58kg	Thor Delta	KSC	229-261,206	34	8,419
29	1965 Nov 6	Explorer 29 GEOS A	175kg	Thor Delta	KSC	1,120-2,269	59.4	120.3
30	1965 Nov 19	Explorer 30 Solrad 8	57kg	Scout X-4	W	671-856	59.7	100.1
31	1965 Nov 29	Explorer 31 DME-A, S-30A	99kg	Thor Agena B	WTR	505-2,833	79.8	119.7
32	1966 May 25	Explorer 32 AE-B, S-6	225kg	Thor Delta	KSC	282-2,716	64.6	116
33	1966 Jul 1	Explorer 33 IMP-D	93kg	Thor Delta	KSC	265,679-480,762	24	38,792
34	1967 May 24	Explorer 34 IMP-F	75kg	Thor Delta	WTR	242-214,379	67.2	6,358
35	1967 Jul 19	Explorer 35 IMP-E	104kg	Thor Delta	KSC	Lunar orbit		
36	1968 Jan 11	Explorer 36 GEOS B	209kg	Thor Delta	WTR	1,081-1,574	105.8	112.2
37	1968 Mar 5	Explorer 37 Solrad 9	198kg	Scout B	W	353-433	59.3	92.4
38	1968 Jul 4	Explorer 38 RAE-A	190kg	Thor Delta	WTR	5,8851-5,861	120	224
39	1968 Aug 8	Explorer 39 AD-C	9kg	Scout B	WTR	670-2,538	80.7	118.3
40		Explorer 40 IE-C Injun 5	70kg			679-2,489	80.7	117.8
41	1969 Jun 21	Explorer 41 IMP-G	174kg	Thor Delta	WTR	80,374-98,159	86	4,906
42	1970 Dec 12	Explorer 42 SAS-A *Uhuru*	143kg	Scout B	S Marco	521-570	3	95.5
43	1971 Mar 13	Explorer 43 IMP I	288kg	Thor Delta	KSC	1,845-203,130	39.9	5,957
44	1971 Jul 8	Explorer 44 Solrad 10	118kg	Scout B	W	433-632	51.1	95.2
45	1971 Nov 15	Explorer 45 S^3	52kg	Scout B	S Marco	233-26,895	3.2	466.8
46	1972 Aug 13	Explorer 46 MTS	136kg	Scout D1	W	492-811	37.7	97.7
47	1972 Sep 22	Explorer 47 IMP-I	376kg	Delta 1604	KSC	201,100-235,600	17.2	17,670
48	1972 Nov 16	Explorer 48 SAS-B	185kg	Scout D1	S Marco	526-526	1	95.2
49	1973 Jun 10	Explorer 49 RAE-B	328kg	Delta 1913	KSC	Lunar orbit		
50	1973 Oct 25	Explorer 50 IMP-J	371kg	Delta 1604	KSC	190,749-244,361	28.7	17,576
51	1973 Dec 15	Explorer 51 AE-C, S-3	658kg	Delta 1900	WTR	155-4,306	68.1	132.5
52	1974 Jun 3	Explorer 52 IE-D *Hawkeye*	27kg	Scout E1	WTR	469-125,569	89	3,032
53	1975 May 7	Explorer 53 SAS-C	195kg	Scout F	S Marco	499-508	3	96
54	1975 Oct 6	Explorer 54 AE-D, S-6	676kg	Delta II910	WTR	151-3,819	90.1	126.8
55	1975 Nov 20	Explorer 55 AE-E, S-6	721kg	Delta II910	KSC	154-3,002	19.7	117.7
	1975 Dec 5	DAD-A, B	Failed to reach orbit					
56	1977 Oct 22	ISEE 1	340kg	Delta II	KSC	337-137,904	28.9	3,441
		ISEE 2	166kg			341-137,847	28.9	3,439
57	1978 Jan 26	IUE	669kg	Delta II	KSC	25,669-45,888	28.6	1,436
58	1978 Apr 26	HCMM	134kg	Scout D1	WTR	560-641	97.6	96.7
59	1978 Aug 12	ISEE 3	469kg	Delta II	KSC	180-1,151,664	28.9	73,702
60	1979 Feb 18	SAGE	147kg	Scout	W	549-661	54.9	96.7
61	1979 Oct 30	MAGSAT	181kg	Scout	WTR	355-562	96.8	93.8
62	1981 Aug 3	Dynamics Explorer A	403kg	Delta II	WTR	559-23,295	89.9	410

(continued)

List of Explorer Missions

			Weight	Launcher	Site	Orbit km	Incl.	Per.
63		Dynamics Explorer B	415kg			298-996	89.9	97.7
64	1981 Oct 6	SME	437kg	Delta II	WTR	538-542	97.5	95.5
65	1984 Aug 16	AMPTE	242kg	Thor Delta	KSC	1,113-49,667	4.8	939
66	1989 Nov 18	COBE	2,500kg	Delta	WTR	888-897	99	102.3
67	1992 Jun 7	EUVE	3,275kg	Delta II	KSC	514-529	28.4	94.9
68	1992 Jul 3	SAMPEX	157kg	Scout	WTR	515-691	81.7	96.8
69	1995 Dec 30	RXTE	3,035kg	Delta II	WTR	565-585	23	96.2
70	1996 Aug 21	FAST	180kg	Pegasus XL	WTR	351-4,165	83	133
	1996 Nov 4	HETE 1	Failed to reach orbit					
71	1997 Aug 25	ACE	752kg	Delta II 7920	KSC	L1		
72	1998 Feb 26	SNOE	132kg	Peagaus XL	WTR	535-581	97.8	95.8
73	1998 Apr 2	TRACE	250kg	Pegasus XL	WTR	599-641	97.8	97.1
74	1998 Dec 6	SWAS	283kg	Pegasus XL	WTR	637-653	69.9	97.6
75	1999 Mar 5	WIRE	270kg	Pegasus XL	WTR	539-594	97.5	96
76	1999 May 18	TERRIERS	123kg	Pegasus XL	WTR	542-554	97.7	95.6
77	1999 Jun 23	FUSE	1,335kg	Delta II	KSC	754-770	25	100
78	2000 Mar 25	IMAGE	494kg	Delta II 7326	WTR	987-45,993	89.9	856
79	2000 Oct 9	HETE 2	124kg	Peagasus XL	Kwa	595-637	2	97
80	2001 Jun 30	WMAP	840kg	Delta II 7925	KSC	L2		
81	2002 Feb 2	RHESSI	293kg	Pegasus XL	KSC	579-609	38	96.6
82	2003 Jan 13	CHIPS	60kg	Delta II 7320	WTR	578-594	94	96.4
83	2003 Apr 28	GALEX	280kg	Pegasus XL	KSC	694-700	29	98.7
84	2004 Nov 20	Swift	1,470kg	Delta II 7320	KSC	584-604	20.6	96.6
85	2007 Feb 17	THEMIS A	126kg	Delta II 7925	KSC	470-87,337	16	1,870
86		THEMIS B	126kg			Lunar orbit		
87		THEMIS C	126kg			Lunar orbit		
88		THEMIS D	126kg					
89		THEMIS E	126kg					
90	2007 Apr 25	AIM	195kg	Pegasus XL	WTR	586-600	98	96.5
91	2008 Oct 19	IBEX	107kg	Pegasus XL	Kwa	86,263 - 258,932	45	12,559
92	2009 Dec 14	WISE	661kg	Delta II 7320	WTR	526-531	97.5	95.2
93	2012 Jun 13	NuSTAR	350kg	Pegasus XL	Kwa	615-633	6	97
94	2013 Jun 27	IRIS	183kg	Pegasus	WTR	620-664	97.9	97.5

KSC = Kennedy Space Center; W = Wallops Island, Virginia; WTR = Western Test Range, including Vandenberg Air Force Base; Kwa = Kwajelein Atoll.

Index

A

Abbey, George, 268
Acceleration, Reconnection, Turbulence and Electrodynamics of the Moon's Interaction with the Sun (ARTEMIS), 229
Active Magnetospheric Particle Trace Explorer (AMPTE; Explorer 65)
  AMPTE 1, 162
  AMPTE 2, 162
  AMPTE 3, 162
Advanced Composition Explorer (ACE; Explorer 71), 178, 180, 192–195, 249, 266
Advanced Solar Coronal Explorer (ASCE), 180, 263
Aeronomy of Ice in the Mesosphere (AIM; Explorer 90), 231–233, 249, 269
Agnew, Spiro T. (U.S. Vice-President), 138, 139
Apollo program, 48, 95, 139, 232
Apollo-Soyuz Test Project (ASTP; Apollo 18), 168–170
Applications Explorer Missions (AEM), 151, 156
Applied Physics Laboratory, Johns Hopkins University, 3, 86, 91, 112, 193
Army Ballistic Missile Agency (ABMA), 8, 9
Astro-H/Hitomi Soft X-Ray Spectrometer (SXS), 259
Atmosphere Explorer (AE), 70, 97, 125, 126, 130, 132, 158
Atmospheric Density Explorer (ADE), 81, 88, 89, 107–109, 132
Auroral Multiscale Mission (AMM), 180, 263

B

B-52, launcher for Pegasus, 181–183
Beacon Explorer (BE), 83, 86, 91, 132
Berkner, Lloyd, 30
Bold Orion interception, 57
Bowyer, Stuart, 169
Brzezinski, Zbigniew, 141
Bush, George H. W. (U.S. President), 177
Bush, George W. (U.S. President), 194, 267

C

Cape Canaveral launch centre
  description of, 14–15
Carter, Jimmy, (U.S. President), 141, 142, 144
Chapman, Sidney, 1
Clinton, Bill, (U.S. President), 177
Cooperative Astrophysics & Technology Satellite (CATSAT), 181, 264
Coronal Physics Explorer (CPEX), 249, 265
Cosmic Background Explorer (COBE; Explorer 66), 43, 66, 163–170, 172, 187, 210–213, 234, 238
Cosmic Hot Interstellar Plasma Satellite (CHIPS; Explorer 82), 214–218, 249
Cosmos program (USSR and Russia), 41, 250
Coupled Ion-Neutral Dynamics Investigations (CINDI), 249, 250, 258, 259, 269

D

Delta, rocket
  description of, 54, 65, 66
Delta II, rocket
  description of, 66–67
Direct Measurement Explorer (DME), 97, 132
Dryden, Hugh, 27–29, 31, 32, 41, 43, 47
Dynamics Explorer (DE)
  DE 1 (Explorer 62), 157–160
  DE 2 (Explorer 63), 157–160

E

Eisenhower, Dwight D. (U.S. President), 7, 8, 12, 15, 19, 20, 27, 28, 31, 34, 37, 46, 47
Energetic Particle Explorer (EPE), 70, 79, 90, 132
Explorer 1, 10, 13, 14, 16–25, 27, 47, 59, 108
Explorer 2, 21–23
Explorer 3, 22–25, 47
Explorer 4, 24–27, 74
Explorer 5, 27
Explorer 6, 27, 45, 51–57, 65, 74, 266
Explorer 7 (S-1A), 57–59, 64, 70, 132
Explorer 8 (S-30), 59–61, 64, 70, 78, 84, 132, 134
Explorer 9 (S-56), 61–65, 81, 83
Explorer 10 (P-14), 54, 60, 64–68, 104, 132
Explorer 11 (S-15), 60, 69, 70, 119, 132, 134
Explorer 12 (S-3), 44, 60, 70–72, 74, 79, 90, 132
Explorer 13 (S-55a), 70, 72, 73, 76, 132
Explorer 14 (S-3a), 66, 72–76, 79, 90, 132, 134
Explorer 15 (S-3b), 66, 73–76, 79, 90, 91, 132
Explorer 16 (S-55B), 70, 76, 77, 87, 117, 132
Explorer 17 (S-6), 77, 78, 97, 98, 125, 132
Explorer 18 (S-74), 78–81, 125, 132
Explorer 19 (ADE-A), 64, 81–83, 89, 107, 132
Explorer 20 (Topsi), 82–85, 97, 132, 134
Explorer 21 (S-74), 80, 85, 92, 132
Explorer 22 (S-66a), 82, 84, 86, 87, 91, 93, 132
Explorer 23 (S-55c), 70, 87, 88, 117, 132
Explorer 24 (ADE), 64, 82, 88–90, 107, 132, 133
Explorer 25 (Injun 4), 82, 88, 89, 132, 133
Explorer 26 (EPE D, S-3c), 74, 90, 91, 132
Explorer 27 (BE-C), 84, 87, 91, 93, 132
Explorer 28 (S-74b), 80, 92, 93
Explorer 29 (GEOS 1), 93–95, 105, 132
Explorer 30 (Solrad 8), 45, 95–97, 132
Explorer 31 (DME-A, S-30a), 97, 132
Explorer 32 (AE-B), 77, 97–99, 125, 132
Explorer 33 (IMP D), 79, 80, 99–102, 104, 114, 125, 132
Explorer 34 (IMP F), 80, 103, 104, 109, 125
Explorer 35 (IMP E), 79, 80, 99, 104, 105, 107, 114, 118, 125, 131, 133
Explorer 36 (GEOS 2), 105, 132
Explorer 37 (Solrad 9), 45, 95, 105, 132
Explorer 38 (RAE-A), 66, 106, 107, 121, 132
Explorer 39 (AD-E), 64, 82, 107–109, 132
Explorer 40 (Injun 5), 82, 89, 107–109, 132
Explorer 41 (IMP G), 109
Explorer 42 (*Uhuru*), 109–114, 120
Explorer 43 (IMP I), 114, 115
Explorer 44 (Solrad 10), 109, 115
Explorer 45 (S³), 116, 117, 132
Explorer 46 (MTS), 117, 118, 132
Explorer 47 (IMP H), 66, 80, 99, 109, 118, 119, 123, 125, 132
Explorer 48 (SAS-B), 119–121, 132
Explorer 49 (RAE-B), 66, 107, 121–123, 132, 133
Explorer 50 (IMP J), 66, 80, 118, 119, 123–126, 132, 266
Explorer 51 (AE-C), 66, 77, 78, 125, 126, 130–132
Explorer 52 (*Hawkeye*), 24, 82, 89, 126–128, 132
Explorer 53 (SAS-C), 128–130, 132, 157
Explorer 54 (AE-D), 77, 78, 125, 130–133
Explorer 55 (AE-E), 51, 77, 78, 125, 130–133, 160, 172, 266
Explorer 56 (ISEE 1), 119, 144–146, 152
Explorer 57 (IUE), 43, 144, 146–151, 154, 170, 172, 205, 266
Explorer 58 (HCMM), 144, 151, 152, 156, 172
Explorer 59 (ISEE 3), 144, 146, 152–156, 194
Explorer 60 (SAGE), 144, 156, 160, 172
Explorer 61 (MAGSAT), 144, 157, 172
Explorer 62 (DE 1), 144, 157–160
Explorer 63 (DE 2), 144, 157–160
Explorer 64 (SME), 144, 160–162, 169, 172, 196
Explorer 65 (AMPTE), 140, 144, 162, 163, 193
Explorer 66 (COBE), 43, 66, 144, 163–170, 172, 187, 210–213, 234, 238
Explorer 67 (EUVE), 140, 144, 168–172, 175
Explorer 68 (SAMPEX), 63, 82, 180, 185–188, 191, 248, 249
Explorer 69 (XTE/Rossi), 187–191, 213, 234, 248
Explorer 70 (FAST), 180, 182, 191, 192, 214, 249
Explorer 71 (ACE), 180, 192–195, 249, 266
Explorer 72 (SNOE), 181, 195–197, 215, 249, 264, 269
Explorer 73 (TRACE), 180, 197–199, 249
Explorer 74 (SWAS), 180, 184, 199–201, 249
Explorer 75 (WIRE), 201, 202, 248
Explorer 76 (TERRIERS), 181, 195, 203, 215, 248
Explorer 77 (FUSE), 170, 180, 184, 204–206, 237, 249

278 Index

Explorer 78 (IMAGE), 180, 206–208, 237, 249, 259
Explorer 79 (HETE 2), 208–210, 249, 256
Explorer 80 (WMAP), 168, 210–213, 219, 237, 249
Explorer 81 (RHESSI), 213–215, 242, 249, 269
Explorer 82 (CHIPS), 214–218, 249
Explorer 83 (GALEX), 213, 218–220, 249
Explorer 84 (Swift), 208, 219–226, 234, 237, 249, 263
Explorer 85-9 (THEMIS), 214, 226–230, 237, 249, 267, 269
Explorer 90 (AIM), 231–233, 249, 269
Explorer 91 (IBEX), 234–237, 249, 257, 269
Explorer 92 (WISE/NEOWISE), 237–242, 249
Explorer 93 (NuSTAR), 242–246, 249
Explorer 94 (IRIS), 246–249
Explorer S-1, 51, 52, 58, 59, 132
Extreme Ultraviolet Explorer (EUVE; Explorer 67), 140, 144, 148, 168–172, 175

F
Farquhar, Robert, 152, 153
Far Ultraviolet Spectroscopic Explorer (FUSE; Explorer 77), 170, 180, 184, 204–206, 237, 249
Fast Auroral Snapshot Explorer (FAST; Explorer 70), 180, 182, 191, 192, 214, 249
Faster, better, cheaper (philosophy, description of), 176, 179–181
Fisk, Lennard, 175, 186
Full Sky Astrometric Mapping Explorer (FAME), 180, 249, 263

G
Galactic/extragalactic Ultra Long Duration Balloon Spectroscopic Terahertz Observatory (GUSTO), 258
GALaxy Evolution Explorer (GALEX; Explorer 83), 213, 218–220, 249
Gemini program, 42, 232
Geodetic Earth Orbiting Satellite (GEOS), 105
 GEOS 1, 93–95
 GEOS 2, 105
Giacobini-Zinner, comet, 153, 154
Glennan, Keith, 31–33, 35, 42, 44, 47
Global-scale Observations of the Limb and Disk (GOLD), 249, 254, 258, 261, 262
Goddard Space Flight Center (GSFC), 22, 31, 41, 44, 45, 179, 202, 250
Goett, Harry, 42, 142
Goldin, Daniel, 176–178, 203, 248, 267

Gravity and Extreme Magnetism (GEMS), 249, 263, 264
Greenbelt, Maryland, 43
Grigorov, Naum, 187

H
Halley, comet, 142, 143, 148, 149, 152, 153, 155, 176
Hawkeye (Explorer 52), 24, 82, 89, 126–128, 132
Heat Capacity Mapping Mission (HCMM; Explorer 58), 144, 151, 152, 156, 172
Hidden Figures, book and film, 44
High Energy Solar Spectroscopic Imager (HESSI), later Ramaty High Energy Solar Spectroscopic Imager (RHESSI; Explorer 81), 213–215, 242, 249, 269
High Energy Transient Explorer (HETE)
 HETE 1, 208
 HETE 2 (Explorer 79), 208–210, 249, 256

I
Imager for Magnetopause-to-Aurora Global Exploration (IMAGE; Explorer 78), 180, 206–208, 237, 249, 259
Imaging X-ray Polarimetry Explorer (IXPE), 254, 257, 258
Injun
 Injun 4, 88–89
 Injun 5, 107–109
Inner Magnetosphere Explorer (IMEX), 215, 249, 264, 265
Interface Region Imaging Spectrograph (IRIS; Explorer 94), 246–249
International Cometary Explorer (ICE), 154, 155, 172
International Geophysical Year (IGY), 1, 6, 7, 23, 24, 29–31, 35–37, 42, 44, 46, 47, 132
International Sun Earth Explorer (ISEE)
 ISEE 1 (Explorer 56), 119, 144–146, 152
 ISEE 2, 144–146, 152
 ISEE 3 (Explorer 59), 144, 146, 152–156, 194
International Traffic in Arms Regulations (ITAR), 209
International Ultraviolet Explorer (IUE; Explorer 57), 43, 144, 146–151, 154, 170, 172, 205, 266
Interplanetary Monitoring Platform (IMP)
 IMP A, 78–81
 IMP B, 80, 85
 IMP C, 80, 92, 93

IMP D, 80, 99–102
IMP E, 80, 104, 105, 118
IMP F, 80, 103, 104, 118
IMP G, 80, 109, 118
IMP H, 80, 118, 119, 123
IMP I, 80, 114, 115, 118
IMP J, 80, 118, 123–125
Interstellar Boundary Explorer (IBEX; Explorer 91), 234–237, 249, 257, 269
Ionosphere Explorer (IE), 83, 84, 132
Ionospheric CONnection (ICON), 249, 254, 255, 262

J
Jet Propulsion Laboratory (JPL), 12–14, 34, 38, 42, 143, 154, 160, 162, 169, 178, 193, 199, 201, 202, 213, 238, 240, 242, 244, 250, 258
Johns Hopkins University (JHU), 2, 3, 86, 91, 93, 112, 192, 193, 200, 204–206, 219
Johnson, Lyndon B. (U.S. President), 28, 33, 134
Joint Astrophysics Nascent Universe Satellite (JANUS), 249, 265
Juno, rocket, 15
Juno 1 rocket, 10, 17, 27, 51, 64
Juno 2 rocket, 51, 61–64, 69, 70
Jupiter, rocket, 8–10, 12, 16, 20, 54, 64, 69, 70, 106, 107, 124, 149, 151, 169, 200

K
Kennedy, John F. (U.S. President), 37–41, 47, 76, 269
Killian, James, 27
Krause, Ernst, 3
Kwajalein atoll, launch site, 182, 183, 208, 209, 234, 242

L
L-1011 Tristar, launcher for Pegasus, 183, 184, 191, 234
Langley Research Center, NASA, 42, 62, 107, 118, 127
Logsdon, John, 269

M
Magnetic Storm Satellite (MSS), 262
MAGSAT (Explorer 61), 157, 172
Manned Space Flight Network (MSFN), 42
Mather, John, 164, 165, 167, 168

McDonald, Frank, 45, 70, 73, 78, 79, 92, 93, 109, 114, 156
Medium Explorer programme (MIDEX), 180, 181, 204, 210, 213, 220, 237, 248, 249, 263
Mendaris, General Bruce, 8, 9, 12, 14–16, 18
Mercury program, 4, 32, 42, 53
Meteoroid Technology Satellite (MTS) (Explorer 46), 117, 118, 132
Microwave Anistropy Probe (MAP), later Wilkinson Microwave Anistropy Probe (WMAP; Explorer 80), 168, 210–213, 219, 237, 249
Minitrack network, 11, 20, 44
Morgan, Mary Sherman, 10

N
National Advisory Committee for Aeronautics (NACA), 27–29, 31, 42, 62
National Aeronautics and Space Administration (NASA), 12, 20, 27–47, 51–58, 62, 65, 67, 69, 79, 84, 86, 88, 89, 93, 95–98, 100, 102, 103, 105, 107, 112, 114–118, 120, 126, 130, 133–135, 174–182, 184, 186, 192, 193, 195, 200–204, 208–210, 215, 216, 220, 224, 226, 232, 237, 240, 244, 248, 254, 259, 262, 263, 266–270
Naugle, John, 34, 45
Ness, Norman, 80, 81, 101, 102, 109, 119, 124
Neutral Ion Coupling Explorer (NICE), 249, 265
Neutron star Interior Composition ExploreR (NICER), 191, 249, 258, 260, 261
Newell, Homer, 3, 4, 6, 31–33, 35, 36, 45–47, 79, 131, 133, 134, 138, 176, 269, 270
Next Generation Sky Survey (NGSS), 180, 263
Nixon, Richard M. (U.S. President), 138, 139, 141, 144, 163, 174, 266
Nuclear Spectroscopic Telescopic Array (NuSTAR; Explorer 93), 242–246, 249
Nuclear tests
    Argus, 24, 25, 27, 76
    Orange, 4, 24, 25, 76
    Starfish, 73–76, 90, 91, 108
    Teak, 4, 24, 76

O
Obama, Barack (U.S. President), 194, 268
Oberg, Jim, 178, 179
Oersted satellite, 157
Office of Naval Research (ONR), 2, 6
Owl mission, 262

P

P-14 (Explorer 10), 54, 60, 64–68, 104, 132
Paine, Thomas, 138, 139, 143
Pegasus, launcher
 description of, 181–182
Pegasus XL, launcher
 description of, 183, 184
Pellerin, Charles, 179, 180
Pickering, William, 3, 12, 14, 15, 18, 19, 29, 31, 42
Prognoz 9, *Relikt*, 168
Project *Reboot*, 156

R

Radio Astronomy Satellite/Radio Astronomy Explorer (RAS/RAE), 106, 121
Ramaty, Reuven, 186, 213–215
Reagan, Ronald (U.S. President), 26, 139, 140, 142, 144
Redstone, rocket, 2–5, 7–10, 12, 15, 24
Roman, Nancy, 107, 112, 113, 123, 148
Rosman tracking station
 description of, 85, 124
Rossi, Bruno, 188

S

S-1A (Explorer 7), 57–59, 64, 70, 132
S-15 (Explorer 11), 60, 69, 70, 119, 132, 134
S-3 (Explorer 12), 44, 60, 70–72, 74, 79, 90, 132
S-3a, b (Explorer 14), 66, 72–76, 79, 90, 132, 134
S-3c (Explorer 26), 74, 90, 91, 132
S-30 (Explorer 8), 59–61, 64, 70, 78, 84, 132, 134
S-55a (Explorer 13), 70, 72, 73, 76, 132
S-55B (Explorer 16), 70, 76, 77, 87, 117, 132
S-55c (Explorer 23), 70, 87, 88, 117, 132
S-56 (Explorer 9), 61–65, 81, 83
S-6 (Explorer 17), 77, 78, 97, 98, 125, 132
S-66a (Explorer 22), 82, 84, 86, 87, 91, 93, 132
S-74 (Explorer 18, 21), 78–81, 85, 92, 125, 132
S-74b (Explorer 28), 80, 92, 93
S^3 satellite (Explorer 45), 116, 117, 132
San Marco launch centre
 description of, 110–111
Satellite Tracking and Data Acquisition Network (STADAN), 42, 89, 91, 98, 107, 109, 115, 117, 119
Scout rocket
 description of, 61–62
Sergeant, rocket stage, 8, 10, 63
Silverstein, Abe, 29, 32, 33, 35, 46, 47
Small Astronomy Explorer, 109, 119, 132
Small Explorer programme (SMEX), 180–182, 185, 186, 191, 201, 203, 213, 218, 231, 242, 246, 248, 249, 256, 263, 268
Smoot, George, 167, 168
Solar Anomalous and Magnetospheric Particle Explorer (SAMPEX; Explorer 68), 63, 82, 180, 185–188, 191, 248, 249
Solar Mesosphere Explorer (SME; Explorer 64), 160–162, 169, 172, 196, 231
Solrad
 Solrad 8, 95–97
 Solrad 9, 105
 Solrad 10, 115
Space Science Board (SSB), 30–33, 35, 36, 40, 47, 60, 119, 134, 140, 143, 153, 175, 210, 267, 268
Spectroscopy and Photometry of the Intergalactic Medium's Diffuse Radiation (SPIDR), 231, 249, 263
Sputnik, Soviet spacecraft, 10–12, 15, 19, 24, 27, 31, 37, 41, 46, 266
Stratospheric Aerosol and Gas Experiment (SAGE; Explorer 60), 156, 160
Student Explorer Demonstration Initiative (STEDI), 181, 195, 203, 215, 264
Student Nitric Oxide Explorer (SNOE; Explorer 72), 181, 195–197, 215, 249, 264, 269
Submillimetre Wave Astronomy Satellite (SWAS) (Explorer 74), 199–201
Sustaining University Program (SUP), 39
Swift (Explorer 84), 180, 208, 219–226, 234, 237, 249, 263

T

Tape recorder, role of, 22, 133
Thor Able, rocket
 description of, 54
Thor rocket
 description of, 82
Time History of Events and Microscale Interactions during Substorms (THEMIS; Explorers 85-9), 214, 226–230, 237, 249
Tomographic Experiment using Radiative Recombinative Ionospheric Extreme ultraviolet and Radio Sources (TERRIERS; Explorer 76), 181, 195, 203, 215, 248
Topsi, 83–85
Total Ozone Mapping Spectrometer (TOMS), 180, 249
Townsend, Marjorie, 112, 113, 120
Transiting Exoplanet Survey Satellite (TESS), 203, 249, 254, 256, 257

Transition Region And Coronal Explorer (TRACE; Explorer 73), 180, 197–199, 249
Two Wide-Angle Imaging Neutral-Atom Spectrometers (TWINS), 249, 258, 259, 269

U
Uhuru (Explorer 42), 109–114, 119, 120, 128
University Explorer programme (UNEX), 180, 181, 215, 216, 248, 249, 262

V
Van Allen, James, 1–4, 7, 10, 12, 13, 18, 19, 22–29, 47, 51, 56, 70, 71, 81, 89, 108, 109, 126–128, 139, 185, 230, 257, 262, 264
Van Allen radiation belts, 24, 56, 230, 264
Vandenberg Air Force Base
   description of, 81
Vanguard, 5–12, 14–17, 20–22, 27, 31, 35, 36, 42, 43, 46, 54, 63, 65, 67, 76, 93, 179
Vigorous Explorer program, 268, 269
Vision for space exploration, 267, 268

Von Braun, Wernher, 2–10, 12–15, 17–20, 29, 31, 36, 46, 54
V-2 Upper Atmosphere Research Panel, 3, 29, 46

W
Wallops Island launch centre
   description of, 61–62
Webb, James, 37–41, 47, 48, 95, 133, 134, 142, 213, 238, 242, 257
White Sands, 2, 3, 5, 163, 238
Wide Field Infrared Explorer (WIRE; Explorer 75), 201, 202, 248
Wide-field Infrared Survey Explorer (WISE, later also NEOWISE; Explorer 92), 66, 202, 219, 237–242, 249

X
X-ray Timing Explorer (XTE; Explorer **69**) also called Rossi, 140, 170, 175, 178, 180, 187–191, 213, 234, 248

The manufacturer's authorised representative in the EU is Springer Nature Customer Service Centre GmbH, Europaplatz 3, 69115 Heidelberg, Germany. If you have any concerns regarding our products, please contact ProductSafety@springernature.com

Printed and bound by CPI Group (UK) Ltd, Croydon, CR0 4YY

23/03/2026

02076657-0003